INVESTING
FROM THE
HEART

INVESTING
FROM THE
HEART

*The Guide to Socially
Responsible Investments
and Money Management*

JACK A. BRILL
AND ALAN REDER

CROWN PUBLISHERS, INC.   NEW YORK

Published by Crown Publishers, Inc., 201 East 50th Street, New York, New York 10022. Member of the Crown Publishing Group.

CROWN is a trademark of Crown Publishers, Inc.

Manufactured in the United States of America

Book design by June Bennett-Tantillo

Library of Congress Cataloging-in-Publication Data

Brill, Jack A.
    Investing from the heart : the guide to socially responsible investments and money management / by Jack A. Brill and Alan Reder.—1st ed.
    Includes bibliographical references (p. 398).
    1. Investments—Social aspects—United States.   I. Reder, Alan.
II. Title.
HG4910.B734   1992
332.6'78—dc20   91-22420
                                                                    CIP

ISBN 0-517-58495-6

10   9   8   7   6   5   4   3   2   1

First Edition

To our families, who enabled this project with their faith and their love.

Contents

Acknowledgments

This book was a collaborative effort from its very inception, for the original idea came from Laurie Fox, our literary agent's right-hand person. Laurie also provided valuable guidance and editing for our book proposal.

Our agent, Linda Chester, with Laurie's assistance, had a number of major publishers excited about the book idea before the proposal was completed. Thanks to Linda and Laurie's groundwork, the industry's interest in the proposal exceeded our wildest fantasies. On the other end of the line, our editor at Crown Publishers, David Groff, with his assistant, Carol Taylor, made the process of wrestling the book into its final form as pain-free as possible. Their enthusiasm at every stage of the game eased our emotional load considerably.

Much of the technical research on investment opportunities and socially responsible investment resources was carried out by two individuals, Hal Brill and Ardie Andrews. Their devotion to the book's purpose is reflected in the excellence of their work. Hal, himself a socially responsible financial adviser, gathered together investment tools never before available in the field. Ardie researched technical material for the appendixes and other sections and provided word processing and correspondence.

We are grateful to a number of players in the socially responsible investing and allied fields for their help: George Gay, chief executive officer of First Affirmative Financial Network, for technical assistance and critical review of the completed manuscript; Ed Winslow and Scott Flora of the same organization for information and graphics assistance; Joan Shapiro of South Shore Bank for information (and inspiration!) on socially responsible banking; Rian Fried and Doug Fleer of the Clean Yield Group and Alisa Gravitz, executive director of Co-op America, for information and encouragement; Peter Kinder for information on the Domini 400 stock index and mutual fund; and Patrick McVeigh of Franklin Research and Development for background information and support. Kent Taylor and Michele André of Taylor's Herb Garden in Vista, California, provided information on agriculture- and pesticide-related issues. We apologize for any others we may have overlooked.

We also want to thank Paul Hawken and Charles Garfield, who cheered the book when it was still in the proposal stage, and Phil Catalfo, for reviewing the final manuscript from a writing standpoint and from the standpoint of a typical reader.

The support for both of us was at its most incredible where it absolutely needed to be—at home. Jack's wife, Sandra Brill, encouraged the project when it was still just a sliver of an idea—despite all the uncertainties of publishing. She also assisted with research and was yet another valued reviewer of the various drafts. While Alan disappeared in his office for most useful hours of the day, Hyiah Reder put her own career as a psychotherapist on hold to raise their daughter, Ariel. As monumental a task as writing a book like this is, Hyiah's task—at her level of dedication—was even more trying. Yet Alan resists the notion that his role in the family system is the more glamorous—anyone who knows Ariel knows that Hyiah's "project" shines the brightest.

INVESTING
FROM THE
HEART

Preface

JACK'S STORY

When I began as a socially responsible financial planner in 1985, I was thrilled to be working for peace and social justice but scared out of my wits about my ability to make a living. For one thing, I wasn't sure where I'd find my clients. Those who cared enough were rarely the same ones with the extra capital to invest. For another, the commissions that brokers earn on socially responsible products are modest, and there just weren't that many products to offer back then—only five mutual funds, for example. It also didn't help that my colleagues in the financial mainstream scoffed at social criteria that they said "limited the universe" of eligible investments and endangered profits.

Times have changed. Today I don't have enough hours in my day to handle all the clients who call me. Socially responsible investing (SRI) is suddenly big business, and even mainstream money managers are paying attention (particularly because many of the socially screened vehicles do *better* than their "unlimited" ones). The number of socially screened mutual funds has more than doubled, I've got a whole array of other socially screened products to offer my clients, and new ones are being readied continuously.

Why the explosion? The massive response of institutions divesting in South Africa is one reason, not to mention the profound results of that effort. The events of 1989 in Eastern Europe didn't hurt, either. Just like that, the cold war was over and the arms race had been supplanted by happier competitions such as the race to develop high-definition television. And, of course, the fact that everyone now realizes that Mother Earth needs our help has created a huge demand for environmentally sensitive investments. That demand is going to grow and grow, because Mother faces a long recuperation.

The bottom line (I love to use that term in this context) is that people really want to make a difference in the world, and SRI gives them that chance—*with interest*. In the lingo of the behavioral psychologist, money withdrawn from companies and government agencies that behave unethically "*punishes* that behavior" and makes it less likely to occur in the future. Money invested in companies and government activities that are socially beneficial "*reinforces* that behavior" and makes it more likely to occur in the future. Researchers (and parents) know that behavioral techniques work.

Of course, none of the above was that clear to me in 1985 when I canceled my role in the military-industrial complex, and my thirty-year engineering career with it. I just knew that I had to get out. At that time I was working for the U.S. Navy, overseeing hundreds of millions of dollars annually in contracts in their San Diego office. This was a job I had reluctantly accepted eighteen years before, after an industrial layoff dropped me out of the peaceful end of the aerospace industry.

Although I had scrupulously avoided taking on weapons-related work up to that point, the navy job did enable me, at times, to act forcefully for the public welfare. Certainly I succeeded in uncovering evidence in many cases that helped convict unscrupulous contractors in the courts. But I also saw repeated instances of high-level officials interfering with investigations of fraud, and the implications—both for taxpayers' pocketbooks and their hopes for a peaceful world—were tearing at my conscience.

I was frustrated, too, with the shamefully excessive military buildup during the Reagan era. I couldn't stop the funding of marginally useful projects and the continued deployment of other inefficient vessels—all to serve the questionable goal of a "600-ship navy." And I certainly had no power to turn around what I saw as a misguided and dangerously aggressive foreign policy. Yet—however reluctantly—my finger was undeniably one of those on the trigger.

By 1985 the moral compromises I had made for a secure living were starting to ruin my health. One night while I was at the symphony with my wife, Sandra, my childhood training about standing up for what I held true flooded back into my consciousness with the meditative power of the music. Two days later I resigned from my job and turned what had been a fifteen-year-long hobby—personal investing—into my sole livelihood. With one twist: I no longer was willing to build wealth for myself or others at the expense of the society I wanted for us.

If you are just now entering the world of SRI, you're picking a far more opportune time to do it than I did. You're entering a thriving industry with even more promising times ahead. The myth that drove me in my days with the navy—that financial security has to come at the expense of ethical compromises—has been dispelled by the success of these investments and the ethical companies they represent. Socially responsible investments are thoroughly researched and brokered by trained professionals and represent some of the most innovative thinking in the financial world. As a class, socially responsible investments are also some of the more prudent investments available in the field. Of course, it is their difference from traditional investments that is the real source of the excitement—and the reason for this book.

ALAN'S STORY

Your present point in time can be traced back to everything that's ever happened to you, but there are certain moments when this is especially clear. When Jack Brill approached me just about a year ago about collaborating on this book, it was one of those moments. It was also immediately obvious that my meeting Jack a year before had been another one.

Providential design notwithstanding, my path to that first meeting gave me no indication of where I was headed. In the late 1960s I had been a student activist in the classical boomer-generational mode, and my countercultural values had survived largely unchanged in the years since. Of course, because I also refused to do work that compromised my social values, I had little money to show for lots of effort.

Then my mother and only surviving parent died, leaving me heir with my sister to a complex portfolio of investments. Suddenly the portfolio was in large part all that physically remained of my folks. They had saved and invested diligently—and just as diligently lived well within their means—to give this gift to their children. The only way left to me to express my gratitude was not to blow it. I began feverishly educating myself on money management principles, but I knew I needed expert help to fulfill my commitment.

Expert socially responsible help, that is. As moved as I was by my parents' drive to extend their loving embrace beyond death, I was uncomfortable with the inheritance's contents. My parents had been unwaveringly ethical models to their children. They had marched against the Vietnam War beside me. They had stood up for civil rights even in the most hostile surroundings. They were caring employers. But like the rest of their generation, they never thought to carry their social values into their personal finances. Business was business. You invested to make money, and then you gave generously to the causes you supported.

I had known about socially responsible investing peripherally for years. I found it intriguing that people with values like mine were actually able to get ahead financially with those values intact. So I went back to a socially responsible investing story that I had clipped from a local alternative publication to find a referral. The only one mentioned was Jack Brill. He came to my home, enthusiastically explained the range of socially responsible investments to my wife, Hyiah, and me, and offered to research the quality, both ethically and financially, of the portfolio. He also immediately adopted us as ex officio children.

The upshot is that Jack helped us transform a portfolio of socially questionable (although financially intelligent) investments into one that made us feel we were investing in a better world. And the returns we built in were still more than respectable in the current marketplace.

Nevertheless, when Jack came to me with the book collaboration idea, I didn't immediately get the fortuitousness of the whole thing. I had long since studied my way to competence in financial planning—partly out of necessity and partly out of my fascination with social investing as a progressive activist strategy. But I approached the proposed book project like any other contract writing job. I told Jack what I thought a book proposal would take in terms of my time and discounted my hourly for a friend. My response was a professional writing estimate, pure and simple.

Then the adrenaline started pumping. The timing for the book was auspicious. The field was expanding almost as fast as the national debt and still had no comprehensive guidebook. The vast baby boom generation—marked both by its social consciousness and its procrastination in planning its financial affairs—was reaching an age where the need to plan was urgent. Somebody needed to get them the word that they could do that and further their social goals simultaneously. Jack and I—the fatherly socially responsible broker and his boomer/ writer/client—were a perfect pair for the job. The bonus was that we had rapport between us like blood relations.

WELCOME TO *INVESTING FROM THE HEART*

No matter how much or how little prior knowledge you bring with you to this book, we have something for you here. If you are starting from square one and need to understand the basics of money management before even thinking about investments, you've come to the right place. If you already know this stuff and are ready to do some serious investing, the "How to Use This Book" section at the end of this preface will direct you where to turn. If you want to use the information in this book in combination with the advice of a professional consultant, we will offer sources of SRI specialists as well as show you how to use the book with a traditional adviser. If you want to manage your own plan, we will direct you to a host of periodical publications designed to help you. We've also provided worksheets to assess and prioritize your needs, financially and ethically.

As for social strategy, you will learn how to keep your money out of the hands of companies and institutions that, by your standards, behave irresponsibly; how to direct your money toward companies and institutions whose actions you support; and even how to "infiltrate" irresponsible companies as an investor so you can lobby for change from the inside.

We do assume, however, that even though the readers of this book are socially concerned, they still want their ethical investments to earn returns at least *equal* to those of traditional investments. Thus, community development investments—for which investors generally accept below market returns in exchange for very focused social gains— are covered only briefly, with referrals to other sources for more information. In addition, we offer only a brief introduction to shareholder activism, the third strategy mentioned above, because it is more a political tactic than a financial one. Again, you will be referred to more comprehensive sources on this intriguing topic.

That said, we're ready when you are to get on with the rest of the book. But, as you'll learn in the early chapters, the best time to start—for both financial and social reasons—is *now*.

HOW TO USE THIS BOOK

We have compartmentalized the information in this book so it can serve your needs no matter what degree of sophistication you bring to the subject matter. Whether you simply want to use the book as a shopping guide to specific socially screened investments or you want to start schooling yourself on basic principles of money management, the following chapter content summaries will direct you where to go.

1: The Socially Responsible Investing Universe

An introduction to the history, concepts, and current state of socially responsible investing (SRI) and money management.

2: Social Screening—A Closer Look

A review of the typical social issues considered by SRI professionals, including typical questions asked of companies and issue-specific resources consulted. The information in this chapter will also assist you in doing your own social screening.

3: A Money Management Primer

The basic principles of money management and investing, including SRI considerations where relevant.

4: A Guide to Socially Responsible Investments— Individual Investments (Except Stocks)

For new and sophisticated investors alike. New investors will learn the typical attributes and financial considerations of various categories of individual investments, as well as the social implications of each category. All readers can ''shop'' from the lists of socially positive or benign investments provided after appropriate categories.

5: A Guide to Socially Responsible Investments—Stocks

New investors will learn how accessible the ''forbidding'' world of stocks really is. A list of socially screened stocks has been provided near the chapter's end for all readers to use as a starter resource for ethical stock investments.

6: A Guide to Socially Responsible Investments— Mutual Funds

New investors will learn the basics of mutual fund investing. All readers will want to turn to the detailed evaluations of socially screened mutual funds and other "shopping lists" of mutual funds relevant to ethical investors.

7: Building Your Financial Plan, Stage I—Balancing Wants and Needs

How to apply the basics of socially responsible money management and investing to your particular circumstances and desires.

8: Building Your Financial Plan, Stage II—Smart Money, the SRI Way

The wrap-up—activating your plan (including how to select an adviser if you wish to use one) plus everyday financial wisdom with an SRI spin.

Appendixes

Lists of companies screened for particular issues, SRI periodical and organizational resources, stock indexes composed of socially screened stocks, resources for high-impact social investing, and a glossary of money management, investing, and social investing terms.

1

The Socially Responsible Investing Universe

WHY SRI? THE RATIONALE BEHIND SOCIALLY RESPONSIBLE INVESTING

Suppose each or your investment dollars—including every dollar in your bank account—could help build the world you want for yourself and your family *and* earn you a healthy return besides? Well, that's exactly what your money *can* do. With every passing day, more individuals and institutions are joining the parade of socially responsible investors, which means that their influence is making ever greater impacts on the marketplace. As of last year, more than $625 billion were invested according to some social criteria, and that number, along with the number of products, is expanding rapidly.

As for returns, the range and quality of investments now screened for social issues is such that you do not have to sacrifice just because you care enough to invest according to your ethical principles. You can do just as well as or better than investors who pay no attention to ethical criteria. *In fact, based on past performance, many of the products available to you as a socially responsible investor do better than the average investments in those categories.* And that impressive record promises to continue as industry turns toward more peaceful pursuits in the post–cold war era. In essence, socially responsible investing puts a whole new twist on the phrase "getting more bang for your buck."

Before we go any further, let's make sure you understand exactly what we mean by *socially responsible investing*, or SRI. First, we use the term *investing* not only in the classical sense of stocks and bonds and so forth, but also as shorthand for everyday money management— bank accounts, checking accounts, bank loans, and so forth. In fact, *socially responsible investing begins with socially responsible banking*. So if you have enough money to maintain a bank account, you have enough to play the SRI game.

The full term, socially responsible investing, simply means money management and investment decisions made according to both financial and ethical criteria.* Not just any ethical criteria, however—*your* criteria. SRI is not about satisfying someone else's political or social agenda. Although most socially responsible investors would agree generally about a broad range of issues, the emphasis in SRI is on your individual ethical stance. Besides, almost all socially responsible investments require you to examine your own priorities, because there are few "purely" ethical investments. With enough information, however, you will probably find several investment opportunities that attract you, both socially and financially. The purpose of this book is to provide you with the data you need to make informed choices.

Certainly, as a socially responsible investor, you will need to process a bit more information than traditional investors. However, that is where the satisfaction of SRI comes in, because it is that extra step that ensures that your money is invested where it is doing both you and the society you live in some good. Besides, consider the alternative. As a wise person we know once said: When you don't vote, you automatically vote for the winner. Everything you do has consequences. Not to consider the social consequences of where your money goes is to endorse those consequences automatically.

Probably for most of your life you have heard the phrase "Money is the root of all evil," and there have undoubtedly been times when it appeared that truer words were never spoken. Greed has certainly been a major motivation behind a lot of the environmental damage sustained by our planet and a lot of the human suffering on the planet as well. But money only has the power that people give it. When investors place their money with social consequences in mind, it becomes a powerful force for good. The large-scale withdrawal of funds invested in South Africa has forced the government there to begin dismantling

---

* The Social Investment Forum, the national professional association for SRI, defines socially responsible investing as "the channeling of personal, community, or workplace capital toward just, peaceful, healthy, environmentally sound purposes and away from destructive uses."

its policy of apartheid. As more and more investment money is directed toward companies whose products, services, and practices contribute to a better world (and withdrawn from those that don't), all business will get the message. Money does talk, after all. So certainly one of the most efficient ways to create the world you want is to invest in it.

LETTING YOUR CONSCIENCE BE YOUR GUIDE— NEGATIVE AND POSITIVE SOCIAL INVESTING

Having defined SRI for you in broad terms, we're now going to reexamine the concept in finer focus, because what is meant by socially conscious investing also depends on whose consciousness is doing the investing. Some people feel it's enough just to keep their money out of the hands of companies whose policies or products are repugnant to them. Others see investing as a form of social action. They look to invest in companies and projects whose purposes they support. Those who research investments according to social criteria use both negative and positive "screens" so that the interests of both types of investors will be served.

A *negative screen* is a sorting that reveals companies that are *not* involved with destructive products, practices, or services considered in the screening. In other words, the social value of these companies is assessed negatively, by what they do *not* do. Examples: a company that does not do business in South Africa; a utility not involved with nuclear energy; a cosmetic company that does not test products on animals.

A *positive screen* sorts out companies whose social value is assessed positively—that is, with regard to these issues considered in the screening, their products, practices, and services make a positive contribution. Examples: a company with progressive minority hiring practices; a company that develops solar energy technology; a company that recycles discarded materials.

All socially responsible investors consider both negative and positive screens. However, the activist investor takes a particular interest in the latter.

Screening also takes into account the broad range of social issues that may be implicated in a single investment. For a stock to be purchased by one of the broadly screened mutual funds listed in chapter 6, it must pass a number of social tests. You'll get a good look at that testing process in the next chapter.

Obviously an issue that concerns one socially responsible investor may be of less or no concern to another. A position that one investor supports may even be one that another opposes. In SRI, you make the choices according to your own set of principles.

YOUR DAILY BREAD—SOCIALLY RESPONSIBLE (AND FINANCIALLY SENSIBLE) BANKING

Socially responsible investing begins with socially responsible banking. You may not think you are in any financial position to invest money, but the fact is, if you have a bank account—even a basic passbook savings or interest-paying checking account—you are investing in every sense of the word. Banks are not just safe places to keep your money. When you make a deposit, you are also *loaning* your bank your money, and it is paying you interest as a return on your loan. Do you realize that if you invested in one of those "high-flying" corporate bonds that we'll be talking about later, the deal would be very much the same? You would loan the corporation money, and it would promise to pay you a specific interest rate.

The social implications of banking have to do with how the bank uses the money you and others lend it. Banks generate income in two primary ways: by investing deposited money and by loaning deposited money out at higher rates than they pay to depositors. Most banks do not invest according to social criteria (even though—as we demonstrate throughout this book—doing so would not necessarily compromise their profits). Most banks do not lend according to social criteria, either.

There are a few banks, however, that specifically serve the needs of socially responsible depositors and investors. And some types of banking institutions, by their very structure, are well suited to the needs of socially responsible individuals even if they don't specifically define themselves as SRI-oriented.

The following section evaluates the social and financial attributes of the various types of banking institutions, in descending order of social responsibility. This will be followed by a listing of specific SRI institutions and other resources to help you find a socially appropriate banking institution in your community.

Credit Unions

Credit unions are not only the most socially appropriate of the bank types, but they offer the best interest rates and lowest-priced loans as well. Credit unions are able to offer the rates they do because they are nonprofit organizations by law, existing purely for the needs of their members. (In fact, any profits generated are distributed back to members in the form of periodic dividends.) Essentially, credit union money circulates as a closed circuit—in other words, money deposited by members is loaned to other members, and so on. Thus, you do not have to worry that your deposits are buying the stock of some corporate polluter or being loaned to some repressive foreign government. And in case you were wondering, your deposits enjoy the same $100,000 of federal deposit insurance as they would in a bank or savings and loan.

In their charters, credit unions define a "common bond" that links their members, and that bonds may define a social purpose. For example, some credit unions serve special populations such as teachers, lesbian and gay people, women, unionized workers, and the like. Some common bonds will be more socially benign: government employees, employees of specific companies, and so on.

Community development credit unions (CDCUs) are a particularly socially proactive form of credit union, designed to help disadvantaged communities improve their lot. Most low-income communities are either poorly served or neglected completely by traditional banks. When banks exist at all in these communities, they mostly *extract* money in the form of deposits and high fees: very little is loaned back to the members of the community, who are usually written off as bad credit risks. (This illegal but still prevalent practice is known as "redlining" a community.) Since, again, credit unions by law lend only to their depositors, a credit union in a disadvantaged community is, by definition, a community bootstrap organization. The fact that credit unions loan at low interest rates and with minimal fees helps, too, as does the fact that CDCUs are locally controlled instead of run by some distant, unresponsive corporation.

One of the biggest drains on a deprived community's precious few resources are the loan-sharking, check cashing, and other exploitive financial services that normally feed upon low-income people who have no other recourse. CDCUs supplant these services so that more of the community's money circulates within it. CDCUs also provide a host of other valuable financial services, including free money man-

agement training. For all these reasons, CDCUs (and socially responsible banks—see following) are a popular investment with religious
organizations, charitable foundations, and corporate charitable efforts.

Becoming a member of a credit union if you don't qualify by its
common bond may take a little resourcefulness, but it is not a major
challenge. Credit unions usually allow members to bring relatives (and
often friends) in as associate members, so see if someone in your
family who is a credit union member can add you. Some credit unions
also have very liberal membership requirements (such as residence in
the community for one year).

Contact information about CDCUs and other credit unions in your
community will be found at the end of this section.

Socially Responsible Banks

These are the commercial banking equivalents of community development credit unions, in that they provide otherwise difficult-to-obtain
services to disadvantaged and modest-income people. They support
goals such as affordable housing, loans to small community businesses
and farms, inner-city rehabilitation, land conservation, and other community development projects.

In addition to funneling depositors' money back into the community in the form of loans, some community development banks seek
charitably inclined ethical investors from outside the targeted community to finance the bank's goals. For example, South Shore Bank
in Chicago, probably the best known of the SRI banks, offers a
rehab-CD loan program designed to finance the rehabilitation of mul-
tifamily housing in Chicago's South Shore neighborhood. The CD
(certificate of deposit) pays an interest rate several percentage points
below competitive market rates. The difference between the interest
rate paid to the depositor and the current market rate goes into a
special housing fund from which low- and no-interest loans are made
to qualified borrowers.

You will find regular market-rate certificates of deposit and other
banking investment opportunities at the typical community development bank as well. As of this writing, there are only a handful of such
institutions throughout the country, with all those we know listed at the
end of this section. If you live in Corpus Christi, Texas, it may be more
convenient for you to do your everyday banking at your local credit
union than at, say, Community Capital Bank in New York City. Still,

you may want to do some investing or depositing at Community Capital to advance its social goals.

Minority-Owned Banks

As a rule, these institutions are more responsive to the needs of minority citizens than other commercial banks (although some minority owners divest in their bank's communities as well). Community-minded minority-owned banks provide mortgages, loans to local businesses, job opportunities, and a variety of other financial services—including credit—to citizens normally excluded from these benefits because of prejudice or lack of credit history. You can support these vital institutions by doing your banking with them and by using their other investment services, such as certificates of deposit, IRAs, money market funds, and the like. In return you will receive the same federal guarantees on your deposits that you receive in other banks, along with market-rate interest on your deposits and other investments.

Savings and Loans

These are not as clear an SRI choice in the last decade or so because deregulation has blurred the distinction between S&Ls and commercial banks. S&Ls used to be much more akin to credit unions in structure in that their primary business was lending deposits back out as home, consumer, and other loans to their primarily middle- and low-income customers. The smaller S&Ls are still more likely to do business primarily within their community as opposed to making large commercial and real estate development loans or outside investments.

Oh yes, the S&L crisis—lest you thought we'd forgotten something. Facilitated by lax federal regulation, this industry has suffered some spectacular institution failures because of irresponsible management and, in some cases, outright fraud. There is no question that the entire industry is feeling the consequences as nervous depositors take their money elsewhere. But most of the consumer panic is just that, panic, because your deposits up to $100,000 are safe in any S&L carrying federal deposit insurance, and the majority of S&Ls do not deserve the bad name pinned on the industry because of a sleazy or incompetent few. Besides, interest rates paid on deposits will tend to be higher, and interest charged on loans lower, than at commercial banks. (Many of the troubled S&Ls started their downhill slide when they invested in junk bonds and made ill-advised real estate development loans in the mid-1980s. The junk bond collapse is detailed on page 133.)

Commercial Banks

These are the worst choice from an SRI standpoint and from a financial standpoint as well, as already indicated. In general, commercial banks are more oriented to the needs of large business depositors and their owners than are other banking institutions. They will generally lend to businesses—and, in the case of the major banks, foreign governments—without regard for the borrowers' social ethics. If you are wary of S&Ls, then you won't like banks much better, because this industry is only marginally healthier than the former.

Again, as a class, small local banks may be an acceptable SRI choice at least in comparison with their larger brethren, because the smaller the institution, the more likely its business is to be community-based.

Actually, you no longer have to guess to what extent your local banking institution invests or divests in its community. Since 1977, under the Community Reinvestment Act (CRA), banking institutions have been required to provide the public with a CRA statement, in which the banking institution defines its geographical community and specifies how it is meeting that community's credit needs. The intent behind the act is to encourage banks to provide credit to their local communities, including—where relevant—low-income communities. Under the act, banks are also required to maintain a file of public comments.

As part of the S&L bailout and regulatory crackdown, the Financial Institution Reform, Recovery, and Enforcement Act (FIRREA) put some more teeth into the CRA, effective July 1, 1990. FIRREA mandates that federal regulators evaluate a bank's CRA statement and requires the bank to publicly disclose the rating one month after receiving it. So ask your bank's head office for its CRA statement (and FIRREA rating if available). One warning: Don't put too much stock in the FIRREA rating. Initial independent social responsibility research reveals the federal regulators' standards to be less than exacting. Check with CANICCOR for an independent evaluation of an institution (see next page).

Resources

American League of Financial Institutions
1709 New York Avenue NW
Washington, D.C. 20006
(202) 628-5624
This organization can direct you to women- and minority-owned S&Ls nationwide (about 50, as of this writing).

CANICCOR, An Interfaith Council on Corporate Responsibility
P.O. Box 6819
San Francisco, CA 94101
(415) 885-5102
CANICCOR is a nonprofit interfaith research and organizing agency focusing on social responsibility in the banking industry. Its primary concerns include financial sanctions on South Africa, third world debt crisis issues, and domestic housing lending in low-income areas.

Credit Union National Association
805-15th Street NW
Suite 300
Washington, D.C. 20005
(202) 682-4200
Credit unions are organized by national region under the overall coordination of this national organization. CUNA will advise you on available credit unions in your community and tell you how their common bonds are defined.

National Banker's Association
122 "C" Street NW
Washington, D.C. 20001
(202) 783-3200
Call for direction to women- and minority-owned commercial banks nationwide (about 120 as of this writing).

National Federation of Community Development Credit Unions
59 John Street, 8th Floor
New York, NY 10038
(212) 513-7191 or (800) 437-8711
This umbrella organization provides technical assistance and services to credit unions throughout the country. NFCDCU will help you locate a community development credit union in your community and help you select investments in CDCUs around the country for your personal portfolio.

Community Development Banks

Ameritrust Development Bank
840 Halle Building
1228 Euclid Avenue
Cleveland, OH 44115-1831
(216) 861-6964

Blackfeet National Bank
P.O. Box 730
Browning, MT 59417
(406) 338-7000
This institution, 94 percent owned by the Blackfoot national tribe, focuses on developing the reservation economy. Approximately 90 percent of loans made go to reservation residents.

Community Capital Bank
P.O. Box 404920
Brooklyn, NY 11240
(718) 768-9344 or (800) 827-6699

The Development Bank of Washington
3614-12th Street NE
Washington, D.C. 20017
(202) 832-2865

Elk Horn Bank & Trust
P.O. Box 248
Arkadelphia, AR 71923
(501) 246-5811
An offshoot of South Shore Bank.

South Shore Bank
71st & Jeffery Boulevard
Chicago, IL 60649-2096
(312) 288-7017 or (800) 669-7725

Banking Institutions of Special Note

First Trade Union Savings Bank
10 Drydock Avenue
Boston, MA 02205-9063
(617) 482-4000
The only bank in the country owned by union pension funds.

National Cooperative Bank
1630 Connecticut Avenue NW
Washington, D.C. 20009-1004
(202) 745-4600
Develops financial services for cooperative businesses.

Self-Help Credit Union
Center for Community Self-Help
413 East Chapel Hill Street
Durham, NC 27701
(919) 683-3016
A national leader among community development credit unions, SHCU loaned $2.79 million in mortgages in 1990, of which 92 percent went to minority households and 64 percent to female-headed homes. Of their $2.56 million in commercial loans that year, 44.4 percent went to minority-owned firms, 37.6 percent went to women-owned firms, and 13.5 percent went to co-ops and nonprofit ventures.

Vermont National Bank
P.O. Box 804-C
Brattleboro, VT 05301
(802) 257-7151 or (800) 544-7108, ext. 2414
More a broadly conceived SRI option than specifically a community development bank. Vermont National Bank invites depositors at any of its thirty-one branches to earmark their funds for socially responsible investments. As of December 1990, $45 million of earmarked funds were invested in loans to Vermont-based affordable housing projects, small businesses, agriculture, education, and environmental protection groups.

Women's World Banking
140 East 40th Street
Suite 607
New York, NY 10016
(212) 953-2390
Provides loan guarantees to banks and other financial institutions for the purpose of promoting entrepreneurship of women, particularly those women generally without access to established financial services.

THE TIGER'S TAIL—THE STRATEGY OF SHAREHOLDER ACTIVISM

For investors who really take their activism seriously, one of the most exciting aspects of SRI is the ability to influence a company's behavior from within. When you purchase a share of stock in a company, you become an owner of that company. As an owner, you have the opportunity at least once a year to elect directors to the company's board and to vote on policy resolutions, because corporations are required by law to hold annual meetings of their shareholders. You also have the right to take part in discussions at the meeting. And, under certain conditions, you can even propose resolutions that will be put before all the shareholders.

It is the latter right that is the most potent, but it does have limits. In 1983 the corporate community persuaded the Securities and Exchange Commission (SEC) to restrict access to the resolution process after being annoyed by Ralph Nader's use of the procedure against General Motors in 1970 and by an ensuing decade of shareholder actions in other corporations. To propose resolutions, shareholders must now either own at least 1 percent of shares outstanding or $1,000 of stock for eighteen months before an annual meeting convenes, and there are other procedural limitations as well as content restrictions. However, persistent activists now pool resources to meet the eligibility requirement, and institutional shareholders sometimes buy "protest shares" to provoke corporations into more responsible behavior. It works, particularly when the corporation fears the resolution will generate bad publicity. For example, the Hospital Corporation of America—America's largest hospital chain—asked its hospital staffs to encourage breast feeding and discourage formula use after the issue was raised in a shareholder resolution a few years ago. As of June 1990, about twenty companies had received shareholder resolutions that year, asking that they submit to an annual environmental audit, a thorough public audit of environmental practices; as of this writing, more than 75 percent of those companies had consented.

As a shareholder, you have one vote for every share of stock you own. Normally you will vote by *proxy*—a short "absentee ballot" sent to shareholders before the annual meeting—rather than vote at the meeting itself. Although most shareholders vote according to the board of directors' recommendations (as stated on the proxy), you should still read the resolutions on your proxy carefully and vote your conscience. Even if the resolution has no chance of passing, it can still embarrass the company into constructive action, as with the Hospital Corporation of America.

If you qualify to propose shareholder resolutions for one or more companies, you can become a particularly effective activist by informing progressive organizations whose causes you support that you will submit resolutions for them. In 1987 People for the Ethical Treatment of Animals (PETA), working through friendly shareholders, introduced resolutions at major cosmetics companies' annual meetings demanding the release of animal test data. The companies targeted included Gillette, Bristol-Myers Squibb, Schering-Plough, and Johnson & Johnson. In 1990 Colgate-Palmolive became the first company to make this data public.* The Council on Economic Priorities reports that Colgate-Palmolive is now a corporate leader in substantially reducing its number of animal tests and/or actively researching alternatives.

Resources

Interfaith Center on Corporate Responsibility (ICCR)
475 Riverside Drive, Room 566
New York, NY 10115
(212) 870-2995
ICCR initiates a number of shareholder actions and is a clearinghouse of information on the subject.

HOORAY FOR THE UNDERDOG! INVESTING IN SMALLER COMPANIES

Investments in companies screened for their small *scale* offer, as a group, more ethical consistency than investments in major corporations. They present some of the most intriguing financial possibilities in the investment world as well. Activist investors often prefer these local and other independently owned businesses—as a group, smaller companies tend to have a far better record of socially responsible conduct than corporate giants, who sometimes act as if they are above accountability. Frequently, smaller companies are also purer in an ethical sense than larger corporations because their activities are not so diverse. (For example, industrial giant Boeing is both a leading solar cell producer and a major weapons maker.) In general, supporting small companies supports job creation (nearly all new American jobs these days are created by smaller companies), local economies, and

* Source: PETA.

product diversity and innovation, since most new American products and services originate with smaller companies.

On a strictly financial basis, the smaller companies are far better candidates for rapid growth than their rich corporate counterparts. The large American corporations tend to be bogged down by massive, stagnant bureaucracies. These days they are also notorious for paying far more attention to short-term profits than to research and development. Thus many of the real innovations are coming from small companies run by a few sharp people with good ideas. Apple Computer, Inc., is an example, as is Windham Hill Records, which began in guitarist Will Ackerman's garage and became a pacesetter of the new age music industry.

Two other growth factors favor the smaller companies. Small, well-run companies often confine their production to a single, innovative product. If that product takes off in the marketplace, it can take the company's shareholders for quite a thrilling—and lucrative—ride. Also, because of their low position on the financial food chain, small, well-run companies are logical candidates for acquisition by bigger ones. That is good news (at least financially) for shareholders in the acquired company, whose stock is usually valued at a favorable price.

Despite the potential for large capital gains, investing in small companies does carry considerably more risk than buying blue chip and big. For one thing, the same acquisition that boosts stock value may wipe out its social value. Suppose Tobacco and Firearms, Inc., acquires Oat Bran Industries, in which you own shares, and gives you its stock in exchange. (True story: In early 1990, Carme—whose personal care products under the brands Mill Creek, Sleepy Hollow, and Jojoba Farms are advertised as "cruelty-free" items not tested on animals—agreed to be sold to International Research and Development Corporation, which conducts animal tests for various industries.)

There's another major caveat. Although smaller companies are far more likely to soar than big ones, they are also far more likely to get very sick or even disappear because they do not have the resources to endure disasters. (Who but an Exxon could have survived the *Valdez* catastrophe? The U.S. government bailed out Chrysler—would it bail out Windham Hill?)

Still, from a financial standpoint, the most potentially rewarding investments are nearly always the most risky ones. That goes with the investment territory. (See the discussion of the risk/reward balance in chapter 3.) However, there is a place in many investment portfolios for this type of investment, and we will discuss this topic later in the book.

HOORAY FOR THE GOOD GUYS! HOW GOOD DEEDS TRANSLATE INTO GOOD BUSINESS

The same stuff that makes for a socially responsible company often makes for a highly profitable company as well. Highly principled companies are likely to bring that level of consciousness to every aspect of the operation, including management of the bottom line. In turn, they are less likely to take unethical risks that can culminate in expensive cataclysms. (For example, after a Texas plant explosion killed seventeen workers, ARCO Chemical agreed to pay $3.5 billion in Occupational Safety and Health Administration fines for nearly 350 instances of willful safety regulation violation.*) They also tend to treat their employees more humanely, leading to increased employee productivity and loyalty. And their concern for their impact on society is often reflected in better customer relations and awareness of the needs of the marketplace. Below are some company case histories that show why socially responsible investing is also smart money!

Ben & Jerry's Homemade

Ben and Jerry's story has been told many times, but it's such an exemplary tale that it bears repeating like a mantra. Ben Cohen and Jerry Greenfield, high school buddies and fellow ice-cream lovers, started cranking out ice cream by hand in a converted gas station in Burlington, Vermont, after completing a five-dollar extension course on the ice-cream business. Today their product rivals any in the world for taste and quality, and their projected net sales for fiscal 1991 are a freezer-cool $91.5 million.

However, the real story is how they handle all that money. Cohen and Greenfield—a couple of 1960s counterculture types who still look and act the part—claim that they are in business as much to be a force for social good as to turn profits, and they back it up:

- B & J say they want to be the most charitable corporation in America. They give away 7.5 percent of their pretax profits (mostly to model social change projects and community events); the corporate average in America as of 1989 was 1.7 percent.
- B & J want to be the best employers in Vermont. None of their employees (including Ben and Jerry themselves) can make more

---

* Source: *Insight* 2/91.

than five times what the lowest-paid employee does. This is the most equitable salary structure of any publicly traded company in the country. Ben and Jerry's distributes 5 percent of its pretax profits to its employees and gives them stock equal to 10 percent of their salary as well.

- B & J want to be an agent of social change. B & J's progressive social activities are too many and varied to do them justice in this quick look. Some from a long and constantly expanding list of examples: their product, the Peace Pop, is specifically designated to generate funds for peace organizations through the "1% for Peace" program. Through socially conscious ingredient purchases, their Chocolate Fudge Brownie ice-cream flavor supports the employment of underskilled people; the Rainforest Crunch flavor supports rain-forest preservation; Fresh Georgia Peach ice cream benefits family farms; and Wild Maine Blueberry ice cream benefits Maine's Pas-samaquoddy Indian economy.
- B & J want to inject a sense of fun into both the corporate and social activist universes. Their "1,000 Pints of Light" campaign awards one thousand pints of light ice cream to individuals who have gone out on a limb to help others. (Past winners include Bob Kunst of Cure AIDS Now and Dr. Patch Adams of the Gesundheit Institute, a model health community.)

As you would expect, Ben and Jerry's corporate citizenship has built a most loyal and expanding customer base. And you thought it was great that you could support social causes by investing money? How about producing social change by eating ice cream?

Wellman, Inc.

Just like saving makes profound economic sense, conservation makes profound environmental sense. Because recycling is the obvious an-swer to our glut of technogarbage, Wellman, Inc.—the world's largest recycler of plastic beverage bottles—has a business concept with an unlimited future. They also have a cheap source of raw materials. Wellman does things like turn plastic bottles and photographic film waste into tennis ball felt, carpets, sleeping bag filler, furniture, and blankets. Revenues in 1990 were $827.8 million.

The economics of recycling plastic by itself almost explains Well-man's impressive profit margins and steady growth. Because plastic is petroleum-based, most plastic companies are forced to weather dramatic swings in oil prices. Not Wellman. They are the only major buyer of the used-plastic product, polyethylene terephthalate (PET), that is their material base. With state bottle deposit legislation on the rise, more PET is available than even Wellman can use. What they do use, of course, stays out of the landfills and conserves oil besides.

Wellman employs two other sensible business strategies that keep them strong. They aim to dominate niche markets, and they keep their production facilities flexible so they can provide the marketplace with whichever of their end product fibers are in greatest demand. As they continue to prosper and grow, they are expanding their plant capacity to recycle greater volumes of PET. They also continue to search out new ways to send back to society—in useful form—that which it throws away.

Stride Rite

Stride Rite has been making children's shoes for some seventy years. Talk about an innocent product! More recently they have ventured into the adult athletic- and casual-shoe markets with their Keds and Sperry Top-Sider lines. The company name is synonymous with product quality.

Stride Rite also has its feet on the ground financially. Its profits and return-on-equity are about twice the industry average, according to a late-1990 report by the SRI advisory newsletter, *Clean Yield*. It had a few tough earnings years between 1974 and 1984 when the whole industry struggled under an assault of cheap, foreign footwear. But unlike most of its domestic competitors, Stride Rite resisted the temptation to take its manufacturing overseas. Proving that profits do not depend on exploiting labor, it pays its workers—31 percent of whom are unionized—higher than the industry average, yet its earnings have risen steadily since 1984.

This company does a lot more for the family than put good shoes on its feet. It was the first public company to open an on-site day-care center (1971) and has since added an intergenerational center to address its concern for elder care. It provides strong parental leave and flexible-hour programs as well.

Stride Rite is a nice place to work in other ways, too. It offers incentive pay, aggressively recruits and promotes women and minorities (according to the watchdog agency in its home state of Massachusetts), and was one of the first companies to institute an internal clean air policy (no smoking in common or work areas).

Stride Rite's charitable contributions, 4 percent of pretax earnings, go mainly to scholarship programs, inner-city public service programs, and its own day- and elder-care programs. The word for Stride Rite is *solid*—solid citizen, solid product, solid business, which is why its stock has been a solid favorite with SRI mutual funds for years.

READING BETWEEN THE LINES: THE SUBTLETIES OF SOCIAL INVESTING

Conventional investors mainly read numbers. Numbers are essentially cut and dried; the issues that socially conscious investors consider *with* the numbers are not. Your broker should be able to provide you with all the information you request to help you with the financial aspects of your investment decisions. However, not all ethical decisions are that obvious, even after you have gathered all the available details.

Example: The developers of trash-to-energy plants champion their process as a viable solution to both our nation's landfill shortage and our overdependence on fossil fuels. Of course, "trash to energy" sounds, on first blush anyway, like the essence of good environmentalism. However, in many communities where these projects have been proposed, citizens and environmental groups have resisted them, charging that the burning process releases dangerous pollutants into the atmosphere. By the same token, there are environmentally sensitive citizens who feel that trash-to-energy plants are better than any of the available alternatives to the trash disposal crisis.

You will want to resolve that issue for yourself before buying stock in a plant development company—and consider one other issue besides. These plants normally sell the energy produced to the local power company. You may want to know whether or not that transaction was conducted in a way that does not exploit ratepayers.

The socially conscious investor also needs to consult his/her list of social priorities in selecting an investment because there are few, if any, perfect companies. The SRI strategy is very much like voting for a political candidate because that person represents most of what you stand for even though he or she takes a few positions that are distasteful to you.

A fictitious example: Amalgamated Automobile has an excellent record of hiring minorities (yay!). On the other hand, it has no female executives (boo!). It makes one of the highest-quality automobiles on the road, its cars giving consumers great value for their money and superb fuel economy (yay!). None of its models are equipped with airbags (boo!).

Of course, not all socially conscious investment decisions are this complex. You may find some that are completely straightforward as far as your values are concerned. The key issue here is *priorities*. Do not compromise on those issues of greatest importance to you, but remember that a little bending is required in almost all aspects of life. If you eliminated everything from your diet that was even a little bit tainted by pesticides or high fat or whatever, you could starve to death. There are also few companies—or even social reform organizations, for that matter—that are not a little bit ethically compromised. Still, you and your broker should be able to find socially responsible investments that you feel good about overall, both ethically and financially. (A worksheet to help you assess your ethical priorities is provided in chapter 7.)

A MOVEMENT WHOSE TIME HAS COME: A BRIEF HISTORY OF SOCIALLY CONSCIOUS INVESTING

Most investments are for the long term, so you are wise to wonder if this SRI thing is just some fad that will leave you and your financial future abandoned down the road. Rest assured. Although socially responsible investing has certainly gathered its primary momentum in the last twenty years or so, its roots go back to the early part of this century. In the early 1900s groups of primarily Christian investors were screening their investments with respect to issues that were critical to them, mainly "sin" issues such as alcohol, tobacco, pornography, gambling, and so on. The Pioneer Fund in Boston, founded in 1928, was one fund developed to serve their needs.

In the late sixties and early seventies, more broad-based socially responsible investing became popular as a result of widespread opposition to the Vietnam War. Students across the country began calling for the divestiture of their university's funds in companies that manufactured napalm, herbicides like Agent Orange, and military hardware. In 1970 a group of Methodists organized the Pax World Fund for investors who wanted to be certain that their profits were not the results of weapons production. The Dreyfus Third Century Fund was started two years later, grouping together stocks in companies notable for their sensitivity both to the environment and to the local community in which they operate.

In the few short years since, the number of socially screened investment products has grown to include twelve broadly screened mutual funds (including growth, bond, income, balanced, and environmental sector funds), two broadly screened money market funds, nineteen precious metals funds screened for South African involvement, seven "sin" funds, over thirty SRI mutual funds outside the United States (primarily in the United Kingdom and Canada), many limited partnerships, various ethical venture capital opportunities, and two socially responsible credit cards, with a number of new investment products being authorized and readied for marketing even as we write. Also available are a number of periodical publications that update investors and financial counselors on socially responsible stock investments. Moreover, concerned investors can now consult a steadily increasing number of brokers and financial consultants sensitive to SRI—some of the major brokerage firms, in fact, now have their own SRI divisions.

The divestiture of funds invested in South Africa raised consciousness throughout the business world about both the extensive support for and the power of socially targeted investment funds. With environmental awareness sweeping the country as we enter the nineties, SRI will certainly become an even more potent force for change in society, perhaps as potent as the environmental movement itself.

INVESTING IN YOUR UNCLE (AND COUSINS GINNIE, FREDDIE, FANNIE, AND SALLY): SRI AND THE U.S. GOVERNMENT

Uncle Sam offers you a number of investment vehicles of his own. These vehicles are among the very safest investments because they are backed with the credit of the U.S. government. They can also be attractive in other financial dimensions. However, because our uncle is engaged in many activities that are on most ethical investors' "bad lists," the ethical implications of some of these vehicles are dicey, to say the least. For example, while U.S. Treasury securities (including T-bills, notes, bonds, and related futures contracts) are popular with regular investors, the socially responsible investor normally avoids these because the money is not earmarked; the same general fund supports our national park system, subsidizes tobacco growers, finances higher education, and buys nuclear weapons.

On the other hand, targeted programs like mortgage-backed securities make a great deal of sense in the SRI scheme of things. For instance, Ginnie Maes—so-called because they are issued by the Government National Mortgage Association (GNMA)—group together federally insured mortgages and then offer investors shares. They are safe because the government guarantees them, they pay relatively high interest rates, and they support something with which few would find an ethical quibble—home financing. Uncle Sam offers investments in other agencies that are relatively neutral in ethical terms, such as the Federal National Mortgage Association (Fannie Maes), the Federal Home Loan Mortgage Association (Freddie Macs), and the Federal Home Loan Banks, all of which support home mortgages and construction; the Small Business Administration, which makes loans to small businesses; and the Student Loan Marketing Association (Sally Maes), which enables institutions to make affordable student loans. These investments will be discussed in more detail in chapter 4.

CLOSER TO HOME: MUNICIPAL BONDS

State and local governments (including the governments of U.S. territories) engage in many activities that socially responsible investors support—activities like building schools, roads, public transportation systems, and housing for low-income families and the elderly. These projects are often funded through instruments called *municipal bonds,* which are essentially loans from the bond purchaser to the government agency undertaking the project. Income from municipal bonds is free from federal taxation in most cases; in addition, municipal bond investment is usually tax-free to investors living in the state issuing the bond. You may have heard the term *double tax-free.* This means federal and state tax-free. Local bond issues often have double (or even triple in a few cases, including local) tax-free status.

Because of the trade-off of their tax-free features, municipal bonds offer lower interest rates than other income vehicles. They generally make more financial sense for investors in higher-income tax brackets. From a social standpoint, there is nothing sacred about these instruments, so you do need to pay careful attention to the projects associated with the bonds you are considering. For example, localities also use bonds to finance things like prisons, whereas some socially responsible investors would rather see the money targeted at the root causes of crime.

As of this writing, there is only one socially screened mutual fund concentrating on municipal bonds—and it's targeted for California residents—so you have to examine the bonds held by other funds for yourself. (You can find this information in a fund's semiannual and annual reports.) Do understand, though, that most municipal bonds are socially responsible by definition because they support infrastructure development: again, schools, parks, libraries, and so on. When considering a bond fund, decide if you can live with the mix of bonds included. If not, examine other funds.

Please refer to chapter 4 for more on the financial attributes of municipal bonds.

CLOSER STILL: INVESTING IN COMMUNITY DEVELOPMENT

If you like seeing the "good works" of your social investments up close and personal, you will want to seriously consider the various strategies for investing in local communities. Community development investments serve vital social needs that even socially screened stocks and bonds do not touch. Through direct infusion of capital, they help provide jobs, housing, employment, business loans, and basic human services to those who have been shunted aside by the workings of mainstream economic institutions. In this way, communities of economically disadvantaged citizens are helped to transform themselves from victim economies to self-supporting economies. This, of course, also helps strengthen social relationships, making those communities much more resistant to social problems.

Community development projects address a broad range of social and economic conditions. Some of those conditions are chronic in our culture, like the handicaps that low-income minority citizens face in entering the economic mainstream. Other conditions are more circumstantial, the result of a community being dangerously dependent on a single, and often external, source of economic vitality. When corporations relocate factories, when bad weather destroys farm crops, or when industries slump or fail, thousands of jobs can be lost at a time and whole communities devastated.

We touched on this subject earlier in our discussion of socially responsible banking. Community development banks and credit unions, along with community-minded minority-owned banks, present a number of investment opportunities that either directly or indirectly extend financial services to disadvantaged people and communities.

Community development loan funds (CDLFs) are another important source of loans for community-based projects. Like community development credit unions, CDLFs are grass-roots, nonprofit corporations that attack poverty at the systemic level. CDLFs pool loans from institutional and individual lenders at below market rates and then lend the funds to various projects that meet their community-based goals. CDLFs, also called *revolving loan funds,* have proven to be powerful agents of social change, particularly in the development of low-cost housing alternatives. Like community development banking institutions, CDLFs also help transform the flow of the targeted community's money, so that instead of it draining *out* through exploitative financial "services" (banks that divest in the community, check-cashing outfits, loan sharks, and the like), more of the money generated in the community stays there and circulates. The excellent track record of CDLFs counters the financial myth that money loaned to low-income borrowers is money lost.

CDLFs have also proven to be extremely reliable investments: as of this writing, no CDLF member of the National Association of Community Development Loan Funds (NACDLF) has ever failed to repay its investors. However, the returns to investors are low, by design— investors generally set the terms for their loans, but there is an upper bound (usually the prevailing money market rate). CDLFs receive a lot of their money at below market rates and some at no interest at all. Clearly, if you want to do your social investing without forfeiting profits, CDLFs are not for you. However, if you really want to *see* the results of your socially committed money, you could hardly do better.

Please note that because this is an investment book and CDLFs are not investments in the strict financial sense of market-level instruments, we are referring you to the following resources for more information on this most intriguing strategy:

Resources

National Association of Community Development Loan Funds (NACDLF)
P.O. Box 40085
Philadelphia, PA 19106-5085
(215) 923-4754

Investing from the Heart Appendix Reference: Some current CDLFs are listed in appendix C (page 374), along with other options for high-impact social investing.

A WORLD OF OPPORTUNITY—INTERNATIONAL SRI

The ethical investment movement has begun to cover the earth. The following list is drawn largely from a review of international social investing carried in the SRI periodical *Insight:*

- **United Kingdom.** As of this writing, investors in the U.K. can now select from more than twenty ethical/environmental mutual funds, and supporting structures and data resources have arisen as well. Some recent examples: A British Social Investment Forum, patterned after the American original, formed in fall 1990; social investors can now consult a Green Index of environmentally responsive U.K. companies; Britain's main financial press organ, *The Financial Times,* has commissioned a major report on SRI due for publication in winter 1991; and shareholder activism is beginning to have an impact on U.K. corporate behavior.
- **Germany.** Three large banks now offer socially screened mutual funds to their customers.
- **Australia.** Social investors have three screened funds to choose from.
- **Canada.** Canadians are investing in six different socially screened funds, one of which—the Ethical Growth Fund—has been the country's top performing mutual fund (as of November 1990) for the last three years.
- **New Zealand.** Two loan funds, one with assets in the billions, attract investors who will accept below market rates to fund projects with positive social value.

Resources

EIRIS (The Ethical Investment Research Service)
4.01 Bondway Business Centre
71 Bondway
London, UK SW8 1SQ
Phone: 011 44 1 735 1351
EIRIS, Britain's first social investment research firm, publishes a quarterly newsletter and an annual performance review of Britain's ethical mutual funds.

EthicScan Canada
P.O. Box 165, Postal Station "S"
Toronto, Ontario M5M 4L7
Canada
Phone: (416) 783-6776
EthicScan provides fee-for-service social screening of Canadian firms and publishes a monthly, The Corporate Ethics Monitor, *which covers the field of Canadian social investing.*

Pensions and Investment Research Center
444 Brixton Road
London, UK SW9 8EJ
Phone: 011 44 1 274 4000
PIRC publishes a newsletter reporting legal and financial issues faced by those British public pension funds that socially screen their investments. The newsletter also covers shareholder activism and some company-specific information.

Note that not all international SRI funds accept U.S. investors, and most that do may involve tax complications for American shareholders.

ANYONE CAN PLAY: SOCIALLY RESPONSIBLE CONSUMING AND CREDIT CARDS

Even if you have no investments, you are voting with your dollars every day with the decisions you make regarding purchases, use of credit cards, *and* the stores where you shop. Socially responsible credit cards are covered in chapter 7. Some other strategies that socially conscious consumers should consider:

■ Buying from independently owned local businesses instead of national chain outlets whenever possible. (This supports local economies.)

- Buying organic produce. (Pesticides and other mainstream agriculture practices are responsible for our dwindling soil resources as well as for producing less nutritious, and often toxic, food.)
- Not buying environmentally destructive or nonbiodegradable products like disposable diapers, garden pesticides, fast foods packaged in Styrofoam, plastic trash bags, and the like.
- Buying environmentally positive products such as low- or no-phosphate detergents, rechargeable batteries, low-flow faucet aerators, low-flow shower heads, compact fluorescent light bulbs, recycled paper products, and so on.

For a more complete guide to socially responsible consuming, refer to the bibliography.

Even with socially responsible investing, there is another side to the coin, of course—the *financial* side. If you are new to investing in general, chapters 3, 7, and 8 will give you the foundation you need to start planning your investments. If you feel ready now to start selecting some socially responsible investment products, proceed directly to chapters 4, 5, and 6.

2

Social Screening—A Closer Look

Many—probably most—ethical investors trust the social screening performed by SRI professionals in their behalf without investigating further on their own. And why not let a huge SRI player like, for instance, the Calvert Group conduct your probe? Calvert commands tremendous resources, both in the financial sense and in terms of social expertise. Calvert's Advisory Council is a virtual *Who's Who* of American progressive expertise, including ex-state senator Julian Bond, Sierra Club Executive Director Michael Fischer, author and alternative economist Hazel Henderson, Rocky Mountain Institute Director of Research Amory Lovins, Interfaith Center on Corporate Responsibility Executive Director Timothy Smith, and *Utne Reader* publisher and editor Eric Utne, among others. These people, and their colleagues in the SRI field, not only share most ethical investors' social concerns, but also know the most telling questions to ask a corporation. Just as crucially, Calvert and its companion ethical investment companies have the financial clout to get those questions answered.

Nevertheless, you may wish to know in detail how the professional social screeners see the issues, and you may also be wondering exactly how your pet issues are addressed in the SRI field. A complete chronicling of each major SRI investment company's and publication's social screening process would fill a book in itself. Since many of the social concerns we discuss are shared throughout the SRI industry, what we offer below is a sort of combination plate of the major issues considered and questions asked by SRI professionals in general. We've also provided a list of sources typically consulted by the pros, so if you wish to conduct your own social inquiries, be our (and their) guest.

If you want to maintain files on the companies in which you plan to invest (a swell idea, in our view) or if you want to screen a company not on our lists, start with these basic ingredients:

- **Annual reports.** Companies are required to issue annual reports to all shareholders, detailing their financial status and previous year's performance. The PR-laden tone of most annual reports could put a pretty face on an oil slick. Still, these documents must contain certain numbers and facts, such as legal proceedings against the company. You can usually determine the composition of the company's board of directors from the report as well. Plus, companies proud of their social programs will usually boast about them somewhere in the report's pages.
- **"10Ks" and "10Qs."** In simple terms, 10Ks (annual) and 10Qs (quarterly) are disclosure reports that a company must file with the Securities and Exchange Commission (SEC).
- **Company proxies, prospectuses, and press releases.**
- **News sources, both in the general and business press.**
- **Industry-oriented publications.**
- **Issue-specific newsletters and other periodicals.** See, for example, publications by People for the Ethical Treatment of Animals (PETA). (Many such publications are listed in the "Resources" section in this chapter on an issue-by-issue basis. Others are listed in appendix D.)
- **Issue-specific organizations.** Again, like PETA (listed throughout this chapter under associated issues).
- **Government agency documents.** These are available from such offices as the Environmental Protection Agency (EPA), Occupational Safety and Health Administration (OSHA), Department of Defense (DOD), National Labor Relations Board (NLRB), and so forth.

Rather than do all original source research (a nearly impossible task for a chapter of this scope), we have relied heavily on screening research already generated in the SRI and socially responsible consumer fields. The following sources were particularly useful: "Asking the Ethical Questions of Companies: A guide to *Insight*'s Social Assessment Ratings," a document prepared by the staff of the SRI periodical, *Insight*; various issues of *Insight* and *Clean Yield* (another prominent SRI periodical); literature published by the Calvert Group and New Alternatives Fund, Inc.; and literature from issue-specific organizations such as Co-op America, AFL-CIO, PETA, and 1% for Peace. To quickly expand your social screening "data base," we highly recommend *Shopping for a Better World,* a concise and reliable pocket guide to socially screened companies put out by the Council on Economic Priorities (CEP). One caveat: Because *Shopping* is consumer product–oriented, the companies listed may trade as stocks under different, parent-company names.

Note that most of the questions in this chapter that appear to be in a "yes/no" form are actually designed to generate a range of detailed responses, from which social screeners will typically rate the company on a several-point sale. For example, consider this question: "Does the company have a history of environmental legislation violations as reported by the U.S. Environmental Protection Agency, other governmental bodies, or environmental watchdog organizations?" Not all violators are villainous. An environmentally concerned company could have some history of violations due to inevitable equipment failures and personnel errors. SRI professionals normally consider the company's history in the context of how the rest of its industry performs in the same area of concern.

EVALUATING INVESTMENTS IN SOUTH AFRICA AND OTHER REPRESSIVE REGIMES

South Africa's apartheid system and brutal enforcement of same has provoked more SRI activity than any other ethical issue. We mentioned in chapter 1 that more than $625 billion has been invested with reference to some ethical screen. According to the Interfaith Center on Corporate Responsibility (ICCR), the majority of that money was invested with the single criterion of eliminating companies maintaining business ties to South Africa.

While certainly not attempting to defend apartheid morally, some Americans—particularly in conservative circles—feel that the presence of American companies in South Africa exerts a positive social influence. Their usual argument is that these companies provide jobs and positive working conditions for blacks. President Reagan's policy of "constructive engagement" epitomized this view. Congress disagreed—it imposed partial economic sanctions in 1986.

Most mainstream analysts feel that those sanctions, in combination with sanctions imposed by other nations, worked as intended. The sanctions created economic hardships that hit South Africa's middle class particularly hard. The middle class in turn pressured the De Klerk government for reforms, and De Klerk has turned heads worldwide with the scope of his responses: Nelson Mandela and many other political prisoners have been freed from prison, the ban has been lifted on the African National Congress and other political parties representing the aspirations of the black majority, and the country's segregationist laws have been repealed. Few today would disagree that the end of apartheid is in sight.

Nevertheless, the most important steps remain to be taken. Black South Africans cannot vote, schools and neighborhoods can still exclude blacks through legal loopholes, and the Population Registration Act—requiring all South Africans to be classified by race and thus the core enabling policy of the apartheid system—is still on the books for South Africans born prior to the repeal. The Bush administration and the European community cancelled their countries' sanctions as a means of "rewarding" the reformist measures Pretoria has taken thus far, but few involved in the struggle to dismantle apartheid permanently supported those actions. Besides, dozens of U.S. cities and states retain sanctions independent of the federal ones. Desmond Tutu, Anglican archbishop of Cape Town and the 1984 Nobel Peace Prize winner, wrote in February of 1991:

> Before I can call for sanctions to be lifted:
> - Schools have to be opened to all races . . . under one education ministry.
> - All political prisoners must be freed and exiles allowed home under a general amnesty.
> - The Population Registration Act has to be abolished. . . .
> - A mechanism needs to be established for negotiating a new constitution that is representative of the people of South Africa and does not allow groups defined by race or ethnicity to veto decisions that are democratically reached.
>
> (Los Angeles Times, 2/6/91, B7)

Arguably, none of Tutu's conditions have been met in full, and most socially responsible investors—including state and university pension funds—continue to avoid investments in South Africa.

Given the moral repugnance of apartheid and the effectiveness of sanctions in producing reforms, the SRI considerations regarding South Africa are more complicated than you might first think. One such consideration is the question of *strategic industries*. Some products (such as computers) help maintain the apartheid system directly or impact the economy immediately (transportation and energy products, for example) while others (such as Coca-Cola) have less obvious effects. Some concerned investors avoid investments in strategic industries only.

Other complexities: A company may sell off its South African operations yet maintain licensing, franchising, or vending agreements with either the new buyer or other businesses so that its products can still be sold there. A company may provide financial services to companies doing South African business. A company many be torn between some shareholders' demands that it leave South Africa and its fiduciary responsibility to the shareholders in aggregate to maintain a profitable operation. A company may publish statements about its South African involvement that seem to satisfy your ethical considerations until you untangle its web of parent/subsidiary relationships. For example, Shell Oil, a U.S. company, is factually correct in stating that it has no South African subsidiaries. But "our" Shell is a subsidiary of Royal Dutch Shell, a major player in the South African economy.

Adherence to "the Sullivan Principles" were once vital to a company's SRI social rating, but no longer. In 1977 Philadelphia-based antiapartheid activist Reverend Leon Sullivan developed a model code of conduct for U.S. companies operating in South Africa. The code, which came to be known as the Sullivan Principles, described a series of nondiscriminatory labor practices as well as a minimum wage. Companies operating in South Africa were asked to sign the code as a measure of good faith. Those that signed, and most did, were then rated annually by an outside evaluation agency as to how successfully they had complied. Black leaders within South Africa, however, including the moderate Tutu, decried the code as a way of making apartheid "comfortable." Ten years later Sullivan agreed, withdrawing his support from the code that bore his name. Sullivan's prescription was renamed the Statement of Principles, and companies are still rated for compliance to them, but most people in the SRI community have deemphasized the meaningfulness of a positive rating.

Many ethical investors simplify the plethora of questions emanating from these gray areas by avoiding investments in companies with any South African connections. There is good justification for such a stance: no matter how you finesse the issues, all companies doing business in South Africa pay heavy taxes there that support the status quo. If you do wish to get into specifics, refer to the questions that follow to guide your inquiry.

South Africa is the only repressive regime that is overtly addressed by the SRI industry. Nevertheless, the major SRI mutual funds do attempt to track U.S. corporate relationships with other repressive regimes through reports by organizations such as the Interfaith Center on Corporate Responsibility (ICCR—see "Resources," p. 41).

One international political situation that American ethical investors have tracked with some diligence is religion-based employment discrimination in Northern Ireland. Unemployment in Northern Ireland is the highest in the United Kingdom, about 16 percent in November 1988 compared with 7.5 percent in the U.K. overall. However, the discriminated-against Catholic minority suffers the British province's economic woes disproportionately—about 25 percent of Catholics were unemployed when the above-cited overall figures were compiled. With the United States being the largest foreign investor in the province, some activist shareholder groups have pressed U.S. companies there to adopt the MacBride Principles, a set of fair employment principles that speak directly to the religious discrimination issue.

If you should be offered an investment opportunity that has international implications and are unsure of the political ramifications, another excellent source of information besides ICCR is Amnesty International, the highly respected watchdog for international human rights. Contact information for Amnesty International is also provided at the end of this section.

Questions to Ask

Does the company operate in South Africa?

If yes, what percentage of its employees are blacks?

What percentage of its management are blacks?

Is the company planning to expand its South African operations?

Is the company part of a strategic industry (computers, energy, transportation) that supports either the apartheid system or the South African economy in general?

Does the company sell to the South African police or military?

If the company has sold its South African assets, does it maintain any licensing, franchising, or vending agreements to sell its products there?

Does the company have a policy regarding South African involvement?

Is the company a signatory to the Statement of Principles? If yes, what is its rating?

What is the company's labor relations record in South Africa?

Does the company operate in other countries ruled by repressive regimes? If yes, what is the nature of its involvement?

Resources

The Africa Fund
198 Broadway
New York, NY 10038
(212) 962-1210
Publishes the Unified List of U.S. Companies Doing Business in South Africa and Namibia.

American Committee on Africa
198 Broadway
New York, NY 10038
(212) 962-1210

Amnesty International
322 Eighth Avenue
New York, NY 10001
(212) 807-8400

Interfaith Center on Corporate Responsibility (ICCR)
475 Riverside Drive, Room 566
New York, NY 10155
(212) 870-2296
Publishes The Corporate Examiner, *a monthly newsletter examining corporate policies and practices with regard to a broad range of social issues, including South Africa. Also publishes* Church Proxy Resolutions, *which covers shareholder resolutions requesting the targeted company's withdrawal from South Africa and other responsible corporate behavior.*

Investor Responsibility Research Center
1755 Massachusetts Avenue NW
Suite 600
Washington, D.C. 20036
(202) 939-6500
Publishes the South Africa Review Service.

Investing from the Heart Appendix Reference: See appendix A under "South Africa," page 347.

EQUAL OPPORTUNITY EMPLOYMENT

When you vote with your investment dollars for equal opportunity employment, you are supporting "progress" more than true equality. Then again, progress is your only option—if you hold out for absolute standards in this area, your money will sit idle for a long time. Although 75 percent of this decade's entrants into the labor force are expected to be women or minorities (according to U.S. Department of Labor projections), few women or minority group members hold top management positions today in even America's most progressive-minded corporations.

If you can accept relative gains for the time being, however, you will notice that some companies are actively engaged in at least partially redressing the balance, and some proceed with their business as if there were no balance to be redressed. The kinds of things that companies can do, and are doing, to rectify past inequities include

- seeking appropriate representation of women and minorities on their board of directors and in top management positions. The Council on Economic Priorities requires a company to have at least two women on its board or among its top officers to earn a CEP middle rating, and at least three to earn a top grade. Because minorities as a group constitute about half as much of the U.S. population as do women, one minority member on the board or among top management earns a middle grade; two earns a top grade. Pitney Bowes is an outstanding example, with three women (one of whom is a minority) on its eleven-member board as of November 1990. As of that same date, 22 percent of Pitney Bowes's upper management are women and 8.5 percent are minorities.*
- running programs to encourage the advancement of qualified women and minorities. CEP recognizes companies such as Avon and General Mills for their mentoring programs, advertising in women's and minority publications, and regular reviews of personnel managers' records with respect to hiring women and minorities.
- emphasizing purchasing from businesses run by women and minorities.
- keeping funds in banks owned by women and minorities or in banks (such as South Shore Bank in Chicago) that serve low-income com-

* *Clean Yield,* 11/90.

munities devoid of or deficient in traditional banking services. CEP notes that Quaker Oats did $150 million of banking with minority banks in 1989.

Equal opportunity, of course, applies to other population groupings besides women and racial/ethnic populations. Workplace discrimination also occurs on the basis of age, sexual orientation, religious belief, and physical condition (weight, physical handicap, or HIV infection). Although you will find the occasional company doing something positive in one or more of these subcategories (such as McDonald's well-publicized program of hiring the elderly), SRI social screens do not generally address these issues. Not for a lack of caring, however—what is usually lacking is either data or the presence of any mature remedial activity to measure. Unfortunately, public awareness in these areas lags significantly behind awareness of the narrowed opportunities for women and minorities, just as with minority issues there is more data available on African-Americans than on other minorities. (One exception: the Citizens Commission on AIDS has developed "Ten Principles for the Workplace," concerning confidentiality, nondiscriminatory hiring, prohibition of mandatory HIV testing, and workplace sponsorship of educational activities. As of the most recent survey [July 15, 1989], seventy-three companies have endorsed the principles.)

This does not mean that you can't screen for such criteria on your own. By combing your newspaper in February and March of 1990, for example, you might have discovered that gay activist groups were accusing Cracker Barrel restaurants, a chain in the Deep South, of covertly continuing an antigay hiring policy that it had retracted publicly, and that Carl's Jr., a fast-food chain in the West, is the object of protests because CEO Karl Karcher has supported homophobic politicians and ballot initiatives.*

Questions to Ask

Is the work force representative of the surrounding population?

Are women and minorities adequately represented on the board of directors and in top management positions?

If yes, is their representation token or an accurate indicator of promotion opportunities for women and minorities within the company?

* *Los Angeles Times*, 3/22/91, A52.

If the company is located in an area where the pool of qualified minorities or women is limited, does it recruit women or minorities for its board from outside the geographical area?

Does the company operate advancement programs for women and minorities?

Does the company provide equal pay for equal work?

Does the company allow the public to inspect its Equal Employment Opportunity Commission (EEOC) records?

Does the company publish an antidiscrimination statement?

Does the company provide family leave, maternity leave, flexible work schedules, day care, and other benefits for working mothers (and working parents in general)?

Does the company provide elder care?

Does the company have a policy against sexual harassment? If so, how does it enforce it?

Does the company protect the privacy of those infected with HIV?

Does the company discriminate against those infected with HIV?

Has the company signed the Citizens Commission to AIDS's Ten Principles for the Workplace?

Has the company signed an Operation Fair Share agreement? (The National Association for the Advancement of Colored People [NAACP], through its Operation Fair Share program, asks targeted companies to sign voluntary agreements to hire and promote blacks, purchase goods and services from minority-owned businesses, and make charitable contributions to black causes.)

Has the company been boycotted for discriminatory practice, and if so, how has it responded?

Resources

Black Enterprise Magazine
The Earl G. Graves Publishing Co., Inc.
130 Fifth Avenue
New York, NY 10011

Citizens Commission on AIDS
121 Avenue of the Americas, 6th Floor
New York, NY 10013
(212) 925-5290

Coalition of Labor Union Women
15 Union Square West
New York, NY 10003
(212) 242-0700

Council on Economic Priorities
30 Irving Place
New York, NY 10003
(800) 822-6435

Families and Work Institute
330 Seventh Avenue
New York, NY 10001
(212) 465-2044

NAACP Fair Share Program
586 Central Avenue South
Suite 10-14
East Orange, NJ 07019
(201) 235-3575

National Community AIDS
Partnership
1726 M Street NW
Washington, D.C. 20036
(202) 429-2820

National Gay & Lesbian Task
Force
1734 Fourteenth Street NW
Washington, D.C. 20009
(202) 332-6483

National Leadership Coalition
on AIDS
1150 Seventeenth Street NW
Washington, D.C. 20036
(202) 429-0930

National Minority Supplier
Development Council
1412 Broadway, 11th Floor
New York, NY 10018
(212) 944-2430

Corporate annual reports and other publications.

Investing from the Heart Appendix Reference: See appendix A under "Labor and Equal Opportunity," page 349.

THE ENVIRONMENT

A 1960s slogan—"If you're not part of the solution, you're part of the problem"—has never seemed more apt than when applied to environmental crises in the 1990s. Our planet's environment has become so severely compromised in so many ways that the long-term survival of human life here is no longer assured. Businesses and consumers must be willing to make radical changes of habit—regardless of cost—or bear responsibility for the destruction. In the following discussion, we highlight some of the general areas of concern and summarize the SRI industry's approach to addressing them.

Ozone Depletion

The ozone layer in our atmosphere, which protects the earth's surface from the damaging effects of solar ultraviolet radiation, is being broken down by chlorofluorocarbons (CFCs) and other human-made chemical gases released into the atmosphere through industrial and household usages. The effects include dramatic increases in the growth of skin cancers and cataracts and destruction of crops and fish populations. CFCs are used as coolants for refrigeration and air-conditioning and as propellants in aerosol sprays. They are also a primary ingredient of sprays used to clean VCRs, computers, and photographic equipment. Unfortunately, no adequate substitute chemical has yet been developed. Several corporations, including Du Pont, have committed tremendous resources to develop replacement chemicals. However, Greenpeace and others have been adamant in pointing out that Du Pont's new chemicals—called HFCs and HCFCs—will still eat ozone molecules, only at a lesser rate than CFCs.*

Global Warming

Global warming may be responsible for the record warm seasons and droughts that have plagued much of the world over the last several years. Life on earth depends upon a delicate temperature balance. This balance is maintained by a blanket of natural gases that allows sunlight to penetrate but traps sufficient heat from escaping to support life. (The gas blanket functions much like the glass in a greenhouse, hence the term *greenhouse effect*.) After decades—actually centuries—of accumulation, industrial gases (primarily carbon dioxide, but also CFCs and other gases) have thickened that blanket, trapping additional heat.

Scientific opinion remains divided over whether the greenhouse effect is responsible for all of the climate abberations popularly attributed to it. However, there is scientific near consensus internationally that it is now high time to take radical preventative steps. In fact, many experts predict that unless corrective action is taken immediately, global warming will raise the earth's temperature several degrees over the next several decades. The climatic effects and effects on various life forms could be catastrophic, threatening much of the earth's food supply and submerging low-lying island nations among myriad other consequences. The primary source of carbon dioxide emissions is the burning of fossil fuels, mainly coal, oil, and natural gas; the primary culprits are factories, automobiles, and electric utilities. Even such consumer conveniences as gas lawn mowers and barbecues exacerbate the problem, which is why some experts predict that both items will be outlawed by decade's end.

* *Greenpeace,* March/April 1991.

Acid Rain

Acid rain, caused by gaseous pollutants transforming chemically in the atmosphere and falling back to earth as acidified rain and snow, is another unfortunate by-product of our dependence on fossil fuel–burning vehicles and power plants. Acid rain damages our vanishing forests and destroys plant and animal life in streams. It even erodes buildings. Clearly, alternative energy development would do more to reduce acid rain (and global warming) than any other industrial response, with more efficient fuel burning and emissions control important secondary strategies.

Toxic Waste

The dangers of industrial toxic waste first gained widespread public attention in the late 1970s when citizens of upstate New York's Love Canal community began, with alarming frequency, to report cancers, birth defects, and spontaneous abortions. The New York State Health Department investigated and found to their horror that massive amounts of toxic chemicals buried beneath the housing development twenty-five years earlier were now leaking from their containers. Most of the community's nine hundred–some families have since abandoned their homes there, and Occidental Petroleum—whose Hooker Chemical subsidiary had buried the waste years before Occidental acquired the company—has settled between $5 and $6 million in Love Canal–related claims.*

The Love Canal episode may be just the tip of a toxic iceberg. Other communities sitting atop waste dumps are facing similar threats, and industries have dumped toxic chemicals in vast quantities directly into rivers, bays, lakes, and oceans. For instance, a congressional Office of Technology Assessment report issued in 1986 estimated that more than three thousand tons of toxic metal wastes entered the Chesapeake Bay annually from companies in Virginia and Maryland. The same report mentioned that commercial catches of striped bass there had fallen from six million pounds in 1970 to six hundred thousand pounds in 1983, in spite of increased environmental regulation over that time span. In 1987 U.S. industries reported dumping 9.7 billion pounds of

---

* Lydenberg, et al. *Rating America's Corporate Conscience*. Addison-Wesley, 1986, p. 322.

toxic chemicals into surface waters.* Of course, much of this type of activity took place before the catastrophic consequences were realized. Several of the early industrial polluters have made sincere efforts since then to find more sensible ways of dealing with their hazardous by-products; others have continued to dump toxics—illegally, in some cases—back into the natural environment.

Industry is not solely to blame, of course. We consumers have become dependent on petrochemical and other industrial products (plastics, detergents, aerosol sprays, pesticides, and so on) whose manufacture produces some of the most pernicious of these wastes. We also do our share of toxic dumping when we pour noxious chemicals down drains or discard them with our household trash. The wastes find their way into groundwater and the ground itself through drains, land-fill seepage, and sewer sludge.

The recognition of the toxics problem has spawned a growth industry of companies that specialize in handling it. Thermo Instruments, a favorite investment of the New Alternatives environmental mutual fund, produces instruments that measure pollution and waste emissions, including nuclear waste. The stocks of Sanifill, a leading operator of nonhazardous landfills, and Midwesco Filter Resource, which produces filter bags for industrial machinery, are both favored by Schield Progressive Environmental Fund, another SRI fund.†

Unfortunately, some of the biggest companies in the waste disposal business—notably Browning-Ferris and Waste Management—have shown themselves to be serious ethical violators and are off the investment lists of most SRI advisers and funds. For starters, Waste Management and Browning-Ferris have been targeted in class action suits regarding price fixing relative to their containerized waste disposal. Waste Management has been lambasted by Greenpeace International and other environmentalists for a subsidiary's decade-long incinerating of toxic waste off the Dutch coast. Both companies have shown leadership in solving some of our country's most perplexing waste disposal problems, and each operates outstanding programs for collecting household recyclables. But their good works do not excuse their ethical lapses, in the view of most SRI professionals.

---

* Cohen and O'Conner. *Fighting Toxics.* Island Press, 1990, pp. 189 and 277.
† Source: *Insight,* 2/91.

Air Pollution

In a word, air pollution means "smog." The primary component of smog is ozone (formed when nitrogen oxide and hydrocarbons combine in sunlight). Yes, that same atmospheric gas that protects us from the sun's ultraviolet radiation does us no good at all when it is formed at ground level. The major sources of polluting hydrocarbons and nitrogen oxide include motor vehicles, utilities, and oil and chemical plants. A recent study suggested that children raised in the smog-choked Los Angeles area develop permanently compromised lung capacities. Smog is also thought to be responsible for extensive damage to pine trees and agricultural crops.

It is easy to point a finger at the industrial causes of smog, but it is to industry that we will turn for the answers as well. Some possible industrial solutions to smog become apparent simply from describing the problem. Lighter vehicles that also run more efficiently would yield greater fuel efficiency and spew fewer emissions into the air. Reformulated gasolines and improved catalytic converters would also help. Alternative fuels are another promising—albeit partial—solution. Ethanol and natural gas burn more cleanly than gasoline. The solution we (the authors) like best—electric cars recharged by solar—is the most utopian-sounding, yet it is apparently technologically feasible within the very near future. (Meeting a market demand created by tougher clean air standards, the major automakers are developing electric cars—although not necessarily the solar-charged variety—for use this decade.)

If utilities relied more freely upon alternative energy generation to replace the burning of carbon-based fuels, we would see significant improvement in our air quality. Hydroelectric and geothermal are two alternative technologies with long histories of successful use. Reformulated fuels constitute another promising strategy—innovators Energen and Burlington Resources mix methane gas from unused coal seams with natural gas for delivery to utilities, thus reducing the quantities of natural gas, a carbon-based product, burned. (Natural gas, by the way, is the cleanest-burning of the petroleum fuels, and some SRI professionals will invest in utilities that emphasize it over dirtier or nuclear sources.) Solar—specifically photovoltaic cells, which convert the sun's rays to electricity, and solar thermal, which creates steam to drive electric-producing turbines—again emerges as the most sensible approach because the source is both clean and inexhaustible. Technologists at Luz International Ltd., the world's primary developer of solar thermal power, assert that the primary power needs of the entire nation (730,000 megawatts) could be met by covering a patch of American desert seventy-eight by seventy-eight miles with a solar thermal field. Energy Conversion Devices is a promising company that produces photovoltaic cells. Boeing is another major photovoltaic producer, although off many SRI lists because of its heavy defense involvements.

Energy conservation is a vital component of any comprehensive clean air strategy. Improved building insulation, more durable and efficient light bulbs, and more efficient electric motors are industrial products that help in this regard. Apogee, which produces energy-efficient glass, is a strong company whose stock is held by many ethical investors. Cogeneration technology enables heavy-energy consumers like hospitals to use the same fuel and equipment for simultaneously producing steam heat, hot water, sterilization, and electricity. Tecogen is a leader in the cogeneration industry, and its stock is one of those favored by New Alternatives Fund.

Waste Disposal

Our nation is choking on its garbage. Many, if not most, of our big cities are running out of landfill space. Much of the problem could be alleviated with a basic two-step strategy: generate less garbage in the first place, and recycle to the greatest degree possible what we do generate. Wellman, a company popular with ethical investors, specializes in recycling nonbiodegradable plastic from soft-drink and other beverage bottles. Instead of eternally reposing in landfills, this plastic begins life anew as mattress stuffing, tennis ball covers, and carpeting.

Not every industrial product or process touted as a landfill crisis solution will pass muster with serious environmentalists. For example, Archer Daniels Midland (overall a favorite of the SRI mutual fund field for its use of cogeneration and other energy-efficient industrial processes and for its development of the gasoline substitute ethanol) produces what they call "biodegradable plastics" by mixing corn starches with the nondegradable petrochemical product. Trash bags and disposable diapers—the latter occupying an estimated 3 percent of landfill space by itself—are among the products available now in so-called biodegradable versions. Unfortunately the plastic part of these products—like any other plastic—does not really degrade; instead it enters the ecosystem as a fine sand after the starch holding it together is eaten by bacteria. Many environmentalists fear this sand will present more of a pollution problem than the regular massed plastic it replaces. The other problem with justifying the use of plastic trash bags and disposable diapers is that not even biodegradable products break down fully in landfills because there is not enough air in those tightly compressed plateaus to permit biodegradation in the first place.

Trash-to-energy plants represent another controversial waste-disposal process. These facilities burn garbage and then sell the energy produced to electric utilities. However, the procedure is far from pollution-free. Toxins are released into the air as ash and can also settle into aboveground water sources. Some companies do burn cleaner than others, of course. New Alternatives Fund invests in those that (it considers) burn more cleanly and are otherwise committed to recycling strategies.

Clean Water

We've already discussed water pollution above in relation to acid rain, toxic waste, and air pollution. Each of these pollution problems is in large measure a result of our society's addiction to petroleum consumption. Pesticide (a petrochemical product) and fertilizer run-offs from nonorganic agricultural methods also poison our waterways, killing fish and river animals and contaminating our drinking water. The Exxon *Valdez* crisis and the spills during the Persian Gulf war illustrate another huge reason why dependence on petroleum products is both irresponsible energy and environmental policy. Accidents happen, and the environmental consequences of oil spills are catastrophic.

Endangered or Overexploited Resources (Including Endangered Animal Species)

Examples of resource exploitation include the destruction of rain forests and the harvesting of old-growth forests and other endangered woods such as teak, koa, and zebrawood.

The overharvesting of ancient forests and the horrifically expanding lists of endangered and extinct animal species are not isolated events. According to Peter Warshall, writing in *Whole Earth Review* (fall 1989), the battle to save "almost every major U.S. endangered species (the Mt. Graham spruce squirrel, the red cockaded woodpecker, the spotted owl) is really a battle over old-growth forest or rivers." So severely have ancient forests—including, among other species, Douglas fir, hemlock, and western red cedar—been logged in the Pacific Northwest that only isolated patches of national forest remain in many regions. Species such as the northern spotted owl seem to require larger home patches of forest than are being left to them.

The negative impact for other wildlife comes not so much from the timber cutting itself, but as a result of the roads built to facilitate the cutting. A Wilderness Society report states that "wildlife disruption, vehicle collisions with wildlife, hunting, and poaching along roads are major factors affecting wildlife populations."* The staggering rate of species disappearance in the Amazon rain forest is even more damning evidence of the catastrophic effects of deforestation and wilderness road building. Over half the world's animal and plant species are native to rain forests, which are being cut down at an estimated one hundred acres per *minute*.†

In the United States, the key to this destruction has been cooperative arrangements between lumber companies and the U.S. Bureau of Land Management (BLM), particularly in the timber-rich Pacific Northwest. BLM has given the companies timber rights at fire-sale prices—a loss to taxpayers, by Wilderness Society estimates, of hundreds of millions of dollars per year. The lumber industry rallies support for its cause by claiming that aggressive harvesting of the national forests is good for employment. However, Jeff DeBonis—a U.S. Forest Service timber sales planner assigned to Oregon's Willamette National Forest who is trying to reform the service from within—points out that "between 1979 and 1989 the timber harvest on federal lands in Oregon increased 18.9 percent. In that same period, employment in the wood products industry dropped 15 percent." Why? Because, as DeBonis points out, at the same time timber companies are appealing to labor to join their fight against the conservationists, they are automating mills to eliminate as many jobs as possible.

Lumber companies are not the only culprits in the deforestation crisis. Much of the Brazilian rain forest has been cleared to pasture beef cattle. Much of the beef in turn is sold to U.S. companies at low prices and ends up as fast-food hamburgers and the like. The Asian rain forest is under attack largely for the value of its exotic woods to the furniture industry, including to a large extent U.S. manufacturers. (Rosewood, teak, mahogany, and ebony are rain forest woods.) Urban sprawl is also responsible for much destruction of animal habitats, one

---

* *Wilderness,* spring 1991.
† Earthworks Group. *50 Simple Things You Can Do to Save the Earth.* Earthworks Press, 1989.

reason many ethical investors avoid most real estate investment trusts (see page 162).

Investing in companies that promote vegetarian products and paper recycling are a few ways of having some positive impact on deforestation. Ben & Jerry's Homemade (who else?) has taken a creative approach to saving Brazilian rain forest. Their Rainforest Crunch ice cream uses nuts from rain forest trees in an effort to make saving the trees as economically viable as cutting them down. Redken is a recent hero to the animal rights community—it used knowledge gained from making its hair-care products to help restore the petroleum-damaged coats of animals mucked over by the *Valdez* spill.

Nuclear Energy

Most SRI investment products are screened for involvement with nuclear energy. The reasons are not exclusively environmental, because nuclear power plants as a group have also proven to be financial black holes for the utilities that built them.

Environmentally, nuclear power plants produce hazardous wastes that nobody wants in their backyard, and a serious nuclear accident can wipe out whole population centers and spread radioactivity worldwide. The champions of nuclear power, including many scientists and technologists, feel that all its bugs are correctable. In a perfect world this may be the case, but it's the real-world nuclear history—stained by Three Mile Island, Chernobyl, and numerous near misses—that worries the SRI community and other nuclear skeptics. You wouldn't want to hear your airline pilot say, "Oops! Well, nobody's perfect," at thirty thousand feet. Nor would you ever want to hear it from a nuclear plant engineer. Nevertheless, safe operation of a nuclear facility requires near perfect operation of a most complicated system by very imperfect individuals and companies. It must also be stated frankly that many ethical investors distrust the politics behind the federal government's continued support of nuclear power, particularly considering the plethora of safe, renewable energy sources that continue to receive only minimal federal attention.

The Valdez Principles

The ethical investment community has responded to the multitude of environmental issues with environmental screening of broadly social investments; with specific environmental investment vehicles such as the New Alternatives Fund, which targets socially affirmative investments in alternative energy, resource recovery, and conservation; and with investment-backed activist strategies. The *Valdez* Principles exemplify the latter. Developed in 1989 by Joan Bavaria, president of Franklin Research and Development (which publishes the SRI newsletter *Insight*), the principles propose a standard by which investors—as well as consumers and employees—can measure a company's responsibility toward the environment. The initial corporate signatories were announced on Earth Day 1990. Bavaria and her colleagues have also spawned campaigns to spread the principles' influence through shareholder resolutions (specifically, asking companies to report on how their policies conform to the principles); divestment (petitioning pension funds to invest only in *Valdez* Principles signatories); procurement (petitioning universities and other large purchasers to favor principles signatories in their purchasing); and employment (petitioning universities to require companies recruiting on campus to inform students if they are signatories).

In sum, *Valdez* Principles signatories pledge to

- minimize the release of environmentally damaging pollutants.
- safeguard wildlife habitats.
- minimize their contribution to global warming, ozone depletion, acid rain, and smog.
- make sustainable use of renewable natural resources such as soil, water, and forests and conserve nonrenewable resources.
- preserve open space, wilderness, and biodiversity.
- minimize waste creation, especially hazardous waste; recycle where possible; and dispose of wastes responsibly.
- use environmentally safe and sustainable energy sources where possible.
- invest in improved energy efficiency and conservation in their operations.
- maximize the energy efficiency of their products.
- minimize the environmental, health, and safety risks to their employees and the community in which they operate by using safe technologies and procedures and by being prepared for emergencies.
- sell products and services that minimize adverse environmental impacts and inform consumers of those impacts.
- restore to whatever extent possible aspects of the environment damaged by their operations and compensate those adversely impacted.

- disclose to employees and the public any aspects of the operation that threaten, and specific incidents that cause, environmental damage or health and safety hazards.
- take no action against employees who report such conditions or incidents.
- commit management resources to implement the principles and monitor and publicly report the company's compliance with them on an annual basis. Such an audit is to include a report of compliance with applicable laws and regulations throughout the company's worldwide operations.
- commit the board of directors and chief operations officer to responsibility for the company's environmental record, including appointment to the board of a member qualified to review that performance.

If you wish to join one of the above-mentioned *Valdez* campaigns or otherwise desire more information on the principles, contact the Coalition for Environmentally Responsible Economies (CERES): 711 Atlantic Avenue, Boston, MA 02111; (617) 451-0927. CERES members include public-employee retirement funds in California and New York, the AFL-CIO, the Interfaith Center on Corporate Responsibility, the Sierra Club, the National Wildlife Federation, and the National Audubon Society. Pension and mutual fund managers within the coalition control millions of shares of major company stocks. The group expects *Valdez* Principles–related shareholder resolutions to appear on the ballots of fifty-four different companies in 1991, including those of IBM, Eastman Kodak, General Electric, General Motors, and McDonald's.*

Environmental Investments

Environmentally concerned investors should carefully review any so-called environmental investments offered to them through advertisements or conventional brokerages. Between 80 and 90 percent of Americans (depending on which poll you read) support environmental protection. With that degree of public support, no business (or politician) can go wrong branding its product as environmentally positive. Much in the spirit of George Bush promoting himself as "the environmental president," the *non*-SRI mutual fund industry in particular

* Source: *Wall Street Journal*, 3/25/91.

has jumped on the bandwagon by offering a host of "environmental funds" that hold the stocks and bonds of major polluters and other investments that would not pass SRI environmental screens. We recommend only two environmental mutual funds, New Alternatives and Schield Progressive Environmental. These investments are reviewed in the chapter on mutual funds.

Questions to Ask

Is the company a signatory to the *Valdez* Principles, and if so, what is its compliance rating? (Note: As of this writing, there are few signatories to the principles, and many of those are not in environmentally strategic industries. Therefore, SRI professionals do not heavily weigh answers to questions regarding the principles in their overall evaluation of a company's environmental standing.)

Do the company's products or services promote energy conservation or greater energy dependence?

Does the company package its products in recyclable containers?

Does the company promote recycling and energy conservation on its own premises?

Does the company have a history of environmental legislation violations as reported by the U.S. Environmental Protection Agency, other governmental bodies, or environmental watchdog organizations? (The occasional slipup is less a concern than consistent violations or a poor standing in relation to others in the same industry.)

If comparative environmental studies of its industry have been conducted, how does the company stand in relation to its competitors?

Does the company have any mechanism for monitoring its own environmental performance?

Is the company monitoring its environmental performance at the board or CEO level?

Does the company use or manufacture products containing CFCs? If yes, has it committed to phasing out those products as soon as an acceptable substitute is found? If the company produces CFCs, is it committed to developing an acceptable substitute?

Has the company moved to reduce emissions such as nitrogen oxide and sulfur dioxide, which contribute to smog and acid rain formation?

What steps has the company taken to reduce or recycle the wastes it produces?

If the company has used Styrofoam packaging in the past, is it moving to replace that packaging with a more environmental-friendly product?

Is the company involved in any way with the production of nuclear energy?

Is the company involved in any way with the production of energy from alternative energy sources?

Has the company actively promoted alternative energy, energy conservation, and efficient use of energy?

Has the company educated its employees and customers on energy efficiency?

Has the company consistently opposed conservation, development of alternative energy, or other environmentally positive measures?

Does the company make contributions to nonprofit environmental protection organizations?

Resources

American Solar Energy Society
2400 Central Avenue   Unit B-1
Boulder, CO 80301
(303) 443-3130

The American Wind Energy Association
777 North Capital Street NE
Washington, D.C. 20005
(202) 408-8988

Center for Science in the Public Interest
1875 Connecticut Avenue NW
Washington, D.C. 20009
(202) 332-9110

Citizens Clearinghouse for Hazardous Waste
P.O. Box 926
Arlington, VA 22216
(703) 276-7070

Citizens Fund
1300 Connecticut Avenue NW
Washington, D.C. 20036
(202) 857-5168

Conservation and Renewable Energy Inquiry/Referral Service
P.O. Box 8900
Silver Spring, MD 20907
(800) 523-2929

The Council on Economic Priorities
(See page 44.)

Earth Island Institute
300 Broadway
Suite 28
San Francisco, CA 94133
(415) 788-3666

Environmental Defense Fund
257 Park Avenue South
New York, NY 10010
(800) 225-5333

Environmental Protection Agency data bases, documents, and publications. *Begin your search for these in the government documents section of a large public or university library.*

Greenpeace
1436 U Street NW
Washington, D.C. 20009
(202) 462-1177
Publishes the bimonthly magazine
Greenpeace.

INFORM Environmental Research & Education
381 Park Avenue
New York, NY 10016
(212) 689-4040

Inner Voice
P.O. Box 45
Visa, OR 97488
Jeff DeBonis's newsletter—see page 52.

National Toxics Campaign
37 Temple Place, 4th Floor
Boston, MA 02111
(617) 482-1477

Nuclear News
555 N. Kensington Avenue
La Grange Park, IL 60525
(708) 352-6611

Public Citizen
Critical Mass Energy Project
215 Pennsylvania Avenue SE
Washington, D.C. 20003
(202) 546-4996
Source of publications on renewable energy sources and technologies.

Rainforest Action Network
301 Broadway Suite A
San Francisco, CA 94133
(415) 398-4404

Renew America
1400 Sixteenth Street NW, #710
Washington, D.C. 20036
(202) 232-2252

Rocky Mountain Institute
1739 Snowmass Creek Road
Snowmass, CO 81654
(303) 927-3128
Research and education organization devoted to fostering efficient and sustainable research usage as a path to national security.

The Sierra Club
730 Polk Street
San Francisco, CA 94009
(415) 776-2211

Solar Energy Industries Association
777 North Capitol Street NE
Suite 805
Washington, D.C. 20005
(202) 408-0660
Publishes a newsletter and guides to the photovoltaic and solar thermal industries.

The Wilderness Society
900 Seventeenth Street NW
Washington, D.C. 20006-2596
(202) 833-2300
Publishes the quarterly magazine Wilderness.

Worldwatch Institute
1776 Massachusetts Avenue NW
Washington, D.C. 20036
(202) 452-1999
Reports on progress towards a sustainable society.

Investing from the Heart Appendix Reference: See appendix A under "The Environment," page 358.

WEAPONS INVOLVEMENT

Many ethical investors avoid investments tied to weapons production as a matter or course, as do most broadly screened SRI mutual funds. However, the weapons issue is subtler than you might first imagine. All but the most extreme peace activists favor arms *reduction,* not *elimination.* Thus, as Elizabeth Judd asks in her book, *Investing with a Social Conscience,* is it really consistent for investors with those sentiments to exclude defense-related investments altogether?

The answer, as Judd herself hints, is a qualified yes—because of that complication first made famous by President Eisenhower: the *military-industrial complex.* This term refers to the cooperative arrangement among the military, the industrial producers of military supplies and weaponry, and political leaders beholden to these industries. Defense contractors obviously have economic incentives for marketing their products in the name of national security. High-level military officers placed in charge of single-product projects—such as a B-2 bomber, for example—often find their careers uncomfortably tied to the progress of these projects. Finally, the Armed Services Committees of both houses of Congress tend to be dominated—for a variety of noncoincidental reasons—by legislators whose constituency's financial well-being depends on robust military spending.

Eisenhower's farewell address will always be best remembered for his warning that the military-industrial complex threatened our democracy. Eisenhower feared that stockpiling, and ultimately using, weapons might someday be driven more by the political/economic gains of a few than the legitimate security considerations of the nation. The daily news abounds with evidence of his prescience. Defense critics often cite the Star Wars (strategic defense initiative, or SDI) project as a classic case study in military-industrial complex politics. The project is near impossible to justify in national security terms when you consider that scientific and technological experts are almost unanimously skeptical that the system will ever work as designed. SDI's advocates have yet even to agree on a clear mission for the project. Yet Star Wars is a major item in the defense budget each year, so bureaucratically and financially entrenched that few expect it ever to go away completely. The 1992 defense budget calls for a $4.78 billion SDI expenditure—a 60 percent increase over the previous year. Another project with a dubious purpose is the B-2 bomber. The B-2, of which we already have fifteen, was designed to drop bombs on the Soviet Union, not exactly a high-priority goal in the early nineties. The 1992 budget asks for four more B-2s—at a cost of $1 billion per plane.

One of the more revealing aspects of the military-industrial complex concerns a dynamic we might term the "defense-contractor/congressional complex." It has become common practice for defense contractors to locate plants in the districts of influential members of Congress. In the 1980s, according to a *Los Angeles Times* story (3/6/91, A1), Los Angeles–based Hughes Aircraft built six factories in the South, each "in the district of a key congressional member of a committee overseeing the firm's business." During the same decade, Northrop built a missile plant in the Perry, Georgia, birthplace of Senator Sam Nunn, chairman of the Senate Armed Services Committee. Lockheed, McDonnell Douglas, and Rockwell also made their appeals to Nunn, according to the story, by opening operations in Georgia. Lockheed clearly hoped to get double duty for its efforts, since its facility was built in Marietta, Georgia, home to House Armed Services Committee (HASC) member Buddy Darden. Similarly, McDonnell Douglas put its plant in the Macon, Georgia, district of HASC member J. Roy Rowland, and Northrop's Perry plant employs many of HASC member Richard Rey's constituents.

So effective is this form of defense industry politicking that Congress will often fund weapons that the military does not want. According to the same *Los Angeles Times* story, HASC Chairman Les Aspin (D-Wis.) nursed legislation through that foisted a truck on the army in which it had little interest. The truck was built in Aspin's district. The Boeing/Bell V-22 tilt rotor aircraft and the Lockheed C-130 transport are cited as other industry/Congress pets for which the military had no affection.

When it comes to explaining why they avoid defense-related investments, some ethical investors feel they need point no farther than this last decade's procurement scandals within the military-industrial complex. While the eighties were a financial bonanza for defense contractors, the industry's ethical image was in smoldering ruins by the decade's end. Giant contractors General Dynamics, Litton, Boeing, and Unisys among others were caught red-handed in unethical (and in some cases illegal) contract procurement activities. Disclosures about "$600 toilet seats" and "$900 hammers" made both the news and Johnny Carson's show with regularity. And those who followed such things closely knew that massive cost overruns were the rule, not the exception, on almost every DOD contract. Avoiding investments in all defense companies because of the actions of a few may not be precise social policy, but many ethical investors find it a powerful way to register a protest.

Part of what makes investing in defense industries an ethical mixed bag are the good things that some companies do. Some of the biggest defense contractors also grade high in the kinds of activities that the SRI community applauds. Nuclear Free America ranks AT&T as one of the leading nuclear weapons contractors in the country (although AT&T disputes the figures). The same company has been acknowledged in recent years by *Black Enterprise* magazine as one of the best places for blacks to work and by *Working Mother* magazine as one of the best places for women to work. AT&T, at its industry's forefront with its pledge to halve its use of ozone-eating chlorofluorocarbons by 1991 and eliminate them in 1994, also won the Council on Economic Priorities' environmental award in 1990. Boeing sometimes does over a billion dollars' worth of DOD business in a year and pleaded guilty in 1989 to criminal charges regarding its early 1980s snitch of classified Pentagon documents. Its Wichita, Kansas, plant is a leading emitter of carcinogenic chemicals, according to the National Resources Defense Council. That same Boeing makes a superior solar cell, is a corporate leader in charitable giving, and purchased $166 million of goods and services from minority-owned businesses in 1988 (not to mention $2.191 billion from small businesses).

Some ethical investors, including some SRI professionals, will "forgive" a company's defense contracting if it is merely supplying such items as food, clothing, or fuel to the military. Some social mutual funds will consider investments in companies doing 5 percent or less of their business through DOD contracting. So make sure that you are comfortable with the specific defense-oriented criteria applied to any investment you make.

The defense industry's aggressive marketing of its products to undemocratic and unstable regimes is probably its most controversial practice, and one that has unquestionably made this a much more dangerous world. Even the industry's staunchest backers were up in arms, so to speak, over the prospect of American troops facing down advanced American-made weaponry in the Persian Gulf war. We survived that conflict with our greatest fears unrealized. Nevertheless, a terrible potential remains unless we can exert controls over some companies' willingness, with the government's blessings, to sell anything to anybody for a price. The discussion below examines these concerns, among others.

SRI and the Gulf War

Throughout the military phase of the Gulf war crisis, a large majority of Americans supported their country's participation in the war effort. Presumably that number also included many who invest according to ethical criteria. Therefore, some readers of this book may regard companies profiting from this war differently from the way they regarded companies profiting from the Vietnam conflict, for example.

That said, some industries certainly are reaping huge profits from Gulf War–related activities. This is not to imply a lack of patriotism or any sort of ill intent on the part of those companies—it is simply a statement of fact. The biggest economic winners during the Gulf crisis seem to be the major oil companies and some select major construction firms. The oil companies scored big when fear-driven bidding prior to the war's start drove crude oil prices to unprecedented levels. Oil prices dropped after the fighting started, but the oil industry will still be sitting pretty if the United States accomplishes its postwar objectives, one of which is to secure the continued flow of cheap crude oil to the West. Oil is still our cheapest energy source (if you don't figure in such "incidentals" as the cost of militarily insuring our access to it). But as the price of oil rises, the price of alternative sources of energy becomes increasingly competitive. Should alternative energy sources become truly competitive with oil, oil conservation will start making financial as well as environmental sense, as will reducing our dependence on oil altogether. That day will surely come, because oil is a finite resource, and many alternative energy sources such as solar are infinitely renewable. It is certainly in the oil industry's interest to delay that inevitability for as long as possible.

According to a February 12, 1991 *Los Angeles Times* story based on a (London) *Financial Times* original report from the previous day, several American construction firms will be among those sharing in what may be $500 billion of contracts to rebuild Kuwait. The firms include the Bechtel Group (the massive privately held company of which ex–Secretary of State George Schultz and ex–Secretary of Defense Caspar Weinberger are famous alumni), Fluor Corporation, McDermott International, and Foster Wheeler Corporation. These four companies and an Italian firm, according to the story, have agreed in principle with the Kuwaiti government to rebuild the occupied country's decimated oil and gas industry. AT&T has acknowledged in a separate report that it is negotiating with Kuwait to rebuild the country's telecommunications network, destroyed during the Iraqi invasion.

Contrary to what many might suppose, the defense industry does not expect to share in the postwar bonanza. Economist Robert Samuelson, writing before the fighting ended, explained that for the most part the war was fought with an arsenal built up to deter a land war in Europe. Since the European scenario is now considered virtually obsolete, much of the military equipment lost is not likely to be replaced. Still, the defense industry and its political backers continue to rely on Gulf war–laced justifications as they push hard for their favorite projects.

Perhaps the touchiest Gulf war issue for ethical investors will be determining which defense contractors helped Saddam Hussein build his ghastly arsenal. The rush of American and other allied nations' corporations to sell weaponry to Iraq has been condemned from every sector of the political spectrum, because it clearly enabled Saddam's aggressive ambitions in the first place. Even arch-conservative Senator Jesse Helms has characterized the sales as driven by "unbelievable greed."

Much of the activity, of course, occurred during the period when U.S. foreign policy favored Iraq in its long war with Iran. This alone might justify the sales in some minds. But the plot thickened in late 1986, according to a *Los Angeles Times* report (largely recapitulated in an NBC-TV report several days later), when the Pentagon became concerned that Iraq was developing weapons of mass destruction at Saad 16, a nine acre complex the Iraqis claimed was devoted to non-military modernization. The Pentagon noted that the site was laid out and outfitted like a nuclear and chemical weapons facility. They made their concerns known to the Commerce Department, which serves as watchdog over national security in matters of private exports. Around this same time, the Justice Department and Customs Service were uncovering evidence of a clandestine Iraqi weapons procurement network with contacts in the United States and Europe. The network was arranging illegal exports of such items as the chemicals necessary to make mustard gas. The Commerce Department largely ignored all such contraindications including the Pentagon's specific objection to exports of weapons-related computers from Electronic Associates Inc. of Long Branch, New Jersey, and Gould Corporation of Ft. Lauderdale,

Florida. Between 1985 and 1990, according to the report, approximately $1.5 billion of high technology and other equipment with potential military applications was shipped to Iraq, some to such suspect destinations as Iraq's Atomic Energy Commission, Ministry of Defense, and air force.

The complicity of some of these companies goes beyond simply making legal, if ethically dubious, sales. In 1985 a number of firms dealing weapons to Iraq established the U.S.-Iraq Business Forum to lobby against any regulations that would restrict their Iraqi trade. For example, the forum opposed congressional efforts to punish Iraq economically for its apparent chemical weapons attack on the rebellious Kurdish population. The Reagan White House sided with the forum, and the punishment effort died.

For the time being there is no way of determining the full list of companies that came to the U.S.-Iraq trade party. To save them embarrassment, the Commerce Department and its congressional oversight committee have refused to make the list public. The *Times* article does mention a few besides those discussed above: Hewlett-Packard, Wiltron (Morgon Hill, PA), and Textronix Corporation, all of which shipped—with the Commerce Department's good wishes—high technology with military applications. An earlier story in the *Times* had disclosed that Honeywell Corporation sold Iraq a manual enabling it to develop a fuel-air explosive device with a destructive power several times that of conventional weapons.

The *Times* story is not without its corporate heroes. When Iraq ordered some special capacitors from CSI Technologies in San Marcos, California, company president Jerold Kowalsky noted that the components are most commonly used in the detonating devices of nuclear weapons. He alerted U.S. authorities immediately. His concerns sparked an investigation that exposed parts of an illegal, international arms procurement network. Consarc Corporation president Raymond Roberts expressed fears to U.S. authorities that Consarc's high-temperature furnaces—which Iraq claimed it ordered for an artificial-limbs factory—would end up in Saddam's nuclear weapons facilities. Commerce approved the export over Pentagon objections, but the National Security Council intervened to delay the shipment. Iraq's invasion of Kuwait, and the allies' total trade embargo instituted shortly thereafter, ended any further processing of the report.

Selling weapons to unstable, repressive regimes has been lucrative business as usual for a long time—not only here, but throughout the industrialized world. Ethical investors who believe in a strong defense and/or supported our Gulf war role may still wish to screen their investments for complicity in arming the itchy trigger fingers of the world.

Firearms Production

The firearms-related investment issues are brief and straightforward. Some segments of the ethical investment community are pacifist either through secular moral conviction or through religious affiliation. Certain ethical investment products screen for firearms involvement to serve these investor populations. Pax World and Working Assets, for example, completely avoid companies that manufacture weapons, including firearms. Calvert, on the other hand, will invest in a weapons manufacturer if weapons account for less than 10 percent of sales.

Questions to Ask

Does the company contribute cash or noncash resources to peaceful causes?

Does the company develop, research, manufacture, or service products used as nuclear, biological, chemical, or conventional weaponry?

Does the company have a policy of not contracting with the Department of Defense?

Does the company have a policy forbidding the application of its products to weaponry?

Has the company converted facilities previously used for weapons-related work into plants producing goods with peaceful purposes (economic conversion)? Has the company taken a position for or against economic conversion?

What percentage of the company's activity is committed to weapons development? Is that portion increasing or decreasing?

Has the company been penalized in recent years for unethical dealings in relation to a Defense Department contract?

Has the company sold technology, chemicals, biological agents, or other products with military applications to Iraq or other unstable, undemocratic regimes? If yes, has the company also lobbied against restrictions on or embargoes against such trade?

Do you feel that the Gulf war was in any way a product of the military-industrial complex? (If yes, you will of course be wary of companies profiting directly from the war.)

Resources:

Center for Economic Conversion
222 View Street
Suite C
Mountain View, CA 94041
(415) 968-8798
Publishes quarterly the Plowshare Press.

INFACT
256 Hanover Street
Boston, MA 02113
(617) 742-4583
Campaigns against nuclear weapons industry and its influence on government policy.

Nuclear Free America
325 E. 25th Street
Baltimore, MD 21218
Publishes a newsletter, the New Abolitionist.

Nuclear Information & Resource Service
1424 16th Street NW
Suite 601
Washington, D.C. 20036
(202) 328-0002

1% for Peace
P.O. Box 658
Ithaca, NY 14851
(607) 273-1919

Department of Defense
and other government publications listing defense- and weapons-related contractors.
Your best source of these is the government documents section of a large public or university library.

Investing from the Heart **Appendix Reference:** See appendix A under "Weapons Involvement," page 362.

LABOR

SRI labor concerns begin with the obvious: wages. In the eyes of most ethical investors, a socially responsible company will be one that offers its employees competitive compensation—keeping in mind, of course, that smaller companies may pay less than large, mature ones (which by the same token cannot compete with the small firm's promise of upward mobility).

SRI professionals tend to prefer companies that offer a competitive benefits package as well, with the centerpiece benefit being comprehensive health insurance. It should be noted, however, that company-sponsored group insurance is not the given that it was a few years ago. Health costs have risen much faster than the inflation rate in recent years, and many companies are reluctant to ante up. Besides, the current trend is for companies to cut employee-related costs to the bone. Not only health benefits, but also Christmas bonuses and even parties have been pared down in company after company nationwide. Soon, 100 percent coverage plans may be a thing of the past; many companies now ask their employees to contribute toward a more limited plan's premium. The social implications of all this are a tough call. Health plans do not improve employees' health, although it was because companies wanted to reduce worker sick days that they first started offering plans years ago. On the other hand, work-sponsored health plans have become the nation's primary source of affordable health insurance, with a great number of Americans who can't get insurance through their work going uninsured. Until, or unless, the federal government or states implement comprehensive citizen health insurance, the problem will grow worse as health costs soar.

SRI pros also look for companies that demonstrate respect for their employees. The company that behaves compassionately toward its employees when bad times erode the bottom line will be especially noted. Also considered favorably are promotion-from-within policies; appropriate communications channels between management and line employees (including appropriate grievance apparatuses); minimal distinctions between upper- and lower-level employees (a preeminent example is Ben & Jerry's Homemade's dictum that no employee, including Ben and Jerry, can be paid more than five times what the least-paid employee earns); company-sponsored education programs for skills improvement and in such subjects as substance abuse and rape prevention; corporate fitness programs, quality circles and other employee participation in company decision making; and commitment to worker safety.

One new safety issue may ultimately reveal much about a company's concern for its work force, because the costs of addressing the problem could be astronomical. The issue involves injuries and illness related to work at computer terminals. The infirmities fall into two categories. The first includes overuse injuries to wrists and elbows and eye strain from long hours in front of oscillating VDT screens. The second category is by far the most worrisome, both healthwise and economically. A far-from-conclusive but accumulating body of research is indicating that overexposure to electromagnetic radiation emanating from VDTs may cause cancer and other serious internal problems. Overuse injuries can be prevented or reduced simply by redesigning workstations. The costs—new or modified furniture and more frequent employee breaks—are relatively minor. But keeping employees at safe distances from VDTs (two feet in front and five feet from backs and sides) means completely redoing floor plans or—in cases where computer workstations are packed together tightly—either eliminating workstations or expanding facilities.

Companies that share the wealth with the employees who help create that wealth also get high marks from social investors. Fund-matching and profit-sharing plans, bonuses for innovative ideas and work well done, and employee stock ownership plans (ESOPs) are among the most common ways of sharing. ESOPs remain controversial in this regard, particularly when they are used to serve the interests of highly paid employees and/or ownership without substantially improving the average employee's comprehensive benefits packages. By the same token, ESOPs usually spur increased productivity and worker loyalty when coupled with increased employee participation in decision making. A study published in the *Harvard Business Review* (Sept./Oct. 1987) showed that ESOP companies that instituted participation plans grew three to four times faster than ESOP companies that did not. However, many company owners install their ESOPs primarily for the significant tax advantages that accrue to them (such as the ability to take out part or all of their ownership value in tax-deferred cash). Owners so motivated sometimes resist increasing employee participation because they fear "letting the chickens run the farm." You the investor may want to discriminate between ESOP companies with participation plans and those without. Our list in appendix A will prove useful in this regard—as a starting place, anyway.

As you might well imagine, the company's response to unionization is a key SRI consideration. Union busters fare poorly in professional social screens. But unionization is not an absolute for equitable employee/management relations. Some companies are so progressive in their internal relationships that unionization never becomes an issue.

Labor issues in particular emphasize a point we made in the first chapter: socially responsible companies tend to be well-run companies overall, with carryovers into profits. Low wages, miserly benefit packages, and other poor working conditions often translate into costly high turnover rates and slovenly productivity. Safety violations and associated penalties and lawsuits are expensive in terms not only of dollars, but also of reputation—not to mention lost time for anyone hurt because of the lapses.

On the positive side, employees who share in company decisions, profits, and ownership feel a sense of pride about their work that pays dividends in productivity, innovative ideas, and loyalty. Skills training makes them better employees. Unusual but highly desirable benefits such as corporate fitness programs enables a company to compete nationwide in recruiting the best talent for the company.

One potential example of how a positive relationship between management and line employees works to the benefit of all: The AFL-CIO's *Label Letter* publication salutes Chicago-based Midas Muffler for resisting the temptation to relocate plants outside the country in search of cheap labor and for its worker participation in decision making; its safety education programs, which reward employees for injury-free performance; and its worker training programs in automation and quality control. The company produced a record number of exhaust pipes and tailpipes (nearly $4 million) in 1990 at a unit cost unchanged from 1982—this despite wage and other overhead expense increases.

Questions to Ask

How does the company's compensation and benefits packages compare with the industry average (particularly with companies of comparable size and maturity)?

How does the company's turnover rate compare with the industry average?

Is the company's labor force unionized? Have there been strikes, boycotts, lockouts, or other labor disputes?

Is the company's industry one that is heavily unionized?

Has the company campaigned internally or externally against unions?

Does the company involve employees in decision making? Does the company otherwise solicit ideas and suggestions from its work force?

What communication channels exist between upper- and lower-level employees, including channels for communicating grievances?

Do employees share in ownership or profits of the company (through ESOPs or profit-sharing plans, for example)? If the company has an ESOP, has it increased employee participation in decision making?

Does the company offer any extraordinary benefits such as a corporate fitness program, tuition reimbursement, subsidized healthful meals, on-site seminars, elderly care, distribution of educational materials, financial/retirement planning, disabled dependent care, on-site or near-site care, recreation programs, and so on?

When a company closes a worksite or otherwise faces a potential layoff, does it offer employees retraining and/or counseling?

Has the company been cited for safety (Occupational Safety and Health Administration—OSHA) violations or unfair labor (National Labor Relations Board—NLRB) practices in recent years?

Is the company on the AFL-CIO boycott list?

Resources

The Council on Economic Priorities
(See page 44.)

National Center for Employee Ownership
2201 Broadway
Suite 807
Oakland, CA 94612
(415) 272-9461

Union Label & Service Trades Department, AFL-CIO
815 Sixteenth Street NW
Washington, D.C. 20006
(202) 628-2131

Investing from the Heart Appendix Reference: See appendix A under "Labor and Equal Opportunity," page 349.

CORPORATE CITIZENSHIP

We use the term *corporate citizenship* throughout this book to refer to a company's community-minded behavior, because it is a simple fact that companies are citizens of the communities in which they operate. For companies that operate in the national and/or international arena, citizenship takes on even larger implications.

Corporate citizenship is a composite of a company's actions in a variety of areas: charitable giving (both cash and noncash), loans of company expertise and resources to solve community problems, lobbying and political contributions, role (constructive or exploitive, if any) in third world nations, willingness to operate within relevant laws and industry regulations, social programs undertaken in its resident community, and consideration of the community impact when faced with a possible facility closing.

The elite among our corporate citizens includes companies that use their wealth-creating ability to further broad agendas for the social good. Ben & Jerry's Homemade, Patagonia, Smith & Hawken, and Pitney Bowes are outstanding examples. We've discussed Ben & Jerry's social contributions all through these pages. Patagonia and Smith & Hawken are B & J's West Coast equivalents. Patagonia, which produces outdoor equipment and clothing, donates about 10 percent of its pretax profits to a broad range of environmental causes as well as the 1% for Peace campaign. Another 1% for Peace supporter, Smith & Hawken—a mail-order marketer of garden supplies, tools, rugged clothing, and related books and gifts—also donates about 10 percent of its profits, primarily to environmental and horticultural organizations, and goes to extraordinary lengths to insure that its products are environmentally sound and otherwise socially responsible. Both Patagonia and Smith & Hawken buy buttons made from Ecuadorian tagua nuts for their clothing lines. The tagua nut is a rain forest product; its purchase helps provide economic justification for the forest's preservation. Pitney Bowes acted to help revitalize an economically deprived area of Stamford, Connecticut, by building new corporate headquarters there.* (Investors should note that Patagonia and Smith & Hawken, although both model socially responsible businesses, are currently privately held—in other words, their stock is not sold to the public. Pitney Bowes stock is publicly traded.)

Corporate giving is the most quantitative measure of a company's social commitment. In 1989 the average company gave away about 1.7 percent of its pretax net income, down from 1985's almost 2 percent peak. The Council on Economic Priorities (CEP) gives its top grade to companies giving 1.2 percent or more and its middle grade to companies giving away 0.7 to 1.2 percent. (Individuals gave away about 2.4 piece of their pretax net income in 1989.) Another way for corporations to give charitably is to match employees' contributions. Quaker Oats contributes $3 for every $1 an employee gives. It should be noted that corporate giving must be regarded differently for large companies with large surpluses of cash and for small companies still in their growth phase, who may be able to contribute more in the way of noncash community assistance than in dollars.

As in other areas of socially responsible corporate behavior, corporate community-mindedness can also make good business sense. Companies that support affordable housing, particularly housing near their facilities, experience returns in their ability to recruit and retain a work

---

* *Clean Yield,* 11/90.

force. Many companies, alarmed by the declining intellectual abilities of the American work force, are backing improved education through Adopt-a-School programs and other strategies. You may want to differentiate between companies that support broad educational goals and those that support only narrow curriculums applicable directly to their industry. (Some company-supported education programs, in fact, seem to exist mainly for the purpose of marketing their products to a captive audience.)

Companies also give by involving retired employees and by supporting employee volunteer efforts through sponsorship of campaigns and work-release permission for participants. No matter what the indicator, however, analyzing corporate citizenship boils down to one basic issue: whether or not the company cares about its effect on the communities in which it operates and on society in general.

Questions to Ask

Does the company lend its expertise and other noncash resources to help solve community problems? Does it make any cash commitments to help solve such problems?

If the company is involved in third world countries, is its relationship there exploitive or constructive?

Does the company market products internationally that have been banned in the United States?

What percentage of its pretax profits does the company commit to philanthropic endeavors? Are these endeavors ones you consider beneficial or harmful to society?

Does the company encourage employees to become involved in community organizations? What form does this encouragement take?

Does the company contribute to political campaigns or political action committees (PACs)? If yes, which ones? Who in the company decides on the recipients and amounts of the gifts? Are the campaigns the company supports ones you consider harmful or beneficial to society?

Has the company attempted to influence employees to contribute money or time to specific political campaigns or vote certain ways in elections?

Has the company codified an ethics policy? Does it distribute a code of conduct to its employees? Does it have a board-level committee in charge of social responsibility?

If the company has closed facilities in the past, did it first search for alternatives? Did it involve the community in that search? Did it offer any assistance to the community in the closing's wake?

Does the company have a history of violations of laws and regulations?

Does the company purchase goods and services from minority-owned businesses?

How is the company regarded in the communities in which it operates?

Resources

Council on Economic Priorities
(See page 44.)

Multinational Monitor
P.O. Box 19405
Washington, D.C. 20036
(202) 387-8030

National Minority Supplier Development Council
(See page 45.)

1% for Peace
(See page 66.)

News articles.

Corporate publications.

Investing from the Heart Appendix Reference: See appendix A under "Product Integrity and Corporate Citizenship," page 365.

ETHICAL TREATMENT OF ANIMALS

Many of our creature comforts—personal care products, household products, medical advances, and dietary pleasures—come to us at the expense of terrible creature suffering. Certainly no sane person supports animal suffering per se, but in certain circumstances such as medical testing to save human life, the ethics become more complicated. Viable alternatives to animal testing and other forms of animal suffering do exist, however—even in medical testing—and it seems clear that concerted research efforts could develop others. Our approach in this section is to explore how some companies have discovered such alternatives, so consumers and investors can make informed choices.

The Council on Economic Priorities publication *Shopping for a Better World* describes federal policies regarding animal testing of consumer products as confusing and contradictory. Animal-rights organizations such as People for the Ethical Treatment of Animals

(PETA) can cite chapter and verse when they claim that companies that cannot find suitable alternatives to animal testing will be covered legally simply by placing warning labels on their products. Corporations that continue to animal test quote the same regulations while defending the necessity of their practices.

PETA estimates that fourteen million animals die each year to test personal care and household products. The most common tests include the Draize eye irritancy test, in which caustic substances are dropped into the eyes of unanesthetized rabbits; lethal dose tests such as the LD50 test, in which animals are force-fed toxic substances until 50 percent of the subjects die; and inhalation and dermal toxicity tests, in which animals are forced to inhale toxic substances or absorb them through their skin.

It should be noted that major companies such as Avon, Benetton, and Revlon ceased animal testing altogether after being pressured by consumers and shareholders in 1988. The Estee Lauder Companies and Clarins of Paris were among those banning animal tests forever in 1990. The Council on Economic Priorities cites S. C. Johnson and Colgate-Palmolive as companies still animal testing that have taken great strides in reducing such tests and/or are actively seeking alternatives to live testing. Church & Dwight—makers of such household products as laundry detergent, baking soda, and toothpaste—have a policy against using the Draize or LD50 tests in any activity. As of this writing, their only animal testing involves brushing rats' teeth to determine toothpaste abrasiveness.*

The point is obvious: Other avenues do exist. Avon, in fact, now employs a system called Eyetex—developed by Ropak Industries (Irvine, California)—which determines eye irritancy by using a vegetable protein that mimics the cornea's reaction to foreign matter. Other alternative methods include the neutral red bioassay—developed by Clonetics Corporation (San Diego)—which computer measures the absorption of a dye added to human skin cells in a tissue culture; testing on corneas from eye banks; other testing on artificial skin grown from single animal or human cells; and testing through the use of sophisticated computer and mathematical models.

Of course, cosmetic companies always retain the option of choosing the far simpler—and eminently sane—approach of relying on healthy natural ingredients and ingredients already determined safe by the

---

* *Insight,* 1/15/91.

Cosmetics, Toiletry, and Fragrance Association. Cosmetics companies such as The Body Shop and Tom's of Maine—both founded on the principle of cruelty-free products—use skin-patch tests on human volunteers when testing is necessary, as does Paul Mitchell Systems. Then again, these companies' formulations do not generally depend on potentially dangerous substances.

At the opposite end of the spectrum are companies that have resisted all pressure to reconsider their dependence on animal testing. The most notorious of these seems to be Gillette, frequent target of PETA-organized demonstrations in 1990. PETA literature is quick to point out that PETA's companion demonstrators included environmentalists protesting the company's pollution record and antiapartheid activists angered by the company's continuing holdings in South Africa.

On the medical front the animal-testing issues are stickier, both ethically and legally. The Food and Drug Administration (FDA) requires that pharmaceuticals be rigorously tested as a condition of approval for general use. Historically such experimentation has relied heavily on animal subjects. Since many of these pharmaceuticals are developed to treat serious illnesses, advocates of medical animal testing argue that it saves countless human lives. Nevertheless, at the very least significant reductions can be made in medical animal testing as well. The PETA literature mentions several effective alternatives. Clinical surveys that combine data from human volunteers, case studies, autopsy reports, and statistical analyses are one useful approach. Health Designs, Inc., has developed a widely used software package called TOPKAT, which predicts oral toxicity and skin and eye irritation. Another alternative, the Ames test, mixes the chemical in question with both a bacteria culture and activating enzymes for the purpose of discriminating between carcinogens and noncarcinogens. CEP notes that the National Cancer Institute reduced its animal testing of antitumor drugs from six million cases in 1986 to under three hundred thousand in 1989 by relying on human cell lines, a task it found more reliable and cost-effective than animal tests.

PETA claims studies have shown that these and alternatives developed for consumer products can save both time and money, sometimes by several orders of magnitude (powers of ten). For example, an Ames test of a suspected carcinogen may cost from $200 to $4,000 and be completed in one to four days, compared with animal testing taking four to eight years at a cost of at least $400,000.

Another sensitive animal rights issue—one particularly uncomfortable for the vast majority of us who eat animal products—is "factory farming." Most of the nation's meat, dairy products, and eggs come from huge industrial farms that maximize efficiency at almost unimaginable cost to the animals' comfort and health. Raised and transported in quarters so confining as to deny even the most basic freedom of movement, and fed growth and other hormones to increase their productive value, food animals are tremendously prone to disease and so are often heavily dosed with antibiotics as "preventive medicine." Veal, taken from calves slaughtered at sixteen weeks, is considered an especially unhealthful meat because the quantity of drugs fed the calves is high even for factory farm-raised animals. The broad *non*ethical implication for the consumer is that the food supply from animal sources is heavily laced with hormones and antibiotics with unknown long-term effects on humans.

We will not address here the additional suffering meat animals incur at their slaughter. Obviously one way that concerned consumers and investors can address the cruelty inherent in factory farming is to support companies that develop vegetarian protein sources. Meat eaters should be aware that health-oriented farms exist that treat animals more humanely (free-range cattle and ground-fed chickens, for example) and feed them organically. In addition, animals raised for kosher meats are slaughtered as compassionately as possible in accordance with Jewish dietary laws.

Questions to Ask

Does the company rely on animal testing for any of its products? If yes, has it recently reduced its dependence on animal testing, or is it actively engaged in researching alternatives? How does the company compare with others in its industry on its rates of animal testing and commitments to alternatives?

Has the company published standards of humane treatment of test or farm animals?

Does the company have a "cruelty-free" policy?

What has the company done recently to improve the quality of life of its test or farm animals?

Does the company sell true "health foods" or vegetarian foods?

Does the company develop alternative, animal-free tests to be used by the consumer products and pharmaceutical industries?

Resources

The Council on Economic Priorities (See page 44.)

Greenpeace (See page 58.)

Johns Hopkins Center for Alternatives to Animal Testing
615 Wolfe Street
Baltimore, MD 21205
(301) 955-3343

National Audubon Society
950 Third Avenue
New York, NY 10022
(212) 832-3200

National Wildlife Federation
1400-Sixteenth Street NW
Washington, D.C. 20036
(202) 797-6800

People for the Ethical Treatment of Animals (PETA)
Caring Consumer Campaign: Corporate Responsibility Project
P.O. Box 42516
Washington, D.C. 20015
(301) 770-7382
PETA publishes lists of companies that test on animals and companies that do not. We have not included these lists because the turnover of company names, particularly on the first list, is too rapid due to intense consumer pressure as well as to the efforts of organizations like PETA. Consult with PETA for information about specific companies, but be aware that companies named on PETA's lists may not trade as stocks under those names, particularly if they are subsidiaries of parent companies.

PRODUCT INTEGRITY

Product integrity—how a company handles its products from conception to manufacture to marketing—is perhaps the most obvious example of social ethics dovetailing with smart business practices. First, well-designed, quality products that are effectively marketed is the bedrock of any good business enterprise. Honda and other top Japanese car manufacturers have given their American competition fits with product design and execution that is light-years beyond what American automakers offer in the marketplace. Second, quality products researched to be safe for consumers and advertised truthfully are less likely to land their makers in court. Lawsuits are costly enough; the goodwill lost with the public when the court actions are publicized can be costlier still. The pesticide Alar had not been completely tested when growers first began using it. The EPA determined that Alar should be pulled from the market as a possible carcinogen in 1989. You didn't have to read the business page to know that 1989 turned into a nightmarish year for apple, apple juice, and applesauce sales.

An SRI review of a company's product integrity addresses subtler issues, also. One such issue involves the product's social usefulness. Tobacco, of course, is the most obvious example of a product conceived, manufactured, and marketed with utterly cynical regard for its social impact. Cigarettes are addictive, lead to fatal diseases and other premature death, and are marketed aggressively to female, minority, and teenage populations. Many SRI investment companies screen their investments for tobacco involvement. How a company uses natural resources also reflects on its product's social integrity. Some ethical investors refrain from investments in companies that create wasteful, disposable products such as disposable diapers, pens, razors, and cameras. Overpackaging is a related concern.

Another of the subtler aspects of a product's integrity pertains to the social values implicit in the company's advertising of it. A socially responsible ad is one that neither promotes racial, ethnic, sexual, or other stereotypes nor encourages negative social behavior. In the late 1960s one tobacco company's ads seemed to make a subliminal appeal to white racist attitudes about interracial sex. The ad campaign encouraged smokers to buy the company's filtered cigarette because "only the white touches your lips." A recent flap of a different stripe has focused on the marketing of $100-plus major brand basketball shoes, a campaign that heavily penetrates the ghetto youth market among others. Critics consider the campaign cruel and cite anecdotes of violent struggles for possession of the prized footwear.

A company can advertise truthfully and appropriately in its domestic advertising and then blow its ethical rating entirely through its campaigns in the third world. One of the Council on Economic Priorities' "shopping alerts" concerns the marketing of infant baby formula as breast-milk substitutes in third world hospitals. The typical campaign, CEP alleges, involves donations of free or heavily subsidized samples to the hospitals with sometimes deadly repercussions in the targeted country's destitute population. The following scenario occurs all too frequently: Because the mother begins using the formula in the early postnatal stage, her breasts stop producing milk. Now dependent on a product she can't afford, she dilutes it to make it last longer, causing malnutrition—plus the water used in the dilution may be contaminated, leading to sometimes fatal baby bottle disease.

United Nations officials estimate that one million lives per year are lost in the third world because mothers are persuaded to give up breast feeding. In 1981 the World Health Organization and UNICEF established an international code for infant formula marketing. Action for Corporate Responsibility (ACTION), under the leadership of chairman Douglas A. Johnson, has targeted Nestlé and American Home Products for violating the WHO/UNICEF code. Although both companies have protested that they are in compliance with the code, Johnson implies that the problem may be more widespread rather than less, with other American companies employing similar marketing strategies. He also notes that Nestlé staffer Thad Jackson, the company's chief counterboycott campaigner, has been appointed by the Bush administration to the U.S. delegation of Codex Alimentarius. Codex is developing worldwide product standards, so Johnson sees the appointment as Nestlé's government-assisted effort to undermine the International Marketing Code. (Nestlé, to its credit, earned top 1990 CEP ratings in minority advancement, charitable giving, animal testing, disclosure of information, and community outreach. However, the company did not respond to CEP's questions on South Africa involvement.)

Social screens for product integrity sometimes address customer relations issues, also. The SRI periodical *Insight,* for instance, evaluates how a company relates to its consumers, whether or not it is forthcoming with information, and how it responds to consumer complaints or defective product. It also considers affordability and marketing to a broad range of people as a positive social value.

Some product issues are product-specific. One such issue centers on pesticides. Pesticides certainly have their advocates beyond the companies that produce them. Farmers practicing conventional agricultural methods depend heavily on pesticides to attack the insects that prey on their crops. But pesticides (which can cause cancer, birth defects, and genetic damage) may cause more problems than they solve. First of all, they pollute our air, food, and water—pesticide production itself pollutes, rain can wash pesticides from farmlands into our sources of drinking water, and pesticide residues adhere to our produce and are difficult to remove without elaborate procedures. Plus, pesticides are sometimes rushed into use before final test results on their human toxicity are in, as with the Alar debacle.

Another problem with pesticides is that the modern agricultural methods that depend on them operate on kind of a law of diminishing returns—that is, the quantity of pesticides needed increases with succeeding generations of crops. Conventional agriculture depends on chemical fertilizers and hybrid seeds selected for such commercially desirable qualities as resistance to mechanical damage at harvest and during transportation; cosmetic appearance; and, most importantly, maximum crop yields. Unfortunately these fertilizers and seed properties come at the expense of nutritional value, flavor, ability to keep during storage, and resistance to disease. Because the plants are weaker in a natural sense, they are more enticing to insects, increasing growers' dependence on chemical pest control. Exacerbating the problem is the fact that the pesticides themselves kill harmful insects' natural predators and provoke the evolution of harmful insects that are immune to the toxins, requiring the use of stronger chemicals in their place.

These are just some of the factors contributing to an interdependence of chemical fertilizers and chemical pesticide control. The societal costs of this interdependence—entrenched through government support—is awesome: besides those costs already mentioned, farmworkers can suffer from chemical poisoning; animals and nontargeted plants also suffer from the spraying; and the health of the soil is terribly depleted. Even the atmospheric ozone layer is attacked, by a byproduct of chemical fertilizer action. Organic farming methods have proven effective in controlling pests without significant economic sacrifice and certainly without the nutritional and other social compromises that conventional methods make.

A more immediately topical product-specific issue concerns coffee beans grown in the right-wing-governed El Salvador, where human rights abuses and military death squad activity have been rampant for years. A boycott against Salvadoran-grown coffee, widely supported by U.S. labor and religious organizations, seeks to pressure Congress to end its military-based aid to the El Salvador government and to pressure that government to negotiate an end to the war. Many members of Congress have voted for reductions or eliminations of aid to El Salvador because of the abuses, although the Bush administration has fought those efforts.

Questions to Ask

What are the social repercussions of the company's products and services on society?

Do the company's products and services make a beneficial impact on society?

What are the environmental impacts of the company's products and services?

Has the company demonstrated a concern for product safety and safety of use? If safety problems have been brought to the attention of the company, how has the company moved to correct the problem?

Does the company advertise truthfully? Do the company's ads avoid stereotypes? What are the other social values promoted by the company's advertising?

Has the company demonstrated pride in quality products and positive customer relationships? Does the company have a toll-free number that customers can call to register complaints or request information?

How have the company's products been rated by objective publications such as *Consumer Reports* and *Consumer Digest?*

Does the company market in the third world? How does the company ensure that its products will be used safely and appropriately there?

Do the company's products use resources efficiently through durable design, resource-conservative packaging, and energy efficiency?

Are the company's products and services priced to be affordable by a broad range of people?

Resources

Action for Corporate Responsibility (ACTION)
3255 Hennepin Avenue S.
Suite 230
Minneapolis, MN 55408-3398
(612) 332-6411
Organizers of the Nestlé and American Home Products boycotts.

***Consumer Digest* magazine**
7373 N. Cicero Avenue
Lincolnwood, IL 60646
(312) 676-3470

***Consumer Reports* magazine**
Consumer's Union of U.S., Inc.
101 Truman Avenue
Yonkers, NY 10703-1057
(914) 378-2000

Co-op America
2100 M Street NW
Suite 403
Washington, D.C. 20063
(800) 424-2667
Publishes Building Economic Alternatives, *a quarterly covering consuming and investing strategies focused on a peaceful, just, and environmentally sustainable economy. Excellent coverage of alternative products and businesses.*

The Council on Economic Priorities
(See page 44.)

Corporate advertisements.

Investing from the Heart Reference: See appendix A under "Product Integrity and Corporate Citizenship," page 365.

3

A Money Management Primer

Beginning on this page and continuing through the rest of the book, we teach you the fundamental concepts for doing your own financial planning. The essentials of financial planning, by the way, are the same for socially responsible investors as they are for anyone else. Here you will learn the rationales behind various money management options and strategies. You can then take that knowledge, plus information on specific investments from the next three chapters, with you to chapter 7. There you will learn how to apply the basics to your own situation and develop a financial plan consistent with your social values.

This is not to say that we recommend going it alone over consulting with a professional financial adviser. We don't, unless you have as much time to study available investments as somebody who makes his or her living at it. However, even if you do rely on professional advice, the information in the rest of this book will enable you to get the most out of that relationship. Understanding the principles of money management will also make your experience of following a professionally designed plan that much more satisfying personally.

Now, if you worry about whether you have the necessary background to digest this material, relax. This book is designed for digestibility—no prior background is assumed or needed. As for the "sophisticated" aspects, such as playing the stock market, you should realize that not even experts predict the market's fluctuations with any degree of certainty, so your beginner's standing is hardly any handicap in that regard.

Please note, by the way, that the terms *money management* and *investing,* both of which we use liberally throughout the rest of this book, are not exactly interchangeable, although we've used them as if they were up to this point. Since you are about to become a financial sophisticate, let's be more precise. *Money management* (also called financial planning) includes investing along with budgeting, saving, tax planning, and insurance planning—in short, making sure that you get the most for your money in terms of current purchasing power and future security. *Investing* refers to various ways of using your money to make more money, either by buying something you expect to increase in value or by loaning your money to someone who promises to pay you interest on it.

WHY MANAGE YOUR MONEY?

Why indeed? Well, try these reasons for starters:
- Your kids' college educations
- Your own financial security in retirement
- Funds for special expenses such as vacations, home improvements, and special celebrations
- Protection against medical emergencies and times out of work
- Properly insuring yourself and your family
- Helping your children financially in their early adulthood
- Providing your heirs with a well-planned estate
- Providing funds for contributions to your favorite causes

While you're at it, don't forget to include these fundamental ingredients of economic life:
- Inflation eating up your financial worth (an average inflation rate of 5.3 percent in the 1980s, as high as 13 percent in the 1970s)
- Credit card and other consumer debt costing you usually 16 to 22 percent over the cost of your purchases each year
- Federal taxes taking probably 28 percent or more of your income every year, with state and local taxes devouring an additional chunk

And, finally, add this "good news" reason to manage money: With socially responsible money management, you'll not only have more money for your needs and wants, but you'll also be supporting a more ethical and livable world while the money builds up.

Yes, there is far more to managing your finances today than earning a decent wage and paying your bills on time. But don't feel bad if bill paying is as far as your current money management goes. Most of your friends and neighbors are probably in the same boat. Fewer and fewer Americans are saving, much less investing, these days, which is a shame *because every American of average means can become financially independent through careful money management.*

Of course, few of those who don't plan ever take the time to figure what their lack of planning costs them. (A picky activity like that would be too much like . . . well, planning.) But they would probably find it sickening to see the actual amount of money that has passed from their lives without qualitatively improving their lives—that is, without buying them *anything*.

Without even considering investments, the way you use such everyday financial mechanisms as savings accounts, credit cards, and checking accounts may be costing you hundreds or thousands of dollars a year. Properly managed, that same money might not only still be in your possession, but might be compounding in an interest-earning account or growing in a stock fund, the seed of an eventual small fortune.

And that is just a piddly reason for managing money. Managing money becomes a truly serious topic when you consider such weighty matters as your kids' education and your retirement. The average *annual* cost (tuition, fees, room and board) of a college education for the 1990–91 school year was $4,970 for a four-year public college and $13,544 for a four-year private college. Based on those averages and a 6 percent rate of inflation, a child born in 1990 will face total expenses of some $58,000 for a four-year degree at a public university and $159,000 for a private school.* (See chapter 7 for more on college cost planning.) As for retirement, you know that a life based on Social Security is not at all secure.

The point, simply, is that the nice things in life, and even necessities like a secure retirement, must be planned for. There ain't no other way (unless you like those one-in-umpteen-million odds for winning your state lottery).

* Source: College Board figures.

THE ETHICS AND EASE OF MANAGING MONEY

If you have avoided financial planning up to this point in your life, the chances are good that you find the subject somewhat intimidating. You have plenty of company. Before we get into the nitty-gritty of financial planning, let's examine some of the common reasons people give for circumventing the "mysterious world" of high finance.

MYTH: *Financial planning is complicated.*
REALITY: Unless you want to get your hands dirty and do the research yourself, financial planning is no more complicated than finding good, socially responsible professional advice and following it. (We'll show you how to find that advice in chapter 7.) Planning is so easy with an adviser, in fact, that after you start you will probably kick yourself when you realize all the money you could have made up until then without hardly lifting a finger. Your SRI adviser will crunch all the numbers for you and clarify your ethical options so that once you determine your needs and the amount of risk you can tolerate, the choices you make will mostly be straightforward.

Planning is not that difficult even if you do all the work yourself. Once you understand the basic principles as presented in this book, you can construct a respectable plan on your own—particularly by including professionally managed products such as SRI mutual funds. Whether or not you use professional help, you do not have to have a "head for business" to earn a businesslike profit on your personal finances.

MYTH: *Investing is not for the faint of heart, because you can lose big as well as win big.*
REALITY: It is absolutely true that you can make more if you are willing to risk more (although even these risks can be reduced by diversifying and diligently monitoring your investments). However, there are lots of investments for the "faint of heart"—ethical, relatively risk-free products whose returns, given time, will build their investors a secure financial future.

What many investors do, as most professionals advise, is build a strong base in the safer products and then put a small amount of money that the investor can "afford to lose" in riskier investments. If you don't feel you can afford to lose anything now, you can build up your financial strength (and courage!) through conservative investments,

never seriously putting any of your money on the line. Properly for-
tified, you can then look again at adding some riskier investments to
your financial picture to increase your chances for real growth. (When
we say "risk," we still mean odds that are weighted heavily in your
favor.) The subject of understanding and determining your tolerance
for risk is addressed immediately following this section.

MYTH: *Investment profits are "dirty" money—money generated by
exploiting labor, the poor, the environment; by war profiteering; and
by other unethical "business as usual" practices.*

REALITY: Obviously the above statement is true about *some*
investments—the very existence of the SRI field, which developed as
an alternative for investors of conscience, is proof of that. But just as
obviously, it is not the whole truth—the SRI field's research demonstrates that there are many companies producing absolutely benign,
and even beneficial, goods and services, companies that also do business in a manner that is respectful of life. Furthermore, many of these
companies are among our country's best in terms of profitability, proving that the profit motive can also be a force for good. Widespread
environmental concern is demonstrating more than any other factor
that there is a market for ethical business, and strong markets generate
strong investments.

UNDERSTANDING AND TOLERATING RISK

We've mentioned risk a few times already in this chapter. It's time to
face the issue squarely. To understand risks in investing, you also have
to understand some companion terms: liquidity, appreciation, taxability, and, most important, trade-off.

Liquidity refers to the ease with which you can convert your investment to cash if you need it for some other purpose. Bank deposits, for
example, are completely liquid. You can get your money immediately
from your bank. Money market funds are also liquid. You can write a
check for your money. Common stocks and mutual funds are liquid as
well, easily salable in the investors' marketplace. Real estate is a
classic example of an investment that is not liquid—you need to find
a buyer for it and then wait out an escrow usually of a month or more.

Appreciation, also spoken of as growth potential, simply means
your investment's ability to grow in value. Some investments, like

certificates of deposit, pay good rates of interest but never grow in value. Other investments, like stocks and real estate, are likely to grow in value over time.

Taxability refers to how much of a cut Uncle Sam or your state and local governments will take from your investment earnings. Some investment income is taxed as ordinary income. Some investments, such as real estate, have a number of technical tax advantages. Some investments, such as most municipal bonds, earn income for you free of federal or state taxes or even both (''double tax-free''). (You may also find ''triple tax-free'' bonds—free of local taxes, too—in your locality.)

Trade-off is not an investment term strictly speaking, but it might as well be. Every advantage in an investment comes at a cost, either in terms of real money, or risk, or liquidity, or growth potential. In other words, those investments in which your principal is the safest are the ones with the most modest returns. Likewise, tax-free investments have more modest returns than comparable taxable investments, so their tax-free status only makes sense for persons in particular financial conditions (see page 323). (For those persons, in fact, the net after-tax return may well beat the return of a comparable taxable investment.) An illiquid investment like real estate usually has great growth potential. And so it goes.

Now, risk. As you have already gleaned from the above, ensuring the safety of your principal (the amount of money you invest) will have its trade-offs. Technically the safest investment is a passbook account at a federally insured institution. Your deposits are insured up to $100,000—as depositors in so many recently failed S&Ls are gratefully aware—and are completely liquid besides in case of an emergency. However, bank deposits pay interest rates that are generally below the rate of inflation, which means that if you had all your investment capital in the bank, your purchasing power would erode over time. Thus, while your principal is safe in terms of number of dollars, its real value is at considerable risk relative to inflation. On top of that, the interest income is taxed.

For these reasons, many investors prefer to move up one category of risk to do their banking, keeping their liquid money reserves in a money market fund. Money market funds pay interest rates that, while taxable,* are considerably higher than what bank passbook accounts

---

* Tax-free money market accounts, invested in short-term municipal bonds, are available and make sense for many in higher tax brackets. Most municipal bonds are inherently socially responsible.

pay yet require no broker's commissions to invest. (By the same token, money market funds pay less interest than a number of only slightly riskier investments, so they are best used as kind of a high-interest bank rather than for pure investing purposes.) Most such funds are also as liquid as a bank account—you can access your money with a check. Because they are not federally insured, money market funds are riskier than bank deposits. On the other hand, no one has ever lost a penny of principal investing in them—a record encompassing hundreds of billions of invested dollars. Therefore many investors will take that history to the bank, literally. (SRI money market funds are covered in the next chapter.)

To cite an example on the other end of the scale, "junk" bonds are an extremely risky category of investment, the trade-off being exceptional rates of return—if the company issuing the bond can make good on them. (Companies offering junk bonds are, by definition, companies with already high levels of debt—they have to offer higher rates of return than safer investments yield in order to attract investors.) Aggressive growth investments such as stocks in newer, promising companies carry significant risks to principal also, the trade-off for their exceptional ability to appreciate.

Determining your tolerance for risk is a matter of combining your personal circumstances with the limits of your emotional comfort zones. Your stage in life, for example, will dictate certain obvious strategies in most cases—primarily safe, income-producing investments if you are retired or nearly so, riskier growth-oriented products if you are younger and upwardly mobile in your career. (See "Stages of Life" section page 93 for more on this topic.) But some young people can't bear the thought of losing principal and will invest more conservatively than others in their same financial situation. Some retirees have an instinct for gambling and want to take calculated risks for the (not cheap) thrill of it. You learn very quickly in this game that peace of mind is worth far more ultimately than potential monetary gain. Of course, we figure that since you are interested in ethical investing, you already know a great deal about the value of peace of mind.

The bottom line on risk is that no investment is 100 percent safe— hypothetically, anyway. But *relative* risk is another thing altogether. Municipalities have been known to default on their bonds; however, examples are extremely rare, so that most muni (municipal) bonds are considered safe havens for your principal. Even ordinary bank deposits and certificates of deposit depend on the ability and will of the government to make good on its guarantees. That is exactly why some fearful souls keep all their money buried in their backyards. They prefer to forgo investment returns and take their inflationary lumps rather than let some stranger touch their money. And if that's the only way they can sleep at night, who are we to argue with them?

The various categories of investment, along with their attributes in terms of risk, liquidity, growth potential, and so forth, will be addressed in detail in the next three chapters.

COMPOUND MAGIC—PATIENCE PLUS COMPOUNDING INTEREST BUILDS FORTUNES

Also related to the subject of risk is the "magic" of compound interest. Compounding interest means that your money increases geometrically—that is, faster and faster over time. After a little time, the numbers really start to get incredible. Investors sometimes risk senselessly because they lack patience. If you are patient enough to allow the mathematics of compound interest to work for you, you can probably achieve your financial goals with minimum risk and maximum peace of mind.

For example, if at age thirty-five you invested $10,000 in an investment paying 9 percent annually and allowed your earnings to compound, your investment would be worth $132,676 when you were sixty-five. To look at it another way, if you began investing $100 a month ($1,200 a year) in the same investment at age thirty-five, you would have $178,290 in thirty years. Relatively safe investments in the 9 to 9½ percent range are widely available in today's market. It is with such building blocks that secure retirements are constructed.

Remember our discussion of trade-offs? You don't get something for nothing in this game. That is as close to an absolute in financial life as there is. This means that virtually every "get rich quick" scheme you will encounter is either ridiculously risky, foolish, or even fraudulent. So, buyer beware. You would probably do better by hauling your cash to Las Vegas, plunking it down on "odd" or "even" on the roulette table, and praying. You would do far better still by being patient and letting compound interest earned in solid, socially responsible investments build your financial independence.

WHEN TO START MANAGING YOUR MONEY

Straight away! Posthaste! *¡Muy pronto!* Or sooner. Since you have glimpsed the rabbitlike "breeding power" of compound interest, you can guess how expensive it is to procrastinate. Using one of our examples above, imagine you had started to invest that $100 per month at age twenty-five instead of age thirty-five. When you retired at sixty-five, your account would be worth $441,950 instead of $178,290. That is an increase of $263,660 for just an additional ten years. You have only so much time before you retire, and that time lessens with every passing day. When it comes to compounding interest, time is your friend. Make it one of your best friends. Start planning *now*.

SAVING AND CONSUMER DEBT

The financial independence that we promise can be yours depends on two important budgeting factors: saving/investing regularly and eliminating consumer debt.

Savings

Recall that our dramatic examples of the magic of compound interest were based on either investing a substantial amount of capital at an early age and allowing it to compound, or investing a modest amount ($100 per month) on a relentlessly regular schedule, starting at the same age. Investments require investment capital. Perhaps you have enough money socked away somewhere—in an insurance "savings" policy, say, or in a savings account—to get your compounding ball rolling. If you cannot find a "chunk" of capital somewhere in your collection of assets, or if the amount you do find is hardly big enough to qualify as a chunk, your most obvious source of capital is through savings. In other words, find a way to live on less than you take in, and invest the difference. And not just once or twice, but regularly, without fail, and without ever hitting up your investments for extra cash.

Does that sound austere and terribly regimented? The fact is, it does take discipline to formulate a financial plan and stick to it. The ultimate rewards, as we've demonstrated, should more than justify the sacrifice, however. Besides, a socially responsible financial plan is not built strictly on delayed gratification. It buys you peace of mind right now to know that your future and the futures of those who depend on you is being secured, and you're investing in a better world for all humanity as well. If you can save and invest, say, just 10 percent a month of your take-home pay, you will be well on your way to the financial independence we have promised you since the start of this chapter.

Consumer Debt

Credit cards (and other simple consumer debt devices) are a financial disaster if you use them to live beyond your means and then pay exorbitant interest rates for the privilege. But they are a great boon to a financially well-managed household. If you use credit cards in place of cash and then pay for your charges completely at the end of each month, you will have had free use of the money for up to thirty days and will never pay a penny in interest—which means your money can be earning up to a month's worth of interest in an investment instead.

So, here's how to get your consumer debt situation under control:

1) Before investing, use your newly saved investment capital to pay off your consumer debt. Yes, we know that seems to go against everything we've said about investing immediately so you can maximize the benefits of compound interest. But consider that your credit cards are probably costing you 16 to 22 percent. Your safe income-producing investments are going to pay you 8 to 10 percent annually in today's market, your conservative stock mutual funds 12 to 14 percent. So your cards are costing you far more than you can make back by investing potential payments elsewhere.

2) After you pay off your plastic, don't cut up all your cards. Save one to keep up your credit rating, to use as a second ID, and to bail you out in emergencies (quick cash, emergency purchases, plane fares, car rentals, and so on). We suggest, of course, that you select your card from one of the socially responsible options covered in chapter 7.

3) Use the card to *make* you some money, by putting everything you would normally buy with cash on your card instead. Buy as many of your daily purchases as your vendors will allow with your card—even your groceries. But keep track of your totals so you are never buying more than you can pay for at the end of the month (after subtracting the amount you are saving for investments).

4) Pay your credit bills on time and completely, or die (at least financially). Using your cards for all your cash purchases really runs up

the totals. If you miss a payment or can't pay the full amount, you will pay some hellacious interest.

5) If you can't control your purchases or remember to pay your credit bills on time, disregard points 3 and 4 and use that card only as suggested in point 2. You will never get anywhere financially paying the rates of interest that credit card issuers charge.

The general rule on consumer debt is simple: If you can borrow money at a lower rate than you can get in the investment market, grab it and invest it. Otherwise, either pay cash or don't buy (unless you absolutely need the item right now). For example, if a car dealer is offering you a 7.5 percent loan on a car you want and were prepared to pay cash for, finance the car and invest the cash savings in something that will pay better than 7.5 percent. You must realize, though, that such low-interest loans are almost never available, because no lender can make profits at those rates.

STAGES OF LIFE AND YOUR FINANCIAL NEEDS

As we indicated earlier, your personal circumstances and stage of life will dictate certain financial choices, assuming you want to follow conventional advice. For planning purposes, your adult life falls generally into two major categories—the earning years and the retirement years. Each has distinctly different, but complementary, attributes that should guide your money management.

Earning Years

We'll presume that those of you who fall into this age group earn enough to support yourself and your dependents, if you have any. We'll also presume that you earn enough to save a little each month, although it may take some rebudgeting to find that little extra. Ideally, you are also in an upwardly mobile position with respect to your career. Your major considerations during this time will include

■ the kids' education. Far better to invest a modest sum now and let compounding interest or appreciation build over time than have to spring for the whole amount in cash when Susie and Bob finish high school.

- establishing a living estate for your retirement and to support your heirs in case of your untimely demise or disablement. The so-called golden years are only golden if you plan for them.
- acquiring adequate life insurance until your estate is in place.
- reducing your debt so that you owe little or nothing when you are ready to retire.
- helping your parents in their later years, particularly if they have not prepared adequately for their own retirement or disability.

Most advisers will suggest that, circumstantially anyway, you can tolerate more risk in your investments in your earning years. You have the earning capacity and the upward mobility to offset any losses you might suffer from riskier investments. Thus they will suggest a significant proportion of growth investments for meeting your long-term goals. (If you are already sweating just with the mention of losses, remember that responsible advisers always earmark the large majority of the average client's money for the safer vehicles. Besides, our overriding advice is that you take only the risks with which you feel comfortable.) Clearly, as you near retirement, you start investing more like a retired person, at least in terms of reduced risk.

Retirement Years

Social Security and Medicare were not designed to take care of all of your financial needs in retirement, and they never will. Think of them purely as supplements to your own basic plan. Therefore, your considerations during retirement remain

- income. To maintain your standard of living after you stop working, your investments still have to replace the income you brought in from your career—yes, minus the small contribution that Social Security makes, but, far more significant, adjusted upward for inflation. Consider also that medical costs are rising much faster than the rate of inflation, and your medical needs—statistically speaking, anyway—may be substantial during this period of your life.
- hedging against inflation. Income vehicles, as we've stated, are vulnerable in times of high inflation because the inflation rate can approach, or even pass, the rate of interest paid by income investments. You should protect yourself against that possibility with some growth-oriented investments even in your retirement years.

Overall, you want to minimize risk during the retirement years, because you do not have the time and earning power to catch up that you had when you were younger. Remember, although you have stopped working, your capital is still on the job—in fact, it has become your primary "wage earner." So you need to guard its life with a prudent investment strategy.

ACCOUNTING FOR INFLATION

Our rather offhanded mention of inflation earlier is not meant to understate the need to take it into consideration. Indeed, we often proceed blithely ahead in our money habits as if inflation didn't exist. We get a 5 percent raise at work and celebrate by buying a new car, not considering that inflation is also running 5 percent. In reality, all we've gained with our raise is some protection against losing ground (which *is* a major accomplishment these days). The car only makes sense after the raise if it made sense before it.

To put this in perspective, consider that inflation has been running recently at about 5 percent per year (5.3 percent average in the 1980s). This means that every dollar you have at the start of the year is only worth approximately ninety-five cents at the end. And 5 percent inflation is modest. Inflation averaged 8.8 percent in the last half of the 1970s. In 1979 the cost of living went up 13 percent, meaning that your 1979 dollars lost thirteen cents each in a year!

In your projections for retirement and major expenses like a college education, inflation will dramatically increase your needs in terms of number of dollars. If you live on $3,000 per month now, you will need $7,960 per month to maintain your current standard of living if you retire in twenty years. Looking at it another way, a $10,000 car today will cost $26,533 in twenty years at that same 5 percent inflation rate. Believe us, there will always be inflation, because the government builds a certain amount of it into its fiscal policy, enabling it to pay off its old debt with cheaper, inflated dollars.

The ever-present reality of inflation is behind the notion of what investment professionals call *real return* or *real interest* rates. The mathematics of figuring real return are simple: Subtract the current rate of inflation from the stated rate of return on your investment. In today's market you should expect a real return of 3 to 4 percent from your safest investments. With 1990's inflation rate of 6.1 percent, you

would thus want to see 9 to 10 percent from your most secure investment vehicles.

Because those rates are easily attainable in today's ethical investment market, perhaps you understand better now why we urge that you
consider keeping only the most minimal amounts (if any) in savings
accounts. The interest on savings accounts rarely outstrips inflation by
more than half a percent. In inflationary times such as the early nineties
threaten to be, those interest rates can fall considerably behind inflation. You may also understand now why financial planners and sophisticated investors—whether socially concerned or not—seem to
speak so cavalierly about risk. Your money is at risk no matter where
you put it, because the risk of erosion by inflation is at least as real as
most investment risks.

THE UP SIDE OF THE ECONOMY'S UPS AND DOWNS

By now we probably have you thinking of inflation as some kind of
corrosive acid that eats all your money and destroys every chance you
ever had to get ahead in life. Banish those thoughts. For the "new"
you, the one who is reading this book and getting a grasp on money
management, inflation is actually a financial opportunity. So is recession. Several types of investments do very well in inflationary times
(for example, real estate and stocks). Recessions are great times to
pick up bargains, such as temporarily depressed real estate and stocks;
optimistic buyers turned huge profits on their positive thoughts after
the stock market crash of 1987.

Part of understanding money management is learning how to profit
by the cyclical fluctuations of the economy. We often envy the well-
to-do because they never seem to suffer even in the toughest of times.
That's planning for you—for investors who understand money, there
are no tough times, just different sets of profitable options.

The one thing you can count on in our country's economics is that
things will go up, and things will go down, but over the long haul the
economy will edge upward. That is because American companies continue to produce wealth. In fact, backing your confidence in this truism
with your investment dollars will help keep it true, because American
industry depends on investment.

The other thing you can count on is the continued viability of socially responsible investing, because American business has gone just

about as far as it can go without at least taking environmental ethics into consideration. Cleaning up the resultant mess, and complying with environmental regulation, will be a major growth industry over the next decade or two.

In chapters 4, 5, and 6 you will learn how specific investment categories perform relative to inflation and recession. In chapter 7 you will learn how to figure inflation into your assessment of your personal financial needs.

LIFE INSURANCE—WHERE DOES IT FIT?

Life insurance is an investment only in the most temporary sense. The purpose of life insurance is to provide monetary benefits to your survivors. It ensures, literally, that should you die prematurely, those who depend on your wage earning will be able to maintain their standard of living (or, should the nonworking spouse raising children die, the surviving spouse will be able to afford child care). Once you have established your living estate, the estate itself should provide for your loved ones. Today you can get much better returns by investing your money than are available through even the best life insurance products. Therefore we recommend that you carry life insurance only until your estate is in place.

The insurance industry can dazzle you and dizzy you with its array of products, but essentially there are two broad categories of insurance: 1) pure protection, called *term insurance,* and 2) protection that includes banking and/or investment features. The latter category breaks down into two main subgroups, *whole life* and *universal life*

Term insurance is a simple wager between you and the insurance company—if you die while the insurance is in effect, you "win" and the insurance company pays your beneficiary. If the insurance company wins its bet that you will survive the term, it keeps everything you have paid. Term life insurance gets its name because it is purchased for a specific period of time, although renewable term insurance allows you to extend the policy for another term. The driving principle behind the premium structure for term life is that premiums go up as you get older (since the statistical odds against the insurance company winning its bet are also rising). Various payment options can keep the cash payments level. (In decreasing term insurance, for example, the payments stay level but the coverage decreases as you age.)

Whole life insurance is essentially term insurance plus a low-interest savings account. The level payment structure includes an overpayment in the early years of the policy; some of that overpayment is put in a savings account to build up what is called the policy's *cash surrender value*. Cash surrender value means you can retrieve the money in the savings account in any of three ways. If you cancel the policy, it will be paid to you in cash (although you then lose all insurance protection). Your other options are to draw on the account to pay your premiums, or to borrow the money at low interest rates; the loan need never be repaid, but the amount borrowed is subtracted from the death benefit.

It is important to note that the cash surrender value, the amount in your account, does not increase the death benefit; no matter how much you have paid into the policy, your beneficiary receives only the face value of the policy if you should die. Thus, cash surrender value is a living benefit only, a kind of forced savings plan for you, the payer. But consider these questionable (to say the least) attributes of banking through an insurance policy:

- You are being charged fees to deposit into your account.
- You have effectively made your insurance company an heir to part of your estate, because no matter how much you have paid in, your heirs will only receive the policy's face value if you die—your insurance company "inherits" the savings account.
- You will be charged interest (typically 4.5 to 8 percent) to borrow your own money.

Many consumers eventually wised up to the dubious "savings" aspects of whole life. They began borrowing from their cash accounts, which returned only about 4 percent annually, and put the money in investments returning market rates. To plug the resultant drain on their resources, insurance companies devised another product, *universal life,* to win back their customers' investment dollars. Universal life policies enable consumers to buy term insurance and make truly valuable investments entirely within each premium payment. Universal life gives you term insurance (on a permanent renewing basis) for part of your premium, extracts a management fee as another portion of the premium, and establishes a cash value fund for you with the rest. This cash value fund is invested for you by the insurance company and pays rates competitive with other conservative income-type investments (typically a floor of 4 to 4.5 percent is guaranteed with an actual return, currently 8 to 8.5 percent, varying with the performance of the company's investment portfolio). Additionally, the IRS allows your money to compound tax-deferred—you only pay tax on money actually paid out. Flexibility is another attribute of universal life. You may increase your premium if you want to invest more or decrease it to about the level of a term policy (plus maintenance fee) if you only want to continue with the protection portion of your policy. You can also draw on your cash value fund to pay your premium.

Note that some insurance companies are now offering an intriguing free rider with various whole life and universal policies that allows the insured individual to draw from his/her insurance value for catastrophic illness or long-term nursing home care. Thus the death benefit is effectively accruing to the insured in life under these conditions.

Okay, let's analyze your options. We'll assume at first that you have no current life insurance policy and want to know which way to go. The traditional advice, and ours, is to buy term insurance, the cheapest form of life insurance, *if*—and this is a big, big *if*—you invest the money you save each payment period as regularly as the insurance company would for you. Whole life and universal life are essentially forced savings/investment plans for those who do not trust themselves to invest regularly otherwise. But you pay a price for that lack of discipline. In the case of whole life, you pay too big a price for most personal situations, for even as a "forced" savings account, it ain't much of an account, as we've explained above. Universal life is a far better deal for most people, but you can earn more than its returns on your own: you can get a 10 percent annual return or so in a good socially screened bond mutual fund, 12 percent in a socially screened conservative stock fund compared to the policy's typical (circa 1990) 8 to 8.5 percent returns. Not only that, but typically you'll pay less in commissions, management fees, and other overhead expenses than you would pay to the insurance company to invest for you.

From an ethical standpoint, term insurance was also the only way to go until recently for one simple reason: Only one insurance company invested premiums in a socially responsible manner, and term insurance is all it offered. Consumers United Insurance Co. (see contact information in chapter 7) offers annually renewable term insurance, but no whole or universal life.

However, as just one measure of SRI's rapid growth in the early 1990s, First Affirmative Financial Network expects to be offering a whole range of ethically invested life insurance products—term, whole life, and universal life—as of October 1991. When this comes to pass, you may want to consider a universal policy as a beginning money management tool. Note that 8 percent of something is far better than 12 percent of nothing. Figure honesty into your financial planning. If you are not going to commit your term insurance savings to another investment on an absolutely regular schedule, get yourself a good universal life policy until your living estate is ready to take over the job of protecting your dependents.

If you already possess life insurance, what should you do? Our best advice is to review your situation with a financial planner who is knowledgeable about both insurance and your financial particulars. Your adviser will probably suggest you leave your policies in place

- if your health is such that it would be difficult to pass a new company's physical
- if you otherwise can't lower your cost-per-thousand of protection
- if your policies contain expensive penalties for cashing out.

In general, however, you will want to restructure your insurance if

- you have multiple policies (Every policy has an administrative fee. Why pay more than one?)
- you have whole life or some variant with a "savings" feature (If you don't trust yourself to invest regularly, use universal life.)
- your policy pays you dividends. (These "participating" policies are simply rebating you a portion of an overcharge, for they usually cost more per thousand over their term than non-dividend-paying policies.)

In chapter 7 we will show you how to figure your life insurance needs.

HEALTH INSURANCE—YOU CAN'T AFFORD NOT TO HAVE IT

Although not everyone need consider life insurance, health insurance is indispensable to any financial plan. Otherwise, accident or disease could wipe out even a considerable estate. Therefore, if you do not receive health insurance from your employer, you should figure in your health insurance costs before budgeting your monthly investment capital.

Another important reason for purchasing health insurance now is to ensure your eligibility in the later stages of life when you may really need it. Suppose you decide not to carry any health insurance at all, reasoning that you are young and healthy and can afford to pay your few expected health costs out of pocket. You may win your gamble in the short term, but if your health should deteriorate in your later years, you may not be able to find an insurer to cover you—at premiums less than the national debt, anyway. If you had been carrying health insurance all along, your insurer would be forbidden by law from disqualifying you when you become more of a risk to them.

Although health care is extremely expensive in this country, health insurance doesn't have to be. Most consumers are unaware of the range of health insurance options, however, and the insurance industry does not go out of its way to enlighten them. So here's the skinny.

Individual vs. Group Insurance

If you are not covered by your company plan, you have the choice of either seeking individual coverage for you and your family or affiliating with an organization that offers group-rate health benefits to its members. If after reading the following section on individual insurance you decide that full coverage makes sense in your case, try like heck to get a group rate from someone. If your normal medical costs are quite modest, you will probably do better with the individual high-deductible policies described immediately below. However, you may want to compare their rates to group-rate insurance anyway.

Individual Plans

Individual health insurance is at its most expensive when you pay for full coverage with a minimal or no deductible. If your health is such that you would normally pay thousands of dollars annually for care even if you had no major medical emergencies, then this comprehensive coverage may be worth it. Otherwise consider buying a policy that covers the normal range of health conditions but has a high (low to mid-four-figure) deductible. This way you are still protected against financial catastrophe. If your out-of-pocket medical expenses plus the cost of your premium is less than the cost of a full-coverage policy, you have won your gamble.

Example: Sam buys a full-coverage policy from a health maintenance organization costing $175 per month, or $2,100 per year. During the first year of coverage, he gets a complete physical ($50), eye exam and contact lenses ($250), several refills of his regular prescriptions ($200 total), and emergency treatment for heat exhaustion ($500). The insurance gladly pays 100 percent of his costs ($1,000 total), because they are still ahead at the end of the year to the tune of $1,100. Sam could have purchased instead a discount policy at, say, $75 per month that would have paid 80 percent of his costs after a $5,000 deductible (with all expenses paid after $6,000). He would have had to pay for his exams, contacts, prescriptions, and emergency treatment in cash, but that $1,000 plus his $900 in annual premiums would have saved him $200 over the cost of full coverage. If he had had no out-of-pocket expenses that year, he would have saved $1,200.

There is a lesson to be learned here. We are so used to having health insurance (often purchased for us by our employer) pay for all our sniffles and boo-boos that it seems outrageous that we could come out ahead by paying out of pocket. But full-coverage health insurance rarely pays for itself except when you suffer terrible health.

Plus, these days many employers are requiring employees to pay at least part of their premiums on the company plan. Company plans usually offer far better rates for full-coverage insurance than you can purchase as an individual, but even there, compare the annual cost of company insurance with high-deductible major medical insurance you can buy on your own. (Be sure to ask your insurance agent to discuss maximum-deductible policies with you; insurance agents do not always bring up those options on their own. Find out also if you get higher-deductible insurance through your company plan, the best of both worlds.)

You can, of course, save even more on your maximum-deductible policy by living a healthy life-style and learning to care for those problems that do not necessarily require medical attention.

Group Plans

Most medium and larger-size employers offer some sort of group plan, and most will pay all or part of the premium. If your employer is even just matching your contribution to your insurance, you probably can't beat that deal financially.

However, if your employer simply offers you the group rate but does not pay it, it is time to shop and compare. If your health is good, you may do better with an individual major medical policy that includes a high deductible. But if the difference is small, consider that the group plan will give you much more coverage for your money.

If your employer does not offer health insurance, see if any of the professional associations, clubs, or other membership organizations to which you belong offer group plans that you can join. If none of them do, consider joining an organization that does have a group plan for its members. The discounted premium you pay on the group plan may by itself justify the expense for membership dues. For example, if you join Co-op America—a socially responsible organization if ever there was one—your $20 annual dues will make you eligible for their group plan with Consumers United Insurance Company, the only socially responsible health insurer in the country. (Note that the above advice changes a little if you have a pre-existing medical condition, in which case individual insurance will be expensive and non-employee group plans may exclude you. In this event, you will likely be better off with your employer's group plan—which by law can't exclude you—even if you have to pay the premiums. In fact, it may be your only recourse.)

Temporary Coverage

If you are not currently covered, we hope the forgoing has impressed upon you the importance of doing something about that right away. However, once you apply for a health insurance policy, it will take several weeks for the insurer to process your application. Insure yourself during the waiting period by purchasing temporary major medical insurance. These policies, available at reasonable fees on a month-at-a-time basis, do not usually require any prequalifying. Your insurance agent should be able to sign you up to a temporary policy or at least direct you to one. Don't spend another day without some sort of major medical coverage!

How to Cover Yourself If You Should Lose Your Job

If you should lose your job, you do not necessarily lose the health insurance that goes with it. Employers with twenty or more employees must by law offer terminated employees continuing coverage on the company plan for eighteen months. As an ex-employee, you will have to pay the premiums, but those premiums should be far less than any full-coverage insurance you could purchase as an individual. You will have to determine if you can do better still on a maximum-deductible major medical policy. Until you make up your mind, however, be sure to purchase temporary insurance immediately to bridge any interim periods between coverage.

RETIREMENT PLANNING WITH YOUR UNCLE'S HELP

Whatever his other faults, Uncle Sam has been trying to teach you the wisdom of planning for retirement for almost twenty years now. In 1974 Congress brought Individual Retirement Accounts (IRAs) and Keogh plans into being so that citizens without employee pension plans could create analogous benefits on their own. In 1981 Congress even gave people on pension plans the right to open IRAs. The tax reform measures of a few years ago have limited the benefits of IRAs for those already on employer-sponsored plans, but those who are not (plus those on employer-sponsored plans who earn below a stated amount) get all the goodies of the old IRAs. (Note that any socially responsible investment can be used for your IRA.)

Nor has Uncle stopped there. He has created an array of vehicles by which you can shelter income from taxation and allow money to compound on a tax-deferred basis. The most important features of the most common plans are detailed in chapter 7, with socially responsible options included.

Before chapter 7, however, we want you to understand why these plans are so helpful to your overall financial plan. Remember our little discussion of ''compound magic'' earlier in this chapter? Can you imagine how much more you might have if you could exclude thousands of dollars from your taxable income each year? Are you in a 28 percent tax bracket? That's $560 per year in tax savings on a $2,000 IRA. For those of you in the 33 percent bracket, the savings are $660 per year in the same $2,000 account. Here's how the numbers might project out for you from the time you open a retirement account to the time you hit the Winnebago trail:

Susan Smart, age thirty, earns $24,000 annually and is in the 28 percent tax bracket. She opens an IRA account in an investment earning 10 percent per year. She contributes the maximum allowable. $2,000, the first year because she wants to pocket the $560 in tax savings. Being a steady, disciplined sort—and no fool besides—she commits to the idea of contributing $2,000 every year thereafter until retirement.

At the end of ten years, Ms. Smart's account will be worth $35,062, plus she will have saved $5,600 in taxes. After twenty years the account will be worth $126,500, and the tax savings will total $11,200. After forty years, when Susan is ready to retire, she will have $973,704 in her IRA account alone to retire on, and she will have saved $22,400 in taxes besides. Yes, she will now owe taxes on the money she withdraws, but she is likely to be in a lower tax bracket at this point than she was in her working years (if the government hasn't messed with the tax bracket structure in the meantime, that is).

Now suppose Susan was married and her husband also funded a 10 percent IRA for forty years. The nest egg doubles, to a cool $1,947,408. Consider this as well: We've run these numbers on $2,000 per year. Under some of the pension plan options outlined later—such as salary reduction, money purchase, and profit-sharing plans—you could set aside several times that.

COMMISSIONS

Commissions are a fact of life in most investing, because most financial advisers earn their living through commissions. Think of commissions simply as a payment for services rendered. Your financial adviser, if competent, is helping you earn a great deal of money and saving you a great deal of time and effort—not to mention peace of mind—in the process. If you choose not to follow his/her recommendations, the advice is free. So just factor the commission you pay into your calculation of returns and be grateful that you can secure your family's future at so modest a price.

Actually there are a variety of investments for which no commissions are charged, including new issues of corporate and municipal bonds and new U.S. Treasury issues (if you purchase them directly from the government). These are the exceptions, however. You will pay commissions to purchase any of these same vehicles on the public market. You will pay a broker a commission to buy or sell stock for you. (Technically you could find a buyer or seller yourself, agree on a price with that person, and complete the transaction without either of you incurring a commission—nevertheless, you can imagine how rare such occurrences are.)

You will also pay a "loading" fee—in other words, a commission—with most mutual funds. These can run as much as 8.5 percent (the maximum allowable by law), although most SRI funds are between 3.5 and 4.75 percent. Most mutual funds charge you at the time of purchase—these are called *front-load* funds. *Rear-load* funds charge you when you sell (more if you sell soon after investing, not at all if you hold the fund for a stated period of years). For competitive reasons, some funds, called *no-loads,* eliminate these charges. Although consumers have been conditioned to expect lessened performance from most discounted items, no-load funds perform just as well on the average as do those carrying loads. By the same token, some of the priciest load funds more than justify their high loads. Performance is the point, not the load size. Besides, no-load funds are not free lunches. All charge management fees, and you will have to handle paperwork that a broker would do for you on a commissioned fund. (Three of the current SRI mutual funds—the Pax World Fund, Dreyfus Third Century, and the new Domini Social Index Trust—are no-loads as of this writing.)

Still and all, the cost of commissions is an important reason for seeing your investments as long-term commitments of your money. Frequent trading and panicked selling can eat up not only your profits, but some of your principal as well. You paid some real money to get into your investment—effectively a significant portion of your first year's returns. The investment would have to perform pretty spectacularly to justify trading out of it in that first year.

OWNING VS. LENDING EQUALS GROWTH VS. INCOME

When you examine your investment options in your money management program, you will normally be considering such factors as safety of your principal, income vs. growth, hedges against inflation, and so on. You probably won't be wondering whether to lend your money or buy something with it. However, in effect, each of your investments will be doing one or the other with your money.

The choice of whether to lend or buy relates quite directly to some of the most critical investment criteria. When you look for growth, you normally look toward owning investments. When you look for safety, you normally lend your money. It's just about that simple (with junk bonds, a risky lending vehicle, a notable exception).

Ownership provides the potential for appreciation of your principal, but the trade-off for that potential is risk to that principal. Ownership is a better bet in inflationary times, because appreciation better offsets inflation's ability to erode value.

Lending investments usually yield dependable returns—many, in fact, pay fixed returns on a monthly or quarterly basis. The trade-off here is accepting more modest gains than are possible with ownership. The major risk with most lending investments is that inflation can overwhelm your gains, leaving you with less purchasing power than when you started (although certainly much more than if you had not invested at all).

Chances are that you have done some of each type of investing already. If you have a savings account at a bank, you have lent that money to the bank. The bank invests that money and pays you interest on the loan. If you are a homeowner, you may have made one of the best owning-type investments available to you.

Typical lending investments also include certificates of deposit, money market mutual funds, U.S. savings bonds, municipal bonds, U.S. Treasury securities, government agency securities, and corporate bonds. The most common ownership investments include stocks, real estate, limited partnerships, commodities, precious metals, and collectibles. Ethical options are available in nearly all of these categories, as you will learn in the next two chapters.

Contrasting a corporate bond with stock in the same corporation provides a useful comparison of lending vs. ownership. A corporate bond is a vehicle by which the corporation borrows money from the general public, in return for which it normally pays a fixed interest rate for a fixed period of time. This lending vehicle will never pay more than the stated interest rate, nor will the value of the bond itself increase if the corporation prospers. Your principal will be returned to you in the exact amount lent when the bond matures (or earlier, if the bond is "called"—see discussion on corporate bonds in chapter 4).

If you own stock in the company, there is no limit on your profits. The stock could increase in value several times. The down side is that the stock's value could plunge so low that your principal shrinks. Dividends, the income aspect of stocks, also rise and fall with the fortunes of the company. (Note that it is possible, although unlikely, that a troubled corporation could default on its bond obligation. Generally, however, a financially strapped company will skip out on its debt only as a last resort, after cutting first its dividends on common stock and then its dividends on preferred stock. Only for "junk bond" holders is default considered a significant risk.)

Chapters 4 and 5 will describe in detail the array of both lending and owning investments available today, including stocks and bonds.

DIVERSIFICATION—THE ULTIMATE PROTECTION

Diversification is the investor's response to the wise axiom "Don't put all your eggs in one basket." Diversification guards your overall finances from being wiped out by any single investment going sour, protects you against fluctuations in the economy, and allows you to take some calculated risks without endangering your overall financial plan for the future.

When we say "diversification," we're actually talking about three categories: diversification of type of investment, diversification of single-issue investments such as stocks and bonds, and diversification through mutual funds.

Again, because of the great array of solid, ethical investments, you can build a completely diversified financial plan without compromising its social responsibility.

Diversification of Investments

When your financial planner suggests that you invest in a "diverse portfolio" of products, it is probably not because he or she is trying to sell you everything in the store. As with life in general, very little is for certain in investing, so a diverse portfolio protects you in the unlikely but possible event that a portion of your investments fail. Assuming your adviser is competent, your portfolio will consist primarily of safer investments. This will enable you to invest a smaller portion in more aggressive growth products, if you can tolerate the risk emotionally and financially. Your portfolio will also contain a balance of liquid and illiquid investments—liquid investments, for ready emergency cash and flexibility to respond to today's rapidly shifting economic picture, and illiquid investments, so you can reap some of the higher rewards available to those who allow their money to be lent or otherwise tied up for long periods of time. Your portfolio will include "hedges" against both inflation and recession—if all your money was in fixed income investments, high inflation would gobble up your net worth. If your portfolio was solely in stocks during a falling market cycle, it could be disastrous if you had to sell stock to generate cash. Depending upon your adviser's sophistication, he or she may suggest a variety of "fine-tuning" diversities as well.

Diversification of Stocks, Bonds, and the Like

No matter how hot the company, no matter how experienced its management, no matter how well placed its products in a skyrocketing economy, stuff happens. If you've got all your money in that company's stock or bonds or even both, you are unnecessarily vulnerable to deep "stuff." If you want to invest in individual stocks or bonds or other individual investments (as opposed to mutual funds), invest in several (five to ten) to spread the risk around

You should also diversify your investments in terms of economic sector—in other words, don't put all your money only into the computer industry. In addition to the fact that industry-specific events can devastate a portfolio too heavily weighted in that industry, general events will affect various economic sectors differently. When Iraq invaded Kuwait in mid-1990, energy stocks jumped. But stocks in retail-based companies took a dive, apparently on the fear that rising energy costs would reduce consumers' ability and inclination to spend.

There is such a thing, however, as too much diversity. Do not invest in so many issues that you cannot keep track of your holdings. Many professionals suggest ten as a maximum. Besides, once you get to a plateau of about five stocks, additional diversity does not buy you that much more protection. For more on diversifying stock investments, see page 216.

Diversification Through Mutual Funds

Diversification of investments overall, and of individual issues like stocks and bonds, assumes that you have a healthy chunk of capital to invest in the first place. If your beginning means are more modest, there is another way—mutual funds. A mutual fund is a company that pools investors' money to assemble a portfolio of securities (usually common stocks and bonds) and/or other investment vehicles and then manages that portfolio for the investors. In addition to the professional expert management (for which you do pay a fee), you get automatic diversity at a far more modest price than if you put together a diverse portfolio with your own money.

Do note, however, that you still have not really diversified your investments if you stick all your money into a single mutual fund. A stock market fund is only diverse in terms of stocks; a bond fund has only bonds. Furthermore, mutual funds are rarely diverse in their objectives. There are conservative stock funds and riskier, aggressive growth funds. There are funds with narrowly defined economic strategies, such as investing exclusively in companies climbing back from near financial disaster. To truly diversify with mutual funds, you should ideally buy into several—offsetting riskier stock funds with conservative ones, for instance. However, if you only have enough now to buy into one fund, start there. (In chapter 6 we'll show you how to get started for as little as $25.) For those who can only afford one investment, a mutual fund is the way to go.

The SRI field offers a number of socially screened mutual funds that perform impressively. Chapter 4 covers mutual funds in detail, including reviews of available SRI funds.

DOLLAR COST AVERAGING—DIVERSITY OF TIMING

Want to know the secret to making millions in the stock market? Buy low, sell high. What could be easier?

Proving the existence of God, for one thing. What's low and what's high in the market is clear only through hindsight. The stock market fluctuates not according to real economic factors, which are more or less discernible, but to market players' perceptions of those factors. What people are thinking is hard enough to nail down in the most stable of circumstances. When they have money at stake, their thought patterns—and the actions they trigger in the stock market—are far too erratic to bet your future on. If you think your special insight into people will up the odds in your favor, feel free to test your theory if you can afford it, but note that even psychics find their abilities deserting them when their own money is on the line.

So how else can you beat the system? With a system of your own—a system called *dollar cost averaging*. Dollar cost averaging means putting a fixed amount of money into the same stock (or mutual fund) at regular intervals, regardless of the price of that investment at the time. In terms of our just concluded discussion on a diverse stock portfolio, that means buying each of the stocks you plan for your portfolio in regular increments rather than all at once.

Let's see how it works in three possible scenarios, a fluctuating market, a declining market, and a rising market.

Fluctuating Market

Suppose you have $300 to invest. Because you're going to dollar cost average, you decide to invest $100 every three months instead of investing $300 in one shot. Having picked out a socially screened stock that sells at $10 a share, you plunk down your first $100 and receive ten shares. Then the market drops, and three months later your stock is at $5 a share. Since your investment interval is up, you invest your second $100. This time it buys twenty shares. (These numbers are rather unlikely in some ways, but they work great for an illustration!) Three months later your stock has bounced back to $10 a share, so your third $100 buys ten more shares.

Let's see how you're doing up to this point:

DATE	INVESTMENT	SHARE PRICE	SHARES BOUGHT
January 1	$100	$10	10
April 1	$100	$ 5	20
July 1	$100	$10	10
Total:	$300	$25	40

Your average share cost: $7.50 ($300 divided by 40 shares)
Average share price: $8.33 ($25 divided by 3)

Do you get what's happened here? The market hasn't done a darn thing, but you have. You have forty shares of stock, worth $400 if you sold today. You're $100 ahead of where you'd be if you had invested all your money initially. If you still like the stock and decide to hang on to it, you're ten shares over where you would have been with a $300 single investment.

Declining Market

Now, you diversify. You pick another socially screened stock and decide to invest $1,500 in it at the rate of $300 every three months. The stock is selling for $25 a share when you start, so your $300 buys twelve shares. But bad times hit the economy and your stock starts to drop in price with the rest of the market. Three months later, at your next "buy" interval, the price has fallen to $15 a share. You buy anyway because you believe in the dollar cost averaging system, your $300 getting you twenty shares.

After a temporary recovery, the skid continues, but with dogged determination you continue to invest $300 every quarter. Here's what might have occurred after five quarters of near continuous decline:

DATE	INVESTMENT	SHARE PRICE	SHARES BOUGHT
January 1	$300	$25	12
April 1	$300	$15	20
July 1	$300	$20	15
October 1	$300	$10	30
Jan. 1 next	$300	$ 5	60
Total:	$1,500	$75	137

Your average share cost: $10.95 ($1,500 divided by 137 shares)
Average share price: $15 ($75 divided by 5)

Do you show a profit? No. Frankly, the market hasn't allowed it. Have you protected your backside? Yes. If you had invested your $1,500 all at once, it would have bought only sixty shares and you would have to sell them all at $25 a share just to break even. Because you dollar cost averaged, you have 137 shares. If your stock recovers only to $11 a share, you will be ahead of the game. History shows that the economy will eventually right itself. Therefore, if you've picked a solid company, the long-term odds are well in your favor.

Rising Market

Let's say you had bought into your second stock in a mostly rising market instead. In that case your $300 would have bought fewer as the price of those shares rose. The new numbers:

DATE	INVESTMENT	SHARE PRICE	SHARES BOUGHT
January 1	$300	$ 5	60
April 1	$300	$15	20
July 1	$300	$10	30
October 1	$300	$15	20
Jan. 1 next	$300	$25	12
Total:	$1,500	$70	142

Your average share cost: $10.57 ($1,500 divided by 142 shares)
Average share price: $14 ($70 divided by 5)

During this period the average share price was $14, but you only paid $10.57. Yes, if you had invested the entire $1,500 when the price was at its lowest, you would have done much better. But the fact is, you did just fine.

That's the point with dollar cost averaging—the *averages* are on your side. To recapitulate: You will get more shares for your money during those intervals when the price is down, and when the price rises again, as it generally will, you will make a profit on those shares. Put another way, buying at intervals practically ensures that you will buy some bargain shares, and the historical upward trend of the market over the long term makes it extremely likely that you will eventually see profit.

By the way, the system applies to purchasing shares in stock mutual funds as well, because these also fluctuate in value and tend to appreciate over time. In fact, most mutual funds—including most SRI funds—allow you to invest via an automatic check draft plan (or some similar feature), by which you direct your bank to debit your account at regular intervals and send the money to the fund. Some SRI stock funds allow regular monthly contributions of as little as fifty dollars. This mode of investment is one we heartily recommend: not only does it ensure dollar cost averaging, but it creates a system of regular savings/investment in the same stroke! Throw in the factor of compounding earnings and you've got yourself a pretty potent formula for creating wealth.

Now the caveats: To begin with, the system only works if you have 1) the cash to buy when the interval comes up, 2) the discipline to buy no matter what the price, and 3) the patience and financial staying power to hang on to your stock while the system is doing its job for you. Those who win big in the stock market anyway are those who do not panic when the market drops and who have invested only money they can afford to tie up until it does complete its job, however long that takes.

The other thing to remember is that with investing, as with everything else, nothing is forever and nothing is absolutely certain. Dollar cost averaging is based on holding investments for the long term as opposed to trying to time your investment activities with the market's highs and lows. The mathematics of dollar cost averaging plus the history of the market weigh heavily in your favor. But suppose your mutual fund changes management and the new team is a loser? Suppose the hot growth company whose stock you started with a few years ago has completed its growth phase? Suppose a new, unforeseen technological development has rendered obsolete the products of another company in which you hold stock? Or suppose your goals change—as you near retirement, say, or as you pull in the reins so you can change careers?

The point is that no matter how sound your original plan and no matter how logical a system like dollar cost averaging, you still need to monitor your investments. At least once a year, review your portfolio and ask yourself if each investment in it is meeting its objective and if those objectives themselves are still current. Think of it as your three-thousand-mile tune-up.

LOOKING AT DIVERSIFICATION AND RISK TOGETHER

Now that you have a good understanding of why to diversify your investments, and of the delicate balance between risk and reward, we'll show you graphically how diversity and risk interface. The "Investment Strategy Chart" (see page 115) illustrates the basic "menu" from which a financial plan is constructed. Not all plans select from all categories, of course. The categories you select from for your plan will be determined by an interplay of your circumstances and your risk tolerance.

Please remember, however, that "risk" is a very relative term. The investments labeled "safe" in the chart are the ones most at risk to inflation. So if you play the investment game too close to the vest, you may end up losing more than if you had "risked."

Investment Strategy Chart

I. SAFE	II. VERY LOW RISK	III. LOW RISK
EE & HH US Government Savings Bonds	Annuities	Balanced Mutual Funds
Gold & Silver Bullion Coins	Collateralized Mortgage Obligations (CMOs)	Growth & Income Mutual Funds
Insured Savings Accounts	High-Grade Municipal Bonds*	High-Grade Preferred Stock*
Insured Checking Accounts	High-Grade Corporate Bonds*	Medium-Grade Corporate Bonds*
Insured Certificates of Deposit	Income Mutual Funds	Public Limited Partnerships, U.S. Government Mortgage-Backed
Life Insurance Policies with Cash Value	Money Market Accounts	Utility Stocks*
U.S. Government Treasury Bills, Notes, and Bonds*	Unit Investment Trusts	
	U.S. Government Agency Bonds*	
	U.S. Government Mortgage-Backed Bonds*	
	Utility Company Bonds*	

(*Continued on next page*)

IV. MEDIUM RISK	V. HIGH RISK	VI. VERY HIGH RISK
Blue Chip Common Stock*	Aggressive Growth Mutual Funds	Collectibles
Growth Mutual Funds	Public Limited Partnerships, Leveraged	Futures Contracts
Index Mutual Funds	Real Estate	Low-Grade, High-Yield (Junk) Bonds*
Public Limited Partnerships, All Cash	Sector Mutual Funds	Oil & Gas Limited Partnerships
Real Estate Investment Trusts (REITs)	Speculative Stocks and Bonds*	Penny Stocks
Tax Credit, Low Income Housing Limited Partnerships		Private Placement Limited Partnerships
		Stock Options
		Venture Capital

*Purchase these investments directly or through mutual funds.

The "Portfolio Design Chart" (see page 117) shows our recommended plan balances keyed to different stages of life. Keep in mind that our recommendations are based on certain general assumptions. We're assuming an average level of risk tolerance. Hypothetically we're postulating a small percentage of what we call very high risk investments for younger, upwardly mobile investors, while in fact there are very few ethical choices in this category. As a rule, socially responsible investors tend to be conservative in their tolerance of risk, so very few high-risk investments have been created to serve them.

We're assuming that younger persons are upwardly mobile and older persons are either planning for retirement or already retired. In fact, age is less relevant than the circumstances we associate with it. For instance, if you are only thirty years old but pursuing an artistic career, you may need to invest like a retired person—protecting your principal in conservative investments, investing for regular income to support your artistic endeavors, mixing in a small percentage of growth investments as a hedge against inflation, and so on.

But no matter what your situation in life, you *can* balance risk, safety, and reward—and you can do it with social responsibility, as we'll show in the next three chapters.

Portfolio Design % of Investment Dollars by Risk Category

RISK LEVEL

	I. *SAFE*	*II.* *VERY* *LOW*	*III.* *LOW*	*IV.* *MEDIUM*	*V.* *HIGH*	*VI.* *VERY* *HIGH*
Age 30 Single	*		5	30	60	5
Age 30 Married No Children	*	5	20	25	45	5
Age 45 Married 2 Children	*	15	40	35	10	
Age 60 Pre Retired	*	70	20	10		
Age 60 Retired	*	85	10	5		

* We recommend holding three to six months normal cost of living reserves in this category before making investments in higher risk categories.

FACTORS FOR INVESTMENT DECISIONS
- Growth to match or exceed anticipated inflation
- Income to supplement salary, business, or other income
- Growth for a specific goal, such as a child's college education, a home purchase, or retirement

DIVERSIFICATION
Target for five to eight different investments—bonds, stocks, real estate, and the like—from categories II–VI above. When concentrating funds in safe and very low risk investments, less diversification is needed.

4

A Guide to Socially Responsible Investments— Individual Investments (Except Stocks)

This chapter begins a detailed look at the various categories of investments, including—where appropriate—listings of specific socially responsible investment opportunities that you may want to consider.

Before proceeding, you should note the following:

■ The price and value of many investments fluctuate daily. Thus, in a book of this type it is only possible to be current in a very general sense. Our financial ratings of specific investments are best used as general indications to take to your broker or financial planner. A professional will be able quickly to do the updated research that will verify whether the investment that you read about here still makes sense for your present situation. Or if you are the do-it-yourself type, you will want to research the investment yourself in daily publications like *The Wall Street Journal*.

The social ratings of an investment will tend to be far more stable. Time passes and things change, though, for the bad and the good. SRI periodical resources listed in appendix D will help keep you current on a company's social manners.

■ Our descriptions of the attributes of bonds, limited partnerships, and all the other categories of investment are necessarily general. But within each category there are investments with wrinkles upon wrinkles upon wrinkles. This is yet another reason for engaging the services of a professional adviser before making a final choice on an investment.

- Our financial advice detailed herein is designed to educate you on the financial attributes of each investment, not stick you with our personal hunches. The advice represents a synthesis of general thinking in the field.
- Unless we say otherwise, the investment income discussed in each category *is* taxable.
- This chapter covers only individual investments, except for stocks, which are covered in the next chapter. Mutual funds are reviewed in chapter 6.
- Please be aware with regard to any investment opportunity mentioned in this book that *past performance is no guarantee or predictor of future performance*.
- Before investing in any vehicle, read the official prospectus and relevant background literature. Be confident of your broker's qualifications—track record, training, and so on—before acting on any advice, and make sure all your questions are answered to your satisfaction. Bottom line, make sure you are comfortable with your investment!

THE SOCIAL SCREENING PROCESS

Different categories of investment require different degrees of social screening. For instance, stocks and corporate bonds require rigorous screening because investing in these securities supports the goals of the companies that issue them. Each company must be screened comprehensively, not only for the type and quality of its products, but also across the broad range of its business activities and other public activities with social implications (such as its lobbying efforts). The company's internal practices (for example, its labor practices in general, its minority hiring and promotion, and its sexual fairness) must be considered as well.

With other types of investments, no screening is really necessary. For example, federal mortgage-backed securities support affordable housing. Indirectly they also promote jobs in the building industry. That is about the extent of the broad ethical implications of these investments. Therefore, when you come upon our lists of, say, unit investment trusts invested in these vehicles, they will not have passed through the same screening process as comparable lists of corporate bonds and—in the next chapter—stocks. The same goes for other investments listed in other socially uncomplicated categories.

We also want to offer a word of caution about the social ratings you will encounter in this and later chapters. We rely heavily on social ratings provided by the dominant professionals in the field. For example, the corporate bonds listed in this chapter are those held by socially screened mutual funds. We do not do any of our own rating of individual stocks or bonds (although in chapter 6 we have made some qualitative judgments about the social merits of the socially screened stock and bond mutual funds). As a group, the professionals in the SRI field care deeply about the social issues they screen for, but there is a limit as to how close any social screener can get to the actual facts of a company's operation. In many cases SRI professionals must rely on secondary sources like news reports and information gathered by activist groups. Plus, company policies and actions change, and the SRI field will not always be the first to know.

Therefore, the best anyone in the SRI field, including the authors of this book, can strive for is about 90 percent accuracy on these ratings. It goes without saying, but we apologize for any oversights in advance.

That said, let's get on with it!

BONDS IN GENERAL

A number of the following categories of investment are types of bonds, including various types of U.S. government bonds, U.S. savings bonds, corporate bonds, and municipal bonds. Each type of bond has its own attributes, but there is a set of terms and concepts common to bonds in general, as defined below.

A *bond* is a type of debt security, a loan by you to a company or government body in return for income. The loan is for a specified term, after which your principal is returned.

Principal, par, or *face value* is the amount of money you pay for the bond if you are its original owner (except with *zero coupon bonds*— see below), and the amount that will be paid back to you when the bond matures.

A *maturity date* is the date the principal is due back to you.

Interest or *coupon* are terms that describe the rate of interest the borrowing entity promises to pay. The interest rate on a bond is still called a coupon from the days when most bond certificates included coupons that the investor clipped and turned in to receive periodic interest payments.

Zero coupon bonds make no actual interest payments. Instead you buy a zero coupon bond at a discount off its face value and redeem the full face value when the bond matures. For example, a twenty-year, 10 percent, $1,000 zero will cost $149. You will receive no interest payments during the life of the bond but will collect $1,000 at maturity, effectively receiving all your interest in a lump sum at the end.

Did you catch the fact that your $149 can become $1,000 in twenty years? This frog-to-prince transformation is dramatic testimony both to the power of compounding interest and to a special quality of zeros. Whereas nonzero bonds pay interest only on your principal, zeros pay interest on your earned interest. Again, take our $1,000, 10 percent bond from the above paragraph. If that was an ordinary bond, you would receive $50 every six months ($100 a year) for the twenty-year term of the bond. If interest rates fell to 8 percent in the meantime, 8 percent is all you would get if you reinvested your interest payments. (And that doesn't count the earning time lost between payment and reinvestment.) The 10 percent zero is automatically reinvesting your bond income at 10 percent no matter what happens to interest rates, as long as the bond is held to maturity.

Thus, zeros are a terrific way to "lock in" a high interest rate. Of course, if you guess wrong about the direction of rates, you are stuck with a below market investment. Some zeros can also be called early, which limits the lock-in capacity of those bonds. (See below for an explanation of calling.)

The U.S. savings bonds you may remember from your childhood were, and remain, zero coupon bonds. Some corporations, municipalities, and brokerage houses issue zeros as well.

As for taxes, although you do not receive your zero coupon interest in actual payments, this "phantom income" will be reported to the IRS each year and you will be taxed for it (unless your zero is in a tax-free form such as a municipal bond).

Most bond certificates contain fine print stating that they are subject to a *call* by the issuer after a specified date, meaning that the issuer can elect to buy the bond back from the bondholder before the bond's term is up. Suppose you buy a twenty-year corporate bond paying 12 percent, and in the next three years the economic climate shifts such that comparable issues are paying only 8 percent. The company issuing your bond will likely buy it back at its *par* value (or perhaps at a premium), and pay you your 12 percent for those three years. If the company still wants to borrow money, it can now issue new bonds at 8 percent. (The truth is that even bonds that claim to be "call-protected" in their fine print may be called in times of steep falls in interest rates if the issuer can find a way around the anticall language, which it often can.)

Clearly, a bond's "callability" is a feature that puts a damper on its guaranteed interest rate. If your 12 percent bond from the example above is called in an 8 percent market, you will be able to get only 8 percent in a comparable investment with the money that is returned to you.

Quality ratings refer to a widely used letter grade system with which a number of financial service organizations—of which Moody's and Standard & Poor's are the best known—rate bonds. The ratings reflect the bond issuer's past performance in making good on its debt, its current financial strength, and its future prospects. These ratings are a crucial factor in determining the investment value of most corporate and municipal bonds. (Federal bonds are not rated, although the creditworthiness of the United States is assumed to be tops. In rare instances, corporate or municipal issues may go unrated as well because the issuer does not wish to pay the rating fee.)

Standard & Poor's rates from "AAA," the safest for an investor, to "D," the riskiest. (Standard & Poor's "D" rating means the issuer's debt is in default, the interest and/or principal in arrears.) Moody's uses a nearly identical system, except that "C"—an income bond on which no interest is currently being paid—is its lowest grade. Other than those differences, Moody's Aaa-Aa-A-Baa-Ba-B-Caa-Ca-C scale is comparable, in gradations anyway, to Standard & Poor's AAA-AA-A-BBB-BB-B-CCC-CC-C-D. Like two schoolteachers grading the same term paper, Moody's and Standard & Poor's do not always see each issue exactly the same way, because of how they weigh their various criteria. However, expect them to largely agree with, if not mirror, each other.

Quality ratings are not the rating service's recommendation to buy, sell, or hold a particular bond. They are simply an evaluation of the safety of the investor's principal. Because of the inevitability of risk/reward trade-offs, the highest-rated bonds offer the lowest interest rates. Lower-rated bonds must offer higher rates to entice investors.

Quality Ratings—
Two Agencies, Two Systems

What you see below is a simplified version of the Moody's and Standard & Poor's systems. Each has its own way of modifying the general ratings. Standard & Poor's sometimes adds a + or − to a letter grade. Moody's sometimes adds a 1. A bond rated A1 or A+ is considered of higher quality than an A.

RATING		
S & P	*MOODY'S*	*DESCRIPTION*
AAA	Aaa	Issuer has extremely strong capacity to pay interest on time and repay principal.
AA	Aa	Very strong capacity.
A	A	Strong capacity.
BBB	Baa	Adequate capacity. This level or above = "investment grade" bonds.
BB	Ba	Speculative capacity. This level or below = "junk" or "high-yield" bonds.
B	B	Low-grade, speculative capacity.
CCC	Caa	Poor-grade, speculative capacity.
CC	Ca	Highly speculative.
C	C	These bonds may have already defaulted on interest. (The lowest Moody's rating. S & P also rates C1 for bonds that have missed interest payments, and D for bonds in default.)

If for any reason you need the money in your bond before it matures, you can sell it through the bond market. The price you get for your bond will depend on the state of interest rates. If rates are up relative to your bond's rate, your bond will be worth less. If rates are down, your bond will be worth more. The reason for this inverse relationship will soon be obvious to you if it isn't already.

Suppose you purchase a $1,000 bond with a ten-year maturity that pays 10 percent interest. Suppose further that five years later you have to sell the bond to raise emergency cash. If interest rates at that time are at 8 percent, buyers who like other factors about your bond (its quality, its maturity date, and so on) will pay you a premium to buy it. In fact, they will overpay to the extent that they are effectively buying a bond paying an 8 percent yield. (In this case, they will overpay you about $112.50.)

Suppose instead that rates five years from now are at 12 percent. To raise cash, you will have to sell your 10 percent bond at a discount such that the buyer is effectively getting a 12 percent yield (in this case, a discount of $89.25).

This is why when you sell your bond before its maturity date, you expose it to market risk. If interest rates are up, you lose principal. If interest rates are down, you have a capital gain.

Zero coupon bonds, by the way, are subject to the same market risk as interest-bearing bonds. Suppose you buy a zero at $5,000 that matures in ten years at $10,000. If you sell it halfway through its term, you won't necessarily get the $7,500 you might logically expect. The market will determine the price.

You may wonder at this point why a buyer would buy a "used" bond rather than a new issue. First, the buyer may want the shorter term of a bond that is closer to its maturity date. Second, if an investor wants a better income flow than he or she can get with a new issue in, say, the 8 percent market of the first example, that investor can buy your 10 percent bond. Figuring in the premium paid for the bond, the investor has the equivalent of an 8 percent bond, but until the maturity date the investor will receive income at a 10 percent rate.

One other factor that will immediately affect the value of your bond in the marketplace is a change in its quality rating. A municipality with an aging population and therefore a declining tax base is an example of an issuer that may be down-rated. A company that takes on tremendous debt to finance a takeover is another example of an issuer vulnerable to down-rating.

The various market factors just mentioned make longer-term bonds riskier from a marketplace standpoint than shorter ones. Why? Because a change in interest rates and its resulting pull on the bond's value makes more of an impact when calculated over, say, a number of years than it does over a matter of months.

You can avoid the risk of losing principal in the marketplace simply by holding your bond to maturity. An excellent strategy is to time the maturity date of your bond with a time you will need the money (for instance, to send son Billy to college).

The remainder of this chapter covers the various categories of individual investments, including bonds. Stocks are covered in chapter 5.

U.S. TREASURY SECURITIES

Most socially conscious investors are uncomfortable with U.S. Treasury securities for ethical reasons detailed in this section. We cover treasury issues here both because we want you to decide for yourself and because we could hardly call this an investment book without covering them.

The one thing that treasury securities have going for them more than most other categories of investment is safety. Treasuries are general obligations of the U.S. government, backed by its taxing power. Whatever you think of the IRS on April 15, its ability to fund the treasury dependably makes U.S. government investment vehicles virtually 100 percent safe.

U.S. Treasury securities include treasury *bills* (T-bills), treasury *notes* (T-notes), and treasury *bonds* (T-bonds). U.S. savings bonds, also issued by the treasury, are covered as a special category following the discussion of T-bonds.

Treasury bills are short-term zero coupon bonds sold at auctions for less than their face value with maturity dates of a year or less. (There are three-, six-, nine-, and twelve-month T-bills.) T-bills bear no interest rates, so the difference between a T-bill's cost and its face value effectively determines the investor's interest. T-bills require a $10,000 minimum investment and are available in $5,000 increments after that.

Treasury notes do bear interest rates but are also sold through auctions. They will sometimes sell for more than their face value if the interest rate is considered an especially good deal in the current economy and less, of course, if the reverse is true. Minimum investment is $10,000, with maturity dates varying from two to ten years.

Treasury bonds differ from T-notes only in their maturity dates, which are ten years or more.

Today's *U.S. savings bonds* make a little more sense financially than the ones you may recall from years ago, which paid interest that was well below market rates. Savings bonds interest rates are now adjusted every six months to 85 percent of the current rate on five-year T-notes. Because savings bonds are zero coupon bonds, the interest earned is included in the face value paid at maturity rather than paid out in installments before that time.

There are two varieties of U.S. savings bonds, series EE and series HH. The former is the most interesting from a strict (non-SRI) financial-planning standpoint, because the interest earned is excludable from taxation if the bonds are used to finance the higher education of the taxpayer, taxpayer's spouse, or taxpayer's dependents. (The IRS phases out this tax advantage at higher-income levels.)

Financial Considerations

Extremely solid and safe, treasury securities pay low interest rates relative to other income investments that are technically riskier. You can purchase treasury securities directly from the government, which will cost you some time in paperwork, or through a broker, which will cost you some commission money. The income you earn from treasury securities is taxed by the federal government but not by state and local governments—a small tax advantage, since the feds take a far bigger chunk of your income than state or local governments anyway.

Ethical Considerations

Ethically, treasury securities are problematic—the money you lend the government goes into the general fund, which pays for nuclear weapons and military advisers in El Salvador as well as toxic-waste cleanup and social services. For this reason, most socially responsible investors avoid these government investments. If you like the idea of including safe government investments in your investment portfolio (and they are wise foundation investments), a number of government agencies offer investment opportunities—discussed immediately below—that you will probably find more to your ethical tastes.

U.S. AGENCY OBLIGATIONS (INCLUDING MORTGAGE-BACKED SECURITIES)

U.S. government agencies offer over one hundred different notes, bonds, and certificates to investors seeking the safety of Uncle Sam's towering presence. Ethical investors will find that this field offers a number of excellent alternatives to U.S. treasury issues. Only slightly more risky than treasuries, agency obligations have the moral advantage of being tied to the activities of a specific agency and often a specific purpose within that agency. Many of these are activities and purposes with which ethical investors have no problem aligning. For example, the Student Loan Marketing Association (SLMA, or "Sally Mae") securities finance loans to college and vocational school students.

The most popular category of government agency obligations is what are called *mortgage-backed securities,* of which the best known are "Ginnie Maes." The Government National Mortgage Association (GNMA) puts together pools of federally insured mortgages and then sells shares in the pools to individual investors. As homeowners pay down those mortgages, Ginnie Mae investors receive a proportionate share of the principal and interest.

Financial Considerations

U.S. agency obligations are only more risky than treasuries in the technical sense. By law, the treasury cannot guarantee agency obligations, but the government's moral commitment to back these issues should be considered just about as good. Besides, sometimes the nature of the issues makes them virtually guaranteed. For example, although Sally Maes are not insured by the government, Sally Mae invests only in student loans that are either federal- or state-insured. At any rate, you will profit from this hypothetical increase in risk, because U.S. agency obligations yield slightly higher returns than do treasuries.

Although Ginnie Maes and other mortgage-backed securities return higher rates than might be expected with investments of such high safety, there is a catch. Unlike other bonds, which pay back the investor's principal in one lump sum at maturity, mortgage-backed securities are "self-liquidating." That is, you the investor are receiving portions of your principal with each payment you receive on your

Ginnie Mae investment. Once paid back, the principal is no longer earning interest, and if interest rates are down in the financial universe, you will have difficulty reinvesting that money at the same rate of return obtained from the Ginnie Mae. No, you can't just plow it back into the original investment, because Ginnie Maes and other mortgage-backed securities require a $25,000 minimum if bought as an initial purchase. (You can buy them for as "little" as $10,000 as *seasoned*—not newly issued—securities on the secondary market.)

Mortgage-backed securities are also vulnerable to fluctuations in the home mortgage marketplace. If mortgage rates are dropping, some homeowners whose mortgages are in your pool may decide to refinance. That means those original mortgages will be paid off, and you will receive some principal sooner than you were expecting it. Again, if interest rates are down, where are you going to get the same blend of return and safety that your mortgage-backed investment gave you? Mortgage securities are also subject to market risk—when interest rates go up, their price goes down.

These are not damning indictments of mortgage-backed securities. All investments have trade-offs. Mortgage-backed securities still offer a good combination of superior safety and decent returns. (See the "Debt Instrument Returns" graph on the next page for a representation of their performance abilities. The peaks in the early eighties are due to extraordinary inflation left over from the late seventies.) The trade-off factors just covered can be largely mitigated if instead of investing in the securities directly, you invest in mutual funds that purchase mortgage-backed securities. Unlike directly owned mortgage-backed securities, the mutual funds will pay fairly steady returns. Like other mutual funds, funds dealing in mortgage-backed securities offer liquidity and a price per share affordable to modestly endowed investors. Several such funds of interest to ethical investors are compared in the figure in chapter 6, page 286. (The trade-off with the mutual fund, of course, is in the form of commission and management fees.)

Another alternative to the unpredictability of mortgage-backed securities is the "collateralized mortgage obligation" (CMO). Essentially, CMOs are bonds backed by a bundle of mortgages of similar terms and interest rates to create predictability for the investor. Your principal will begin returning to you with your interest payments after a stated time period and will be returned to you completely by another stated date. In between those times, you will receive relatively level payments. These attributes make CMOs an intriguing SRI vehicle for

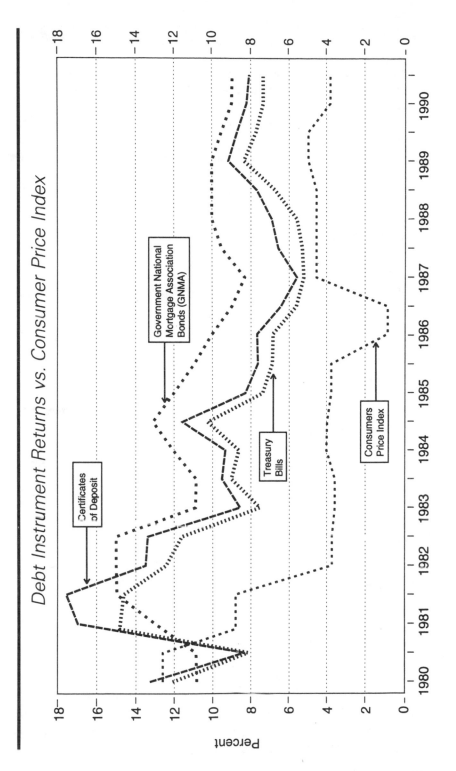

Debt Instrument Returns vs. Consumer Price Index

Certificates of Deposit

Government National Mortgage Association Bonds (GNMA)

Treasury Bills

Consumers Price Index

129

retirement or education planning because you can time the return of principal for when you really need it. CMOs are also available in a zero coupon–type format.

Unlike treasuries, which can be purchased at banks, U.S. agency obligations generally must be purchased through a broker, to whom you will pay a commission. The income is subject to federal tax but not always state or local. You can follow the progress of most agency obligations in *The Wall Street Journal.*

Ethical Considerations

As you can imagine, not all U.S. agency obligations serve purposes that are worthy in the eyes of the average socially responsible investor. The following issuers are among those you might want to consider for ethical investing purposes.

- Small Business Administration (SBA) guarantees portions of loans made to small businesses.
- Federal National Mortgage Association (FNMA, or ''Fannie Mae'') finances home loans and loans for low-income and senior housing.
- Federal Home Loan Banks (FHLB) finances home loans by member savings and loan associations.
- Federal Home Loan Mortgage Corporation (FHLMC, or ''Freddie Mac'') finances home loans and loans for low-income and senior housing.
- Student Loan Marketing Association and the Government National Mortgage Association (already mentioned).

CORPORATE BONDS (INCLUDING JUNK BONDS)

Corporations issue corporate bonds as a means of borrowing money from the general public, to whom they pay interest. Usually sold in $1,000 denominations, most ''corporates'' pay a fixed interest rate over a fixed term. The interest is normally paid every six months, with the principal returned at maturity.

As with all investments, trade-off factors affect the interest rates paid on corporate bonds. One of these factors is length of term—bonds with longer terms pay higher rates. Another factor concerns whether the bond is backed by specified assets of the corporation or is a *debenture,* backed only by the corporation's general credit and care for its reputation. Debentures are riskier for the buyer and thus pay more interest.

One of the biggest trade-offs in bond interest rates is the bond's quality rating, since the rating reflects an expert opinion of the creditworthiness of the issuing company. Obviously, the less likely a company is to make good on its obligations to its creditors (say, because it is already indebted to the hilt), the more interest it will have to pay to entice investors. Similarly, a company in solid financial shape with an impeccable record of coming through for its investors will find plenty of more conservatively inclined investors for the comparatively modest interest rates it offers.

Remember that quality ratings are *not* a recommendation to buy, sell, or hold a particular issue. Your decision should be based on your financial condition and tolerance for risk. For example, many investors with a gambling instinct have enriched themselves in recent years by wagering on *junk bonds*. Junk bonds are corporate bonds issued by companies whose debt level is so high as to be considered potentially unhealthy. (Technically, junk bonds are those rated Standard & Poor's BB—Moody's Ba—or below. Standard & Poor's triple B or Moody's Baa bond issuers are considered to have adequate, although not strong, capacity to pay interest and repay bond principal. These bonds, and those rated higher, are popularly known as "investment grade.")

Despite the implications of the name, junk bond issuers are not necessarily "junk" companies. In early 1991 Standard & Poor's dropped Chrysler Corporation's bond rating to junk level. Although Chrysler was then in worse straits by far than the other domestic automakers, both General Motors and Ford saw their ratings drop a couple of notches, too—from A to AA-minus (GM's lowest ever). Automakers traditionally do poorly in bad times, and there were big drops industrywide in sales and production in the wake of the Gulf war and a persistent recession.

The interest rates offered on junk bonds can exceed those offered on a Standard & Poor's AAA-rated bond by as much as a tempting 5 percent. If you are among the tempted, realize what you are getting yourself into: according to Moody's economist Jerome Fons, junk bond defaults are expected to total 11.5 percent in 1991, the highest rate since the Great Depression year 1933. The 1990 default rate was 8.8 percent, at that time the highest in twenty years.*

As fixed income investments, most corporate bonds have no growth

* Source: Reuters story in *Los Angeles Times*, 8/14/91, D4.

potential when held to maturity (instead of being sold in favorable conditions in the secondary marketplace). That is the trade-off for their guaranteed interest payments. The exception to the rule is the *convertible* bond. Convertible bonds can be exchanged for a specified number of shares of the issuing company's common stock (or, in some cases, the stock of another company in which the bond issuer has holdings). Clearly, as a company's stock rises in value, so does the value of its convertible bonds—making the bond a potential growth investment. The trade-off for that advantage is a lower interest rate on the bond than is paid for comparable nonconvertibles.

Financial Considerations

Corporate bonds usually pay higher interest rates than either money market funds or certificates of deposit—junk bonds, of course, pay considerably more. Although they can appreciate in the secondary market given fluctuations in interest rates, bonds other than junk bonds are sought primarily by conservative investors interested in safety and guaranteed income.

Junk bonds are for speculators only. A safer way to wager on the bond market is to buy investment-grade bonds at a discount when you sense that interest rates are headed downward—if you are right, your bonds will appreciate when the rates drop. If you are wrong, you can still hold your bonds to maturity and redeem the face value, so your down side is protected (as long as the company does not default, that is—but then again, you buy high-quality bonds to diminish that possibility in the first place).

Remember that, unlike stocks, regular corporate bonds give you no equity in the company. The company could triple in size and your bond will still pay you its 10 percent interest and $1,000 (or whatever) at maturity. So, safety—the ability of the company to make good on its promise to pay you interest and return your principal—should be your primary concern when buying bonds. This does not mean that you should buy only triple-A bonds, because you may be able to tolerate a little more risk in return for a better rate. It does mean that you should consider the bond's financial worth strictly on the basis of its rating, interest rate, and length, not on the potential of the issuer's earning capacity.

A company's potential is a factor with convertible bonds, however. Clearly, a convertible bond issued by a newer, growing company can give you a particularly lucrative investment if the company takes off. Then again, these companies are riskier guarantors of their debt.

The risks associated with your bond being called have been discussed in the general bond section (page 120).

Ethical Considerations

When you buy a corporate bond, you are helping that corporation achieve its goals. Are they goals you support? Your ethical considerations with corporate bonds are exactly as they would be if you were buying stock in a company. Your considerations will encompass its environmental record and other ways of doing business, the nature and quality of its products, with whom it does business, employee practices, and the like. One way to get up-to-date ethical information on companies issuing bonds is to consult the various SRI periodicals listed in appendix D that evaluate companies for stock investment purposes. The shortcut method is to find out which corporate bonds the various SRI bond and money market mutual funds are buying. Our "Listings" section on page 136 lists those held by SRI mutual funds as of December 31, 1990. Consider also our list of suggested stocks in chapter 5; some of these socially screened companies issue bonds (bear in mind, however, that a company's presence on a stock list says nothing about the financial attributes of its bonds).

You should be aware that the utilities industry (power, water, gas, and the like) as a whole is a major issuer of corporate bonds. These bonds are generally considered extremely sound financially because ratepayers will always buy power no matter what the condition of the economy or the utility itself. You may be concerned, however, with what type of power the utility is selling. In particular, many ethical investors avoid investing in electrical utilities involved with nuclear power. Nuclear plants, in addition to being ethically controversial, have often proved to be financial black holes. Refer to appendix A for a list of utilities associated with nuclear energy.

The ethics of junk bonds have been complicated, and to a large extent soured, by their recent history. The 1980s was marked by a flurry of leveraged buy-outs and unfriendly acquisitions. These activities, which were financed largely with junk bonds, had serious social consequences, including the elimination of thousands of jobs. The following summary of that history—essentially the story of the rise and fall of Michael Milken and his investment company—will help you understand not only the ethical compromises that many junk bond investments represent, but also some recent events that did horrific damage to the American economic landscape.

First, some terminology. In a *leveraged buy-out,* a group of investors that often includes company managers purchases all of the outstanding shares of a company's stock from the public; an unfriendly acquisition occurs when one company makes an offer to purchase all the shares of stock of another corporation at a price so beneficial to the shareholders that the target corporation's directors are duty-bound to recommend its acceptance.

These aggressive maneuvers require huge amounts of money. No problem, maintained Michael Milken, an executive in the Beverly Hills office of the Drexel Burnham Lambert investment firm. Milken believed that high-risk, high-yield bonds were a vastly underrated financial vehicle that made almost any financial ambition possible. A brilliant man of almost messianic persuasive abilities, Milken convinced his seniors at Drexel Burnham as well as a host of drooling clients that junk bonds could finance leveraged buy-outs and unfriendly acquisitions of previously unimaginable scope. And that is why in the mid-1980s we saw, for example, such seemingly unthinkable acquisition attempts as Mesa Petroleum—net worth $500 million—going after Unocal, and Farley Industries—with $6 million in earnings—bidding about $1.5 billion for Northwest Industries.

It was usually out of the question for the (relatively) pint-size aggressors in this game to finance their pursuits through conventional institutions like banks for any (or a combination) of the following reasons: they were already heavily in debt or they lacked a credit history or they were simply too darn small. Milken and Drexel Burnham's solution—borrowing the money from the public through junk bonds—seemed ludicrous on the surface. The junk bond market—then almost exclusively the playground of individual speculators—was not nearly large enough to support undertakings of this scope. But Drexel Burnham, convinced that Milken was really on to something, promised to underwrite the bonds, thereby creating an active market for them.

Meanwhile, banks and especially savings and loan associations were losing multitudes of depositors to the money market funds, which offered much higher interest rates, good liquidity, and a level of safety that, if not guaranteed, was still acceptable. They knew they had to offer higher interest rates to compete, but the only way to make those pay was to dramatically up the yield of their own investment portfolios. Under the spell woven by Milken (and aided by the relaxed regulatory atmosphere of the Reagan years), they became convinced that junk bonds were a big part of the solution to their troubles. Sure, junk bonds were risky, but the risk was insured by the public, through deposit insurance.

The greed that powered the game also unraveled it. Dennis Levine, one of Drexel Burnham's new hotshot executives, was hauled in by the SEC for trading fifty-four stocks on the basis of illegal insider information. Although Drexel Burnham was not implicated directly, Levine cooperated with the government by fingering a number of others, including another big-time insider trader, Ivan Boesky. Boesky, a Drexel Burnham client, then fingered Milken. Milken, it seems, had buoyed his commanding position in the now burgeoning junk bond market with a number of market manipulations, insider deals, and technical violations. Milken is currently serving a ten-year prison term, has agreed to pay $600 million in fines, and as of this writing has been sued for $6 billion by federal bank regulators for his contribution to the devastation of the S&L industry. He has also been banned for life from the securities business. Drexel Burnham, now in bankruptcy proceedings, has been sued for $6.8 billion by the same agencies. Without Milken and Drexel Burnham to underwrite them, the junk bond market has collapsed, making junk bonds a much more dicey investment than in the pre-Milken days (although select issues may be appropriate for the right well-heeled investor).

The social fallout of the junk bond financing fiasco extends far beyond the S&L crisis and the shaky condition of the banking and insurance industries (which also invested heavily in them) To pay the high interest to their junk bondholders, companies involved in leveraged buy-outs and unfriendly acquisitions usually trimmed all but their most profitable operations, with multitudes of jobs lost in the process. The casualty list also included lots of ordinary investors who bought into junk bond mutual funds on the theory that a fund's diversity would protect against bond defaults while still returning high yields. When the junk bond market collapsed, it was not uncommon for junk fund investors to lose forty or fifty percent of their principal. Junk bond hysteria damaged the environment as well, with junk-financed Pacific Lumber shifting from selectively harvesting to clear-cutting its redwood forest to pay off its debt.

And the corporate raiders were not the only ones closing plants, slashing jobs, and cutting research and development to build up cash.

Their targets, the major corporations, were forced to adopt a similar strategy to defend themselves against takeovers. Essentially, their best defense was to buy back their stock from the public, and that too required awesome amounts of cash. This resulting, junk bond–spawned obsession with short-term profits is widely recognized as a major reason that U.S. business is falling behind its international competitors, who invest more in the research and development that leads to long-term growth.

Obviously there is nothing a priori unethical about a heavily leveraged company offering a bond to the public. But we hope you better understand now why you will have to look long and hard today to find a junk bond that is both a socially benign and financially astute investment.

Listings

The following corporate bonds were held by at least one broadly screened SRI mutual fund—as indicated—as of December 31, 1990. The list is a short one, because although there are hundreds of broadly screened companies, most of them have relatively low debt ratios. Therefore few such companies issue bonds. The bonds of those that do are among the higher-quality bonds available and thus are not as heavily traded on the secondary market.

Explanation of Codes

CB = Calvert Social Investment Fund: Bond Portfolio
CM = Calvert Social Investment Fund: Managed Growth Portfolio
WA = Working Assets Money Fund

12th Street Historic Rehab. Assoc.—CM
AEF Shady Point—WA
Anchor Savings Bank—CB
Bell South Savings—CB
Beneficial Corporation—CB
Citizens & Southern Bank—CB
Corestates Capital Corporation—CB
Great Western Savings & Loan—CB

Hershey Food—WA
MASCO—CB
Maytag—CB
McGraw Hill—CB, WA
Michigan Bell Telephone—CB, CM
Mountain State Bell—CB
PaineWebber CMO Trust—CB
Piedmont Aviation—CB
Pitney Bowes Credit Organization—CB, WA

Quaker Oats—WA
Safety Kleen Corporation—CB
Seamen's Capital—CB

Security Pacific—WA
U.S. Air Equipment Trust—CB
U.S. Wind Power—CB

MUNICIPAL BONDS

Like corporations, state and local governments also borrow money from the general public through bond issues that pay competitive rates. You already know municipal bonds well because they are the bond issues on your state and local ballots. Many socially responsible investors find municipal bond investments tremendously satisfying because the bonds have a direct and beneficial impact on the municipalities that issue them. "Munis" finance undertakings like school and library construction, water projects, elderly housing, and parks, for example.

Bond rating services grade municipal bonds for credit quality much as they do corporates (yes, there is such a thing as junk munis), and a bond's interest rates vary with those ratings and the bond's length of term. But there are important differences between "munis" and corporates. The most notable of these is that the interest earned by munis is exempt from federal taxes. If the purchaser lives in the state issuing the bond, the bond will normally be exempt from state taxes as well. (These circumstances are referred to as "double tax-free"; "triple tax-free" bonds, free of local taxes, are offered in some municipalities, too.) Munis' tax-free status attracts middle-income and above investors who can benefit from the break—for someone paying 28 percent federal and 7 percent state tax, for instance, an 8 percent double tax-free muni is equivalent returnwise to a 12.3 percent corporate. Of course, for that precise reason municipalities can market their bonds at lower interest rates than can corporations, so these bonds make sense only for high-tax-bracket buyers who can benefit from the tax break.

Although municipal bonds as a group are considered safer from defaults than corporates, there have been notable exceptions. In 1983 the Washington Public Power Supply System (WPPSS) defaulted on nearly *$9 billion* of bonds issued to construct five nuclear power plants. These bonds had originally carried a Moody's rating of Aaa and a Standard & Poor's AA. The lesson here, we feel, pertains more to nuclear power plants than to muni bonds in general—not only are nukes a senseless health, environmental, and safety risk, but they have proven to be a terrible risk financially as well. WPPSS was financed by industrial revenue municipal bonds (see below); as such, they were to repay investors from income generated by the project. Given the legal, political, and financial problems that typically dog nuclear plants, the risks should have been foreseen by all concerned.

As a rule, muni bonds are backed either by the taxing authority of the issuing city or state or by the revenue generated by the project the bond is financing, and that backing is almost always sufficient. However, you can buy bonds that are insured by a third party against default. Insurance will usually raise a bond's rating to triple-A status, even if the issuing municipality is not otherwise considered to be of that quality. Then again, you will pay in the form of reduced interest rates (normally about ½ percent) for your increased peace of mind. Most municipalities choose not to insure their bonds because of the cost.

One final note: The fixed income promised by municipal bonds has the same catch as with corporates, because most munis are callable.

Financial Considerations

Although municipal bonds are normally issued in $1,000 denominations (as are corporates), they are generally sold in lots of five—a fact that, combined with munis' tax-free interest trade-off, makes them almost exclusively a product for substantial investors. A formula provided in chapter 7 will help you determine whether a tax-advantaged investment like muni bonds makes sense for your financial circumstances. You can also invest in municipal bonds through mutual funds or unit investment trusts, which lowers the ante to a level most middle-income investors can afford.

Should you be considering muni bonds for your IRA or Keogh or similar retirement plans, stop. It makes no sense to accept the lower interest rates that tax-free munis pay, when any bond—including higher-paying corporates—will compound tax-free in your retirement account. Besides, you *will* pay tax on any money withdrawn from your retirement account, even if it was earned with tax-free investments.

As for the safety of muni bonds, it is the opinion of many bond experts that the "junkiest" munis are considerably safer from default than the worst corporate junk bonds. This is because municipalities are less likely than corporations to be overloaded with debt, giving them more flexibility to reduce costs in tough times. Of course, this is not the same as saying that municipalities never default. Some have in the past, some will in the future. So analyze carefully any potential muni investment before purchasing.

Do be aware that the muni bond market includes numerous quality bonds that pay high yields like the junk munis but have no rating simply because the issuer decided not to pay for one. These municipalities compete with rated municipalities by offering higher rates on their unrated issues. You can uncover these little gems by going to a bond-specialist brokerage firm that does its own rating of unrated issues. Firms such as Franklin and Nuveen, which often include unrated municipal bonds in their tax-free bond mutual funds, do internal ratings of unrated issues as well. Your individual broker should also be able to access information about high-quality unrated bonds.

Although quality seems historically to be less of a real issue with munis than with corporates, do be advised that a recession increases the chances that a marginal municipality will have to default on some of its obligations. Recessions are definitely not the time to be taking a flyer on shaky companies or governments.

Note also that the value of municipal bonds is subject to the same fluctuation in the marketplace as are other bonds.

Ethical Considerations

In terms of ethics, muni bonds offer some of the simplest choices available to socially responsible investors. Investing in school construction or a housing project for the elderly appeals to many ethical investors, who will also find other bond issues, such as toll road or airport construction, at least socially benign. But there is always the odd bond issue of a more controversial nature—if you are an antinuker, the WPPSS tale related previously is a sober reminder on that score. Consider each bond on an individual basis, or examine the holdings of a municipal bond mutual fund for your first culling. (You can find the holdings of a mutual fund listed in the fund's semiannual and annual reports.) By the way, just before this book's publication, a new socially responsible municipal bond fund—the Muir California Tax-free Bond Fund, stressing education, environmental protection, and affordable housing bonds—became available, double tax-free for California investors.

Note that many ethical investors are distressed that the taxes they pay, particularly federal income taxes, are used to finance activities that repulse them—weapons buildups, support to repressive regimes, and so on. Some such investors will buy municipal bonds—usually through mutual funds or unit investment trusts—even when they are not financially well off enough to benefit from muni bonds' tax-free status, accepting the lesser returns as the price of their moral commitment.

COMMERCIAL PAPER

The term commercial paper refers to short-term loans made to corporations to help them meet short-term cash needs. An offer of commercial paper is not some sheepish move by a company that is running a little short this month—rather, it is an economical way of generating cash for special needs such as increased production schedules.

Buying commercial paper is a game for big girls and boys. Commercial paper is usually sold in $100,000 denominations. The terms are normally thirty, sixty, or ninety days.

Financial Considerations

If you have the resources to play, commercial paper is likely to pay more interest than short-term government issues of comparable term. Money market funds, by the way, invest heavily in commercial paper. Mutual funds also use them to park cash in between purchases.

Ethical Considerations

As with corporate bonds, this investment is as ethical as is the company to which you are loaning the money. Check the ethical reviews of individual companies in SRI publications such as *Clean Yield* and *Insight* (see appendix D, our list of screened bonds on page 136, and our list of stocks in chapter 5). SRI mutual funds and money market funds do screen the commercial paper they buy.

CERTIFICATES OF DEPOSIT (CDS)

When you buy a bank's certificate of deposit, you are agreeing to loan your money to the bank for a specified term. The bank will reward you for your willingness to tie up your money by giving you a significantly higher interest rate than you could earn on its regular liquid passbook savings accounts (which typically pay from 4½ to 5¼ percent, depending on your part of the country). Your money will also be protected by the same $100,000 of federal deposit insurance that regular depositors receive on their savings. (Do be aware that as we write this, the federal government is considering several revisions to the protection levels offered by federal deposit insurance.)

CD rates vary from week to week, depending upon whether your bank thinks market interest rates will rise or fall. Rates will also usually vary slightly with the length of term—the longer the term, the higher the rate—a trade-off for reduced liquidity. Many institutions offer somewhat higher rates for large minimum deposits—say, $10,000—as well. With most CDs, the rate on the day you purchase the CD becomes the fixed rate for the life of your deposit. You are gambling, of course, that your CD purchase is locking in a high rate at the peak of the interest cycle. If interest rates should rise, you will be caught with a below market investment. You can always withdraw your money before the CD's term expires, but you will pay a substantial interest penalty for doing so.

Again, most CDs carry fixed rates, but the competition for investment money always creates new variations to attract investors, and the CD market is no exception. Some institutions, for example, offer "liquid CDs," which allow you to make a certain number of withdrawals each month with no interest penalty. These are not really CDs, for they have no fixed term, and the interest rate fluctuates with the market. Effectively, they are high-interest savings accounts, with a high minimum deposit that must be maintained. Interest rates are generally lower than those offered on true CDs and on money market funds, but then they are more liquid than CDs and technically safer than the money market because they carry federal deposit insurance.

Your bank may also offer a variable rate CD. Here's an example of how the variable rate works: You agree to accept a starting rate, say, 8.1 percent that is slightly lower than the 8.3 percent that fixed rate CDs are offering that week. In return, the bank will raise the rate on the CD if interest rates go up in the marketplace, at the same time guaranteeing that your rate will never drop below 7.9 percent. It's a gamble on your part of .2 percent of interest for a chance at higher rates.

If your bank has a sudden need for cash, it may also offer a "hot rate'' CD that week. ''Hot rates'' are like CDs on sale, with slightly higher interest rates than normal.

Financial Considerations

When interest rates are peaking, CDs are one of the safest (although certainly not one of the most lucrative) ways of earning a fixed income. Your primary risk, as with all fixed income investments, is that inflation will gobble up your gains. Even in the best economic circumstances, your gains will be modest relative to most other income-producing vehicles, which is the price you pay for your deposit insurance. (CD interest rates do tend to run slightly higher than money market fund rates, however—the main trade-off here is loss of liquidity, without penalty.)

Depending on whether interest rates justify it, a CD purchase can be an excellent place to ''park'' money that you do not need right away but will need shortly after the CD's term expires—for a down payment on a house, for example. Make sure, though, that you can truly do without the money for the full length of the term before committing it to a CD. The interest penalty you will have to pay for withdrawing early wipes out the CD's investment value. If there is a chance you would need to tap into your CD, a money market fund or a ''liquid CD'' makes a more sensible parking place.

Also remember that there is no ''national'' rate paid on CDs. Rates will vary from institution to institution and from locality to locality. Generally speaking, savings and loans will pay higher rates than banks, and credit unions will pay higher than either because they are nonprofit. (You must be eligible for membership in a credit union, however, to partake of its benefits.) Your broker may be aware of out-of-state CDs that pay more than anything you can find locally, even considering the small commission you may have to pay him/her to place you in one. In many cases the out-of-state institution will pay the broker a commission for placing the CD without passing on the charge to the investor.

Perhaps the biggest risk on CDs is the rollover provision. That is, if you do not instruct the bank otherwise when your CD's term expires, the bank will automatically ''roll over'' your money into a new CD of the same term as the last one (which could be two years or more) and at *current* rates, which could be significantly lower than your just expired CD carried. Your bank will send you a notice that your CD will be expiring soon. The note will have a deadline date, after which time your money will be rolled over into a new CD. There will probably *not* be a tear-off return form that says "Send me my money." And if you happen to be out of town when the CD expires, good luck in retrieving cash from your rolled-over CD without a penalty. Monitor your CD expiration dates more carefully than you would your mother's birthday. Mom will forgive your oversight. Your bank may not.

Final caveat: Read the fine print on your CD agreement. Know its terms, limitations, and options. Pay particular attention to how the interest is paid. Some CDs pay only "simple interest"—a one-time payment on your principal. In other words, your interest will not compound. Those CDs that do compound may do so only monthly or quarterly. By comparison, most money market funds compound daily. Obviously, the more frequent the compounding, the higher the return.

Ethical Considerations

You have no control over how your bank invests your deposits, and—with rare exceptions—banks do not consider socially responsible criteria when they make their investment decisions. Nor will you be able to track the contents of your bank's investment portfolio, which can change daily. Nevertheless, some banking institutions are better from an ethical standpoint than others. We have already mentioned the financial advantage of purchasing your CD at a credit union. In addition, credit unions are usually the best place to shop for your CD from an ethical standpoint. Credit unions steer clear of the complications of corporate and international involvements, because their money flow is pretty much person to person. Deposits are loaned out to members; because credit unions are nonprofit institutions organized specifically to serve the banking needs of their members, members are also distributed the institution's excess profits. If you are not eligible for membership in one of your local credit unions by virtue of your work or government employment or whatever, find out if a relative who belongs can add you as a family member.

Another ethical option is minority-owned banks. Minority ownership does not guarantee an ethical lending philosophy—some minority-owned banks divest in their communities as thoughtlessly as do bigger institutions. Make sure that the bank from which you purchase your

CD does not. (Minority-owned banks and community development credit unions—credit unions organized as bootstrap institutions for disadvantaged communities—are covered in more detail in chapter 1.)

Within the banking community, there are a handful of institutions known for their social consciousness. All of those we know about are listed in chapter 1. Consider also that the smaller an institution is, the more likely it is to meet your ethical standards. This is because the smaller banks will tend to generate their income primarily by lending within the community they serve, as opposed to maintaining extensive stock portfolios and lending to corporations and international governments.

Do not entirely rule out savings and loan institutions, either. Although few Americans will be able to say "S&L" without snorting for years to come, the industry's reputation for being speculation- and fraud-ravaged was based on the actions of only a small minority of S&Ls. Traditionally, S&Ls have loaned their money primarily to depositors for home, auto, and personal loans. They are more middle- and lower-class-oriented than commercial banks, which aim toward serving the needs of large businesses and the people who run them. So in theory, anyway, S&Ls are less likely to be involved in activities that you disapprove of than are banks (although the same deregulation that led to the S&L scandal blurred the distinctions between the institutions).

Listings

See the listings of ethical banking resources on page 17 for issuers of socially responsible CDs.

MONEY MARKET MUTUAL FUNDS

Money market mutual funds are one of the most popular and useful forms of investment for the small investor. To tell the truth, many small investors hold far too much of their money in money market funds, a legacy from the mid-1970s when the first such funds paid solid double-digit interest rates. But don't get us wrong—money market funds have a part to play in almost every financial plan. In fact, many investors rely on money market funds instead of banks to hold most of their daily expense money—quite wisely, we feel.

Like other mutual funds, money market funds are pools of investments managed by professionals, with shares in the pool sold to the public. Also like other mutual funds, money market funds allow small investors to play a game otherwise reserved for big players—in this case, very big players. Commercial paper, for example—which constitutes a portion of most money market fund portfolios—requires a $100,000 minimum investment. Money market funds are typically invested in other short-term lending investments as well, including treasury bills, certificates of deposit, and short-term government agency bonds. Some money market funds are invested only in treasuries to attract investors seeking that extra edge of safety. Tax-free money market funds, including double (federal and state) tax-free, are invested primarily in short-term municipal bonds and other short-term municipal lending investments. As a trade-off for their tax advantages, these funds will pay interest a couple of percentage points less than the taxables.

Your shares of a money market fund are immediately liquid; in fact, most money market funds will issue you a checkbook so you can access your money as easily as if it were in a bank. The only difference between money market and bank checking accounts is that the former often require checks to be written in large minimum amounts—say, $250. There may also be a limit on how many checks you can write per month before paying a per-check fee. Opening deposits are often in the $1,000 range, with subsequent investments frequently a minimum of $250 or more.

Financial Considerations

Money market funds usually pay interest rates that are slightly below the certificate of deposit market. Then again, money market funds are far more versatile investments than CDs. CDs are great for locking in a good interest rate when you think interest rates are about to drop, but if you need that money before the CD's term expires, you will pay back much of your earnings in interest penalties. Your money market deposits are as close to your pocket—without penalty—as the vendor that accepts your check. No, money market funds are not as safe as CDs, technically, because they are not insured. But they are considered virtually risk-free in the investment industry—no one has ever lost a dollar in a money market fund, and no one is ever likely to because of the nature of the instruments in which money market funds invest.

Another advantage over CDs is the fact that money market interest is paid daily and can be reinvested without charge at the current rate. This is one of the two reasons so many investment-wise people do most of their banking through the money market. The second reason is that money market funds tend to beat the interest rates paid on bank pass-book accounts by anywhere from one to several percentage points. (In the wildly inflationary mid-1970s, money market funds were paying in the 12 to 13 percent range.) Thus, money market funds are ideal checking accounts from which to pay your big-ticket bills such as mortgage and car loan payments. In fact, the more bills you can pay from a money market account the better, because with daily com-pounding your money will keep working for you until the day you write a check for it. Money markets are also an ideal place to ''park'' money while waiting to commit it to a more long-term investment or to put the proceeds from the sale of your home while waiting to buy a new one.

The primary disadvantage of a money market fund is that the interest rates paid are low in comparison with most other vehicles described in this book. Thus, money market funds are best used only as advised above. Many individuals delude themselves that they are just parking money in their money market fund while waiting for the right invest-ment opportunity to present itself. Unfortunately they never get around to committing the money to a higher-paying investment, an oversight that gets increasingly expensive with the passage of time.

Ethical Considerations

Very simply, there are two broadly screened *taxable* socially respon-sible money market funds: Working Assets and the money market portfolio of the Calvert Social Investment Fund. No other taxable money market fund can be considered even approximately a socially responsible fund because of the holdings in treasuries and unscreened commercial paper. Besides, the portfolios of money market funds turn over so quickly that you could never keep track of the varying social implications.

Both Working Assets and the Calvert money market portfolio are modest performers. They usually pay slightly below average interest rates in comparison with unscreened funds. Nevertheless, they do perform respectably, better than a significant number of unscreened funds. Plus, the Calvert and Working Assets management groups are among the most ethically committed in the industry. The Calvert fund is part of a mutual fund family, an advantage over Working Assets if you have investments in other Calvert mutual funds. (See page 240 for more on fund families.)

Tax-free money market funds are not screened per se but, as with other municipal bond investments, are socially responsible because of the types of projects that municipal bonds finance. As an ethical investor you may object to a bond's targeted project in rare instances, as we mentioned earlier in our discussion of munis. However, unlike other municipal bond investments, you will not be able to evaluate or track tax-free money market funds' holdings because their portfolios turn over too rapidly. Besides, there are hundreds of these vehicles available—see the Sunday business pages of any major newspaper.

Listings

Both Working Assets and Calvert's money market funds are reviewed in detail, with other SRI mutual funds, in chapter 6.

For the reasons stated earlier, we haven't evaluated the tax-free products for you. This is an investment where you just trust that the ethics are mostly positive—in fact, proactive. If a tax-free money market fund is part of a mutual fund family in which you are otherwise invested, that is the one to pick (see earlier discussion of Calvert's fund). If you are not invested in mutual funds, just go for the fund paying the highest rate—again, look in your Sunday paper. By the way, remember that tax-free funds are designed for those who can use the full tax advantage.

ANNUITIES

Annuities, created and regulated by federal or state authorities, have enjoyed great popularity recently because they offer tax-deferred compounding just like the retirement plan options discussed in the previous chapter. And they have one significant advantage over IRAs and other retirement plans—there is no limit on how much you can contribute to your annuity each year.

Annuities are *not*, however, the pot of gold at the end of the rainbow—particularly for small investors. Like any other investment vehicle, they have their advantages and disadvantages and certain populations for whom they are better suited than others. Although we think annuities are a nice fit for certain circumstances, we do not feel that the average person should consider an annuity as his/her *primary* retirement plan. When you fully understand an annuity's attributes, you will understand why.

In essence, an annuity is a payment stream generated by an investment. You give the insurance company either a lump sum or periodic amounts to invest, and the money grows tax-"free" (tax-deferred) until it is withdrawn.

Annuities fall into two broad categories, *fixed* and *variable*. A fixed annuity pays a fixed payment amount at the time you begin withdrawing the money. (The actual amount will depend upon the type of payout you choose—see the discussion that follows.) The insurance company also guarantees your principal, but note that "guaranteed" does not mean "insured." If the insurance company goes bankrupt, there is no federal deposit insurance to bail you out. What's more, your fixed annuity could be attached by the company's creditors because it is part of the company's general account. Finally, the term *fixed* is something of a misnomer, because the rate of return is not really fixed—it is adjusted to market conditions at the insurance company's discretion, at intervals ranging from several months to several years. Thus, the long-term performance of the market could affect the amount you ultimately receive.

Many investors find variable annuities more comfortable investments than fixed products for a number of reasons. First, control. When you have a fixed annuity, the insurance company decides how it's going to invest your money to generate your payments. With a variable annuity, you choose how the money is to be invested from a variety of options, normally including stock, bond, or money market mutual funds. (Some offer more exotic choices like gold, international, and real estate funds as well.) This brings in the second comfort zone—familiarity. If you are used to investing in mutual funds, a variable annuity will seem very homey—it is just like investing in a mutual fund family, including the ability to transfer your investment from one fund to another. The third comfort zone is safety. Unlike fixed annuities, variable annuities are held in a custodial account, separate from the insurance company's general assets. If the insurance company goes under, your annuity won't go with it.

Variable annuities are not risk-free, however. Your variable annuity, being a mutual fund, carries investment risk. It will fluctuate in value with the performance of the fund. If you choose a risky option like an aggressive growth mutual fund, the risk could be high—although your chances of spectacular gains will increase commensurately.

Mutual funds and variable annuities both charge operating and management fees, but variable annuities usually tack on a couple of extra charges. The largest of these—up to 1.25 percent per year—is ostensibly for "mortality and risk" but in fact is mostly a profit item for the insurance company. With these extra charges, the total annual fees charged variable annuity holders run between 1.5 to 3 percent for most products. You may also be charged a set fee every time you elect to switch your investment from fund to fund.

The management and other fees will not seriously degrade your investment. However, both the insurance company and the government expect you to keep your money in your annuity for a long time. If you decide—or circumstances force you—to pull out early, you will be leaving a serious chunk of your money behind. Most variable (and fixed) annuities charge "back-end surrender charges" (meaning you will be charged if you withdraw) as high as 8 or 9 percent in the first year, although the charges drop as the years pass. Then there are the federal withdrawal penalties of 10 percent that you will pay if you withdraw from your annuity or other retirement plan before age 59½ (with certain exceptions).

Now that you've mastered the differences between fixed and variable annuities, we're going to throw another distinction at you—*qualified* and *nonqualified* plans. Qualified plans are those that can be used in conjunction with a tax-advantaged retirement plan such as an IRA, 401(k), 403(b), Keogh, and so on. Nonqualified plans cannot.

The significance of this wrinkle is the effect it can have on your payout. Keeping in mind that most people invest in annuities for the purpose of guaranteeing lifetime income after retirement, let's look first at your payout options on a *qualified fixed* annuity:

1) You can *annuitize* your payout schedule, which essentially means to set a payment plan in concrete (say, $400 a month for life starting at age sixty-five, payments to end at death. This is called a straight life payout—see payout options below). With a fixed annuity, an annuitized payout will also be fixed—in other words, the payments will all be the same size. Annuitizing your payout also opens up another field of choices. Depending upon your insurance company, you may be allowed to select from two or more of the following:

- straight life annuity, which pays you for life but ends payments when you die;
- life annuity with installments certain, which pays you for a specified minimum number of years, so that if you die within that time, the balance will be paid to a beneficiary;
- joint and survivor annuity, which pays you fixed payments for the number of years chosen or for as long as the second person named lives, whichever is longer;
- refund annuity, which pays you for life or until the payout equals your total contributions—if you die before that balance is reached, your beneficiary receives the refund;
- a lump sum withdrawal.

Individual companies may offer twists on these general categories, not every company will offer every choice, and the choice you make may affect the size of your regular payment.

2) Because your annuity is qualified, your insurance company may also offer you the option of a *systematic withdrawal plan,* which permits you to withdraw according to an IRS minimum distribution formula. (Not all insurance companies with qualified annuities offer this option, but most do for competitive reasons.) On this plan, you have to continue withdrawing minimum amounts according to the IRS table once you begin your withdrawals, but you do not have to annuitize.

3) You can withdraw your money irregularly, as you would money in a bank passbook account, subject to insurance company rules and penalties about maximum withdrawals in any one time period. In other words, since this is your money in the annuity, the insurance company can't prevent you from taking it out. However, they can charge you fees if you take too much at one time.

4) You can leave the money in the annuity for a beneficiary.

(Do note that whichever payout option you choose, you will have to begin withdrawing by age 70½. This federal requirement only applies to qualified plans, although nonqualified plans leave you at the mercy of the insurance company's own rules—not necessarily an advantage!)

If yours is a *nonqualified fixed* annuity, you will (again depending on the company) have all the above options, except for the systematic withdrawal plan.

If yours is a *qualified variable* annuity, you will have the same options (depending upon the individual company) as with a qualified fixed plus one additional: instead of fixed payments, you can choose variable payments that change with the performance of your investment.

If yours is a *nonqualified variable* annuity, you will not have the option of a systematic withdrawal plan. Otherwise, all options available on qualified variable annuities should apply.

Despite the incredible complication of all the options, an annuity can be a fine investment for the right person. To see if that person is you, read on.

Financial Considerations

Tax-free compounding is the main benefit all annuities share. However, every variety of tax-advantaged retirement plan compounds tax-deferred as well. Before committing money to an annuity, answer the following questions:

Have I already invested as much as I can in other retirement plans such as an IRA, 401(k), 403(b), or Keogh (some of which may themselves offer annuities as investment options)? In most cases your contributions to these other plans can be excluded from your income for tax purposes. Your annuity contributions cannot. So an annuity makes most sense as a supplemental retirement plan after you've socked away the full $2,000 per year allowable with an IRA and taken advantage of other retirement plan options available to you. The money you save in taxes can then be invested in an annuity and compound tax-free.

Is there any significant chance that I'll need the money invested within the first few years? If your honest answer is yes, invest your money in something other than an annuity. The surrender charges applicable in the first few years of most annuities will eat up much of your gain. Plus, the IRS will charge you 10 percent on earnings withdrawn before age 59½, in addition to the ordinary income taxes that you will pay on withdrawals made at any age. (Note: The costs of an early pullout can be reduced by choosing an annuity product with no surrender or sales charges. Several are currently available, some of which offer low fees as well.)

Is my tax bracket high enough to warrant this type of investment? The main advantage of an annuity is the tax-deferred compounding. The lower your tax bracket, the less of an advantage this is to you.

Do I want growth or income from this investment? If after answering the above questions you decide that an annuity makes sense for your circumstances, your answer to this question will help you choose between a fixed and variable annuity. A fixed may make more sense for income seekers because the fees are usually lower than those charged on variables. (We'll address the safety aspect of fixed annuities in a moment.) If you seek growth, a variable annuity's stock fund is a nice way to fly because you will be deferring taxes on all your gains. With 100 percent of your gains compounding (instead of, say, 67 percent if you are in the 33 percent tax bracket), you will be so far ahead of the game after a number of years that the fee difference will be negligible.

Now let's look at some of the other choices facing you with annuity products:

- **Investment type (variable annuities only).** The trick here is not to select too conservatively. Yes, you are probably counting on this money for retirement, and that consideration mandates a certain prudence. But you don't want to pick so conservative a vehicle that you lose ground to inflation, especially considering the fact that the annuity's fees will also chip away at your returns.
- **Payout options.** The first question is whether to annuitize or not. We generally advise not to, because you give up flexibility. Once you sign an annuitization schedule, it is usually irrevocable. The insurance company likes this option because it wants to play the classic life insurance game with you, which means betting on your death (for instance, if you die before the "pot" is empty in a straight life annuity, the company pockets the balance). You can win the bet as well as lose it, but remember that the insurance company is betting according to well-researched mortality tables and, like a Las Vegas casino, has tipped the bet in favor of the "house."

    Is there ever a good reason for choosing annuitization? Some people think peace of mind is a good reason. Some individuals choose to annuitize fixed payments for the assurance that the annuity will handle one or more of their fixed expenses when they retire (a $1,000 fixed payment could handle a $1,000 house payment, for example).

    For qualified plan holders, the nice thing about the systematic withdrawal system is that you get a steady payment stream without locking in a payment size. Since life—and the economy in particular—can be so unpredictable, flexibility is usually a good idea when you can work it in. A systematic withdrawal plan allows you to increase the payment size as needed, although the company may penalize you for withdrawing more than a certain maximum percentage (often 10 percent) of your investment within any one year.

If you are the ultra-self-reliant type, you may want to direct your own withdrawal schedule without committing to any plan, an option available to you with all annuity types. However, you will have to enmesh yourself in the insurance company's red tape every time you want some money, which is not the most convenient way to provide regular retirement income if that is your goal.

■ **Qualified vs. nonqualified annuities.** Obviously, take the more flexible qualified plan. But don't just assume that the company offering you a qualified annuity offers the systematic withdrawal plan with it. Ask.

■ **Fixed vs. variable payments (variable annuities only); straight life vs. life with installments certain vs. refund annuities . . .** Take up the fine points of these questions with a financial adviser who knows your circumstances.

If you do decide that a fixed product is your cup of tea, pick your insurance company carefully. Many companies invested heavily in junk bonds during the Milken/Drexel Burnham heyday, when high-yield bonds seemed like easy money. To identify a healthy insurance company, look up its A.M. Best rating. A.M. Best is an independent agency that evaluates the financial strength of insurance companies, much like Standard & Poor's and Moody's evaluate the creditworthiness of bond issuers. See page 303 for more on this topic.

No matter which annuity product you pick, go into it with your eyes wide open. Compare products for differences in sales charges, management fees, withdrawal fees, other costs, and (in the case of fixed products) interest rate guarantees. In fact, our number one recommendation is to *seek the advice of an outside consultant who has no vested interest in selling (or not selling) you an annuity before making any of the complicated decisions that this product requires.*

Ethical Considerations

If you're reeling from all the options upon options in this crazy annuity business (and it is even more complex than we've presented it), you may actually find it refreshing to discover that you have relatively few socially screened annuities from which to choose. Consumers United

Insurance Co. offers a no-load fixed annuity. If you are a teacher and on a 403(b) salary reduction plan through TIAA-CREF (Teachers Insurance Annuity Association—College Retirement Equity Fund), you can select a socially screened variable annuity called the Social Awareness Fund, which is invested in a socially screened stock mutual fund. State Bond Mutual (SBM) Life Insurance offers a fixed annuity that is about 75 percent invested in Ginnie Maes, with the balance in private mortgages and commercial paper. The company assures SRI brokers that it will maintain a socially responsible profile for this product. Hartford Insurance checks in with their Socially Responsive Fund managed by the Calvert Group, available only through employer pension plans. This is a qualified variable annuity, and the systematic withdrawal option is available. Lincoln National Life Insurance Company's contribution is their own Social Awareness Fund, available in both qualified (with systematic withdrawal) and nonqualified variable annuities. The Calvert Group has filed the necessary documentation with the government to open a new variable annuity, which should be available to the public by the time this book is published.

Calvert is also managing a socially responsible option available through the variable annuities of several major insurance companies including Aetna, Metropolitan Life, ITT–The Hartford Insurance Group, Mutual of America (MOA), and Mutual of New York (MONY). This option, a "mirror image" of Calvert's Managed Growth Portfolio (see page 258), is available through employer-sponsored pension plans which utilize the above insurance companies. It is also available to individuals, by request.

That's it. Every other annuity that we know of is unscreened, although more screened annuities should become available as the SRI field grows. Contact information for the annuities just mentioned is provided below:

Listings

Be sure to ask for the socially screened product.

Aetna
151 Farmington Avenue
Hartford, CT 06156
(800) 243-2390

Consumers United Insurance Company
2100 M Street NW
Washington, D.C. 20063
(800) 255-4432

ITT–The Hartford Insurance Group
Hartford Plaza
Hartford, CT 06115
(203) 547-5000

Lincoln National Life Insurance Company
1300 South Clinton Street
Fort Wayne, IN 46805
(800) 348-1212

Metropolitan Life
1 Madison Avenue
New York, NY 10010
(212) 578-2211

Mutual of America (MOA)
666 Fifth Avenue
New York, NY 10103
(212) 399-1600

Mutual of New York (MONY)
P.O. Box 4830
Syracuse, NY 13221
(800) 487-6669

State Bond Mutual Life Insurance
1-106 North Minnesota Street
New Ulm, MN 56073
(800) 333-3952

LIMITED PARTNERSHIPS

Limited partnerships allow investors to profit from direct participation in large operations—from commercial real estate to nursing homes to wind energy farms—without having to put up big-league capital and without any of the day-to-day hassles of running the thing. Nor do limited partners (investors) face major financial "exposure"—their liability is usually limited to the amount invested. The business is managed by *general partners* with unlimited liability who share in the ownership, make all the decisions, and receive a management fee as compensation for their troubles. (In many limited partnerships, the general partner is actually a corporation.)

There are both *private* and *public* limited partnerships. Because private partnerships tend to be riskier, more speculative investments oriented to wealthier investors, they are limited to a maximum of thirty-five "nonaccredited" partners (investors who meet a lower-level net worth eligibility test). The number of accredited partners, those meeting a higher-level test, is not limited. Private limited partnerships typically require a minimum investment of $25,000 or more. Public partnerships often have thousands of partners with minimum investments typically of $2,000 to $5,000. Public partnerships are more closely regulated than are private. They are required to file their investor's prospectus with the Securities Exchange Commission (SEC), and the general partners must demonstrate substantial net worth. Because of their more public nature, and because the general partners are more likely to have a track record in similar ventures, public partnerships are easier for investors to evaluate than are the private variety.

Whether public or private, most income-oriented partnerships endeavor to pay its limited partners a steady income for the life of the partnership; when the partnership is dissolved and the assets sold, limited partners share in those profits as well. Not all partnerships are income-driven, so depending upon the nature of the partnership's endeavor, it could be some time before the investor sees returns. In a development project, for instance, the partners' dollars go to buy raw land; only after buildings are built and then either rented, leased, or sold will money be generated to distribute back to the partners. (Some partnerships distribute profits only after all the assets are sold.)

Unfortunately, limited partnerships as a category of investment still suffer from a bad reputation earned prior to the 1986 tax reform. Before 1986 the tax laws were so favorable for limited partners that a multitude of partnerships were created strictly as tax shelters, with little merit on their own as businesses. In numerous other cases, perfectly legitimate businesses failed because the market became saturated with a flood of tax shelters. (A miniwarehouse isn't worth much even as a tax shelter if there are twelve of them in the same little community.) Today, however, many limited partnerships make excellent sense financially in spite of the continuing public skepticism.

American Retirement Villas III is a good example of a limited partnership that is sound financially and potentially appealing to socially responsible investors. A real estate investment that addresses both the need for affordable housing and the issue of elderly care, this partnership promises income and growth to its investors. Income for this ten-year program is generated through revenues in the form of rents; growth results from appreciation of the land and buildings.

Although your broker can get you any stock or bond on the market, he or she will probably sell only a select number of partnership programs. Brokers are required to personally research and qualify—through a process called "due diligence"—each partnership they offer clients. Therefore a large brokerage will usually offer a greater selection of qualified limited partnerships than will a small firm or an individual broker. In any case, prepare yourself to shop around some between brokerages until you find a partnership you like.

Two traditional drawbacks of limited partnerships—for small investors, anyway—have been their illiquidity and the high minimum initial investment (usually $2,000 to $5,000). Those problems are mitigated to a great extent in *master limited partnerships*. Like mutual funds, master limited partnerships sell shares priced for the smaller investor; as liquid as mutual fund shares, master limited partnerships are traded on major stock exchanges. The latter advantage is particularly noteworthy—in a regular limited partnership, it might be tough to get your money out before the partnership dissolves. In the master form, you can buy or sell at any time. The much-ballyhooed "Own a share of the Boston Celtics!" campaign was accomplished by converting the parent corporation to a master limited partnership. Not all MLPs are single endeavors like the Celtics; some pool several partnerships.

Financial Considerations

The 1986 tax reform removed most, although not all, of the tax advantages accruing to limited partners. Today's investor has to evaluate a partnership strictly on its risk/reward basis. That starts with its merits as a business. Do the general partners know what they are doing? What is their track record with similar ventures? And is the management fee a fair one?

Of the money that you invest in a limited partnership, chances are 15 to 30 percent of it will go to pay management expenses and broker's commission. If you should sell your interest in the partnership any time prior to its natural completion, you will almost certainly feel the effects of that 15 to 30 percent. Therefore, before investing in a limited partnership, you should be quite sure you can hang in for the duration of the project.

One factor that could bear on the riskiness of your investment concerns whether the general partner is investing only cash raised from the limited partners or is borrowing additional funds from a lender. Leveraged partnerships—those that invest borrowed capital—are riskier.

There are exceptions to the rule that limited partners are financially liable only for the amount of their investment. Some partnership agreements warn that limited partners are subject to "cash calls," meaning the general partner can require them to contribute cash to the partnership if the endeavor runs short. Some cash call provisions are limited to a certain amount per period of time or total amount of money; some are open-ended. Be sure to examine your partnership agreement for cash call provisions. Such a provision will make your investment more difficult to evaluate financially, particularly if the provision is open-ended.

Of special interest both ethically and financially are some limited partnerships that offer federal and/or state tax credits because the partnership has been formed to build and operate federally backed low-income housing. Even though most such partnerships make no payments to their investors, the tax credits can figure out to an effective 14 to 17 percent annual yield over a ten-year span. See our listings of limited partnerships at the end of this section for specific examples and contact information.

MLPs often market themselves on a promise of high rates of return. It is a promise often kept, partly because the entity of a partnership—unlike that of a corporation—is not itself taxed. Thus, more of the earnings can be passed through to the investors. Some partnership organizers even guarantee a certain return for a number of years. The danger with some MLPs, however, is that in their efforts to keep the limited partners smiling, the organizers send too much of the earnings their way, starving the business for cash. Check out that possibility in any MLP you are contemplating as an investment. Also assure yourself that the general partners have an ongoing interest in seeing the venture succeed beyond its ability to attract investors.

Ethical Considerations

Obviously the main ethical question here is the nature of the business in which the partnership is engaged. Some pursue business goals that are quite compatible with SRI.

Listings

The partnerships listed here have been selected by the authors as appropriate examples for a generally socially responsible investment strategy. Many of the specific offerings will not be available by the time you read this book. Limited partnerships have a finite life and are rarely open to new investors for more than a year after inception. However, companies that offer successful limited partnerships will often introduce clones to the marketplace after the original partnership sells out. Therefore use these ratings as a general indicator of promising investments. If the partnership listed is no longer offered, perhaps one similar in financial and social value will be. (Of course, the ratings given here may not apply to a similar or cloned partnership offered by the same general partner.)

The financial rankings below are courtesy of *Stanger's Partnership Watch,* December 1990 (Robert A. Stanger & Co., 1129 Broad Street, Shrewsbury, NJ 07702-4314; 908-389-3600). Stanger's risk rating is self-explanatory. "Share value," the first rating, refers to the percentage of your investment that goes directly into the venture itself, as opposed to paying management fees, commission, and so forth. In other words, the higher the percentage, the better for you. For the partnerships listed here—except for those listed under the heading "Government Mortgage Properties"—the following scale was used to determine share value:

Highest = 83.5% or higher
High = 79.0%–83.4%
Above average = 74.5%–78.9%
Average = 70.0%–74.4%
Below average = 65.5%–69.9%
Lowest = 65.4% or lower

For those partnerships listed under the heading "Government Mortgage Properties," the share value scale is as follows:

Highest = 82.0% or higher
High = 77.5%–81.9%
Above average = 73.0%–77.4%
Average = 68.5%–72.9%
Below average = 64.0%–68.4%
Lowest = 63.9% or lower

Socially Responsible Limited Partnerships

Low-Income Housing Tax Credit

American Tax Credit Properties III (also invests in historic preservation)
Merrill Lynch Capital Markets   (800) 288-3694, (212) 888-7444
1 Liberty Plaza   Share Ranking: Above average
165 Broadway, 3rd Floor   Risk Ranking: Medium
New York, NY 10006-1231

Boston Capital Tax Credit Fund II
Boston Capital Services, Inc.   (617) 439-0077, (800) 632-3642
313 Congress Street, 5th Floor   Share Ranking: Average
Boston, MA 02210-1231   Risk Ranking: Medium

Boston Financial Qualified Housing Tax Credits V
Boston Financial Securities, Inc.   (617) 439-3911
101 Arch Street   Share Ranking: High
Boston, MA 02110-1106   Risk Ranking: Medium high

City-Tax Credit Partners
NASD Network   (213) 670-1444, (800) 448-2484
6001 Bristol Parkway, 2nd Floor   Share Ranking: Average
Culver City, CA 90230   Risk Ranking: Medium

1990 Federal Tax Credit Partners
Enterprise Securities Group   (800) 633-2747
c/o AI Housing Inc.   Share Ranking: Average
129 South Street   Risk Ranking: Medium high
Boston, MA 02111

Freedom Tax Credit Plus Program
Shearson Lehman Hutton   (800) 831-4826, (212) 421-5333
625 Madison Avenue   Share Ranking: High
New York, NY 10022   Risk Ranking: Medium high

Gateway Tax Credit Fund, Ltd. II
Raymond James & Assoc., Inc.   (813) 573-3800, (800) 237-4240
880 Carillon Parkway   Share Ranking: High
St. Petersburg, FL 33716   Risk Ranking: Medium high

National Tax Credit Investors II
PaineWebber, Inc.   (213) 278-2191
9090 Wilshire Boulevard   Share Ranking: High
Suite 201   Risk Ranking: Medium
Beverly Hills, CA 90211

Prudential-Bache Tax Credit Properties
Prudential-Bache Securities   (800) 535-2077, (212) 214-1782
1 Seaport Plaza, 33rd Floor   Share Ranking: Above average
New York, NY 10292   Risk Ranking: Medium high

Waterford Tax Credit Partners
NASD Network   (313) 562-5005
22005 W. Outer Drive   Share Ranking: Average
Dearborn, MI 48124   Risk Ranking: Medium

WNC Housing Tax Credit Fund II

NASD Network
3158 Redhill Avenue
Suite 120
Costa Mesa, CA 92626-3416

(714) 662-5565, (800) 451-7070
Share Ranking: Average
Risk Ranking: Medium high

Senior Housing and Retirement Facilities

American Retirement Villas Properties III

ARV Capital Corporation
245 Fischer Avenue
Suite D1
Costa Mesa, CA 92626

(714) 751-7400, (800) 624-0236
Share Ranking: Average
Risk Ranking: Medium

Historic Building Preservation

Diversified Historic Investors 1990

Delaware Securities & Investment,
Inc.
1521 Locust Street
Suite 500
Philadelphia, PA 19102

(215) 735-5001, (800) 468-4017
Share Ranking: Average
Risk Ranking: Medium high

Historic Preservation Properties 1990 Tax Credit Fund

Boston Bay Capital, Inc.
50 Rowes Wharf
Boston, MA 02110

(617) 330-7700
Share Ranking: Average
Risk Ranking: Medium

Manufactured Housing

Uniprop Income Fund III

NASD Network
280 Daines Street
Suite 300
Birmingham, MI 48009

(313) 645-9261, (800) 541-7767
Share Ranking: Above average
Risk Ranking: Medium

Windsor Park Properties 7 & 8

Kensington Financial Inc.
120 West Grand Avenue
Escondido, CA 92025

(800) 821-4715
(800) 821-3736, CA
Share Ranking: Average
Risk Ranking: Medium

Government Mortgage Properties

Capital Mortgage Plus
IDS Financial Services, Inc.   (212) 421-5333, (800) 831-4826
625 Madison Avenue, 9th Floor   Share Ranking: Above average
New York, NY 10022   Risk Ranking: Lowest

NYLIFE Government Mortgage Plus
NYLIFE Securities   (800) 332-5774
51 Madison Avenue, Room 1700   Share Ranking: Average
New York, NY 10010   Risk Ranking: Lowest

Wingate Government Mortgage Partners II
Kidder Peabody/Continental   (617) 574-9000
Wingate   Share Ranking: High
10 Hanover Square, 9th Floor   Risk Ranking: Lowest
New York, NY 10005

Krupp Government Income Trust
Krupp Securities Corp.   (617) 423-2233, (800) 669-5787
470 Atlantic Avenue   Share Ranking: Above average
Boston, MA 02210   Risk Ranking: Lowest

Health Care

Common Goal Healthcare Pension and Income Mortgage Fund II
Healthcare Securities, Inc.   (800) 822-3863
P.O. Box 11269   Share Ranking: Average
Baltimore, MD 21239   Risk Ranking: Medium low

REAL ESTATE INVESTMENT TRUSTS (REITS)

REITs make it convenient for small investors to buy into large real estate investments. Some REITs invest in real property (land and buildings), some invest in mortgage securities, and some buy both. A REIT investment combines attributes of limited partnerships and stock ownership. Like a limited partnership (some of which also invest in real estate), a REIT is a sophisticated investment that is managed by experts. Not only is the investment hassle-free, but as with most limited partnerships, your liability is limited to the amount of your investment.

Unlike a limited partnership, however, which costs at least several thousand dollars to enter and is usually illiquid besides, a REIT is traded on the various stock exchanges, so you can buy shares for comparatively few dollars and have the same liquidity you would have with any stock.

Financial Considerations

As real estate investments, REITs return both income and appreciation. Additionally, the share value itself can appreciate in the market. Of course, shares that can appreciate in value can also depreciate, a risk factor that can cost you principal, particularly if you have to unload shares in an emergency.

We have already discussed the risks inherent in mortgage-backed securities. The main risk in most large real estate investments (apartment and commercial buildings) concerns whether the rents/lease payments collected will support the costs of keeping the building (mortgage payments and maintenance expenses). Excessive and long-lasting vacancies are the primary threats to these types of investments, because just about everything else is in the investor's favor. Property tends to appreciate over time at a rate that exceeds the inflation rate, often by plenty. And you can bet that rent and lease arrangements in a professionally managed investment will increase regularly to keep pace with inflation. There are also tax advantages (deductible mortgage interest, building depreciation, maintenance expenses) to this type of investment. You will not benefit from these directly on your tax return, because your distribution will be taxed just like stock dividends. However, the venture itself will benefit, increasing the worth of your investment.

Ethical Considerations

A number of REITs, such as those invested in mortgage-backed securities, have no substantial negative implications ethically. Others invest in projects whose goals (such as affordable housing, health care, and elderly care) ethical investors support. Such projects make up a relatively small proportion of available REITs, however. The majority fall into the general categories of residential and commercial real estate development, or "paving paradise," as the old Joni Mitchell hit put it. Few ethical investors would want to associate with such projects without first determining whether they are likely to be carried out with sensitivity to the surrounding community's wishes and needs.

Listings

The marketplace in REITs is too dynamic for us to provide any meaningful ratings of specific vehicles, particularly in a book of this type. Remember, REITs are traded and perform like stocks, which are also too much of a moving target to evaluate comprehensively on these pages.

What we have provided instead is a first-stop shopping list of REITs, currently available as of this writing, whose investment holdings might appeal to ethical investors. Consult current issues of the *Value Line Investment Survey* and evaluations in Moody's and Standard & Poor's publications for up-to-date financial reviews. Telescan is a useful computer data base package for the same purpose.

Sponsors of Real Estate Investment Trusts (REITs)

Health Care Industry

American Health Properties
11150 Santa Monica Boulevard
Los Angeles, CA 90025
(213) 477-9399

Beverly Enterprises
155 Central Shopping Center
Fort Smith, AK 72913
(501) 452-6712

Community Psychiatric
24502 Pacific Park Drive
Laguna Hills, CA 92656
(714) 831-1161

H & Q Health Care
50 Rowes Wharf
Boston, MA 02110
(617) 574-0500

Health Care
1 Seagate Street
Suite 1950
Toledo, OH 43604
(419) 247-2800

Health Care Properties Inc.
10990 Wilshire Boulevard
Suite 1200
Los Angeles, CA 90024
(213) 473-1990

Health Equity Properties
915 West 4th Street
P.O. Box 48
Winston Salem, NC 27102
(919) 723-7580

Health & Rehab Properties Trust
400 Center Street
Newton, MA 02158
(617) 332-3990

Health Vest
P.O. Box 4008
Austin, TX 78765
(512) 343-5234

MEDI Trust
128 Technology Center
Waltham, MA 02154
(617) 736-1505

Nationwide Health Properties Inc.
35 N. Lake Avenue
Suite 540
Pasadena, CA 91101
(818) 405-0195

Universal Health Realty Income Trust
367 Gulph Road
King of Prussia, PA
(215) 265-0688

U.S. Health Care Inc.
980 Jolly Road
P.O. Box 1109
Blue Bell, PA 19422
(215) 628-4800

U.S. Government-Backed Mortgage Trusts

CRI Insured Mortgage Association
11200 Rockville Pike
Rockville, MD
(301) 468-9200

Krupp Government Income Trust*
Harbor Plaza
470 Atlantic Avenue
Boston, MA 02210
(617) 423-2233

UNIT INVESTMENT TRUSTS

Unit investment trusts are like mutual funds in most respects. As with mutual funds, professionals pool investors' money to assemble a portfolio of investments for those investors. The difference is that while mutual funds are continuing enterprises, their portfolios changing constantly as well as expanding when new investors buy in, unit investment trusts are static. The portfolio stays the same; once the shares are

* Pending change to REIT from limited partnership.

sold, that's it. Plus, most unit investment trusts invest in term investments; once the terms expire, the investment is over and the principal is returned. Because a unit investment trust is a known quantity from inception to expiration. you know what your return will be throughout its duration. That too is a notable difference from mutual funds, where returns vary with changes in the portfolio.

The vast majority of unit investment trusts are invested in municipal bonds. You will also find unit trusts invested in such vehicles as corporate bonds, U.S. government issues, Ginnie Maes exclusively, and so on. Again like mutual funds, unit investment trust portfolios are put together according to stated goals. For example, a corporate bond trust might contain high-grade bonds with staggered maturity dates. A municipal bond trust might contain only bonds from a single state so investors from that state would enjoy "double tax-free" benefits.

Normally, unit investment trust shares are sold in units of $1,000, with a sales charge—typically 4.5 to 5 percent—assessed on purchase. You can generally choose whether you want to receive monthly, quarterly, or semiannual income from your trust or reinvest the proceeds. There is an active secondary market for unit investment trusts, so they are quite liquid.

Financial Considerations

Note that a unit investment trust's $1,000 "entry fee" (some charge as much as $5,000) is a more substantial commitment than many mutual funds require. (Some mutual funds accounts can be opened for as little as $25 as part of a monthly check withdrawal plan.) Of course, for your $1,000, you are getting a diverse portfolio within an investment category—as you would in a mutual fund—so in that respect, a unit investment trust still makes sense for many modestly endowed investors. Most unit investment trusts are invested in low-risk fixed income vehicles, but be sure to check the trust's investment philosophy before assuming that.

Consider also that you are paying a chunk of commission to get into your unit trust. Therefore you should look on this as a long-term investment.

Other risks and benefits depend on the type of vehicle in which the trust invests. For example, those invested in bonds will fluctuate in value on the secondary market according to the direction of interest rates. Municipal bond trusts, with their tax-free or double tax-free attributes, make sense financially only for those positioned to take advantage of those benefits.

A terrific option for retirees who pay high taxes and other investors who want safe, tax-free income are unit investment trusts invested in insured municipal bonds. You will give up about half a percentage point in interest rate to get the insurance, but if you can't afford to risk principal, the cost is certainly worth it. In return, you have a steady income stream at a good rate (when you figure in the tax benefit), liquidity (because of the marketability of UITs), and safety of principal (because the bond is insured against default). That combination is hard to beat, and the investment is proactively ethical as long as all the bonds held pass your standards. These investments are also known as insured municipal investment trusts.

Ethical Considerations

Although there are no true socially screened unit investment trusts as of this writing, those UITs invested in such U.S. agency obligations as Ginnie Maes would meet most ethical investors' social criteria. So would most unit trusts invested in municipal bonds, although you should examine the holdings of the trust before investing. Just as with municipal bond mutual funds, the unit trust portfolio may contain bonds financing projects that you cannot support. In general, however, municipal bonds make beautiful ethical investments, because they are socially proactive.

By the same token, it is unlikely that a unit investment trust invested in corporate bonds will pass any SRI type's ethical test. And, again, most socially conscious investors stay away from any portfolio containing U.S. Treasury issues because treasuries finance the nasty things our government does as well as the good things.

Listings

The following companies offer unit investment trusts in vehicles— such as municipal bonds and Ginnie Maes—acceptable to most ethical investors. We have not provided performance ratings of unit investment trusts because these investments are effectively "prerated." The quality rating of municipal bonds will be stated in the trust's prospectus, and the interest rate paid will be a direct reflection of the trust's collective quality. For example, a trust paying higher-than-average interest rates will be invested in riskier-than-average bonds. (Ginnie

Maes are considered equivalent to AAA-grade bonds—bonds of the highest quality.)

When you do go to select a unit trust from a sponsor, consider that unlike mutual funds and other investments with fluctuating values, unit investment trusts are known quantities as to their rate of return, risk level, and term. Most any available unit investment trust is likely to be competitive with comparable products at the time the trust is formed, although you should check before assuming this. As we just mentioned, the risk level of a municipal bond trust is a direct reflection of the rate of return: simply select the risk/return balance with which you are most comfortable. As for term, select a term that returns your principal when it is most strategic for you to have it back (when your child starts college, for example).

To ethically screen the holdings of a municipal bond or government bond unit investment trust, refer to the list of holdings in the trust's semiannual or annual report.

The following listings are all sponsors of unit investment trusts.

Tax-free and Insured Tax-free Municipal Bond Trusts

Independent Sponsors

Advest, Inc.
1 Commerce Plaza
280 Trumbull Street
Hartford, CT 06130
(800) 733-1194

J. C. Bradford & Co.
330 Commerce Street
Nashville, TN 37201
(800) 251-1060

Clayton Brown & Associates
300 W. Washington Street
Chicago, IL 60606
(800) 621-0325

Glickenhause & Co. (New York trusts only)
6 East 43rd Street
New York, NY 10017
(800) 642-1055

Kemper Capital Markets
120 S. Riverside Plaza
Chicago, IL 60606
(800) 343-7999

John Nuveen & Co., Inc.
333 West Wacker Drive
Chicago, IL 60606
(800) 351-4100, IL
(312) 917-8138

Van Kampen Merritt
1001 Warrenville Road
Lisle, IL 60632
(800) 225-2222

Brokerage Firm Sponsors

Bear, Stearns	PaineWebber
Dean Witter Reynolds	Prudential-Bache
Kidder Peabody	Shearson Lehman Hutton
Merrill Lynch	Smith Barney, Harris Upham

GNMA Bond Trusts

Clayton Brown & Associates	Kemper Capital Markets
Shearson Lehman Hutton	Merrill Lynch
Dean Witter Reynolds	

Other Government Agency Trusts

Freddie Mac Trusts—Merrill Lynch
FNMA Trusts—Kemper Capital Markets

PRECIOUS METALS

Precious metals are one of the oldest means of storing wealth. Even today, people of means in the world's more unstable societies will often hold precious metals and jewels to preserve whatever wealth they can in the event of political or economic collapse. Gold particularly is ideal for this purpose, because it is the traditional base of many countries' currencies and is highly liquid as well, salable worldwide at market prices.

In stable countries like the United States, precious metals—again primarily gold—are a popular buy during inflationary times for one simple reason: They tend to inflate in value at the same rate as the economy. Why? During an inflation, the price of stuff goes up, and gold is "stuff."

Despite our country's historical stability, we see a rush toward gold and other precious metal buying every time there is widespread concern for the republic's economic or political health. It is one of the topsy-turvy truths of the investment world that what is bad for the nation as a whole can be profitable for certain markets. In fact, the worse it gets, the better the goldbugs seem to do. When the United States called up its military reserve units after Iraq invaded Kuwait, gold shot up about $45 per ounce.

You can invest in gold and other metals as bullion; as coins issued here and in other countries; through stock in mining companies; as special precious metals portfolios put together by various brokerage houses; through mutual funds of mining stocks; or through commodities futures contracts.

Financial Considerations

Investing in gold as a bet against our nation's stability runs counter to everything we have already said about building wealth through investing in America's history of consistent growth. There is no denying, however, that gold (and to a lesser extent other precious metals like silver) is an excellent protection against inflation. There is also profit to be made, however cynically, on the panic of others in troubled times. For these reasons, precious metals deserve consideration for the smaller, more speculative portion of your portfolio. Do understand, though, that in times of low inflation, there may be little demand for your metals holdings, driving down their value.

The gold market is also among the most emotional of financial markets—panic drives it like almost no other. This means its value fluctuates wildly. It is the last place you would want any money that you need to tap for an emergency. On the other hand, like other investments in "things," gold investments do tend to at least keep pace with inflation if held for the long term.

As for the different ways of getting into the market, your best choices are stocks and precious metals mutual funds. Playing with commodities futures contracts is like playing with fire, and you are likely to get burned unless you are already a major player in the metals business. The financial problems with owning bullion or coins—that is, the actual *stuff*—is that you have to pay to store and insure something that is not earning interest and that is not otherwise creating wealth (as stock does). If you do decide to purchase the real thing, check out the company you buy from thoroughly. When the economy went blooey in the early seventies, a lot of gold buyers discovered that their dealers had sold them "stored, insured gold" that somehow could never be located on the shelves.

Ethical Considerations

A lot of gold mining takes place in South Africa, so beware if you are buying into precious metals mutual funds or stocks. Some precious metals mutual funds are screened for South African involvement (these are listed in chapter 6). South African Krugerrands are also among the most popular coin investments. Jerry Falwell ran a "Buy Krugerrands" campaign to show support for the regime there—one of the only examples of socially directed investing with a conservative agenda. Obviously, most ethical investors avoid Krugerrands instead of seeking them out. In the eyes of most in this country, the Chinese government is a repressive regime as well, yet its Panda coin is also grabbed up by goldbugs.

COLLECTIBLES

The chances are excellent that you have already dabbled in this market, as a child. And if you are one of the very few who hung on to your drawers full of baseball cards and Superman comics, you are probably well aware that they may now be worth thousands to adult collectors. So, by the way, may your old furniture, knickknacks, and the like—all that stuff you were about to unload in your next garage sale.

But don't count on it. That is the bottom line with collectibles, which also include more culturally rarefied items such as paintings, other fine arts, and antiques, as well as rare coins and stamps. Besides the fact that every collectible market has its quirks, most collectibles are attractive only to a restricted and idiosyncratic market in the best of circumstances. In other words, you can't always be sure that someone is drooling to acquire your 1963 Willie Mays card, whereas you will always be able to market your shares of Calvert Social Investment Fund.

Realizing a profit on your collectible is also complicated by a factor known as *spread*. Spread is the difference between an item's net buying price and its net selling price. When gold is at $350 an ounce, you can buy it or sell it for $350. Only the dealer's commission will affect the spread, which should be negligible in any case. However, most collectibles are purchased at retail stores, where their price includes a markup likely to be 100 percent or more. So the antique armoire you buy for $1,000 may have cost the dealer only $500. The value of the piece could increase a full 100 percent, and the dealer will still only pay you $1,000 to buy it back.

Financial Considerations

Like precious metals, collectibles are a ''stuff'' investment, as a hedge against those times when the price of stuff inflates. Collectors bank on the principle that time increases the value of tangible assets, which it often does. Still, collectibles make up one of the most speculative and, therefore, riskiest categories of investment. Not only are values difficult to determine and growth potential difficult to predict, but collectibles can be troublesome to unload if you suddenly need cash. This is a game that should be played only by the very wealthy if the intent is to play for serious money—and then only with money the investor can afford to be without. You can increase your chances of gain by beating the retail spread—that is, by buying at auctions and privates sales or bargaining retail prices down—but the odds may be long in any case.

As with all speculative investments, growth can also be astronomical. Baseball cards, in fact, have been very hot for the last couple of years. So have rare coins, particularly silver coins. When the Hunt brothers of Texas tried to corner the silver market several years ago, silver prices skyrocketed to some $50 per ounce, and many people holding silver coins melted them down to recover the silver. That hysterical flurry only added to the rarity of some silver coins, increasing their value.

Gold and silver coins are hybrid investments, having value both as collectibles and as precious metals. Special cases aside, however, the best general advice for serious investors and dabblers alike is to collect only those things that you would enjoy owning anyway, because whether you intend to or not, you may be admiring them for a long, long time.

Ethical Considerations

By and large, purchasing collectibles is an ethically benign activity. But not always. For example, if you collect pre-Columbian artifacts, you may unknowingly be contributing to the illegal plundering of Mayan ruins in Central America. Those rare exceptions aside, buy and enjoy.

VENTURE CAPITAL

Venture capital, basically, is what it sounds like—capital provided to finance business ventures for a price. For reasons explained below, the businesses and projects seeking venture capital are almost always small and without a significant track record.

Businesses without a track record of performance and creditworthiness are extremely risky investments. So venture capitalists will usually exact a lofty price for their faith—either a high rate of return, a percentage of future business, shares of private stock, or some combination of the above. (Certainly, if the business had a track record and other creditworthy resources, it could raise money at far less cost by borrowing from a financial institution, issuing bonds, or obtaining financing through some other more public channel.)

Venture capital is frequently raised through private limited partnerships. The Academy Award–winning movie *Rain Man* was financed through a venture capital private limited partnership—Star Pictures II—which also backed *A Dry, White Season,* the antiapartheid film about South Africa starring Marlon Brando. Many of the Disney/Touchstone films have been backed by Silver Screen partnerships.

Financial Considerations

Obviously, providing venture capital is one of the riskiest investments you could make, although the potential for spectacular gain should be equally great if you have chosen your investment well. Only the very well-to-do should make venture capital investments to any significant degree (although if you are young, upwardly mobile, and on the nervy side, you might consider allocating, say, 5 percent of your financial plan to such vehicles). Of course, the fact that so many of these investments are made through private limited partnerships will eliminate all but the well-to-do anyway.

Ethical Considerations

The ethics of venture capital obviously depends on the project it finances. Many earth-friendly industries have spawned via venture capital.

Listings

The following are a few firms that offer socially screened venture capital investments.

Calvert Social Venture Partners
1715 18th Street NW
Washington, D.C. 20009
(202) 462-5449

Global Environment Fund L.P.
412 N. Coast Highway
Suite 345
Laguna Beach, CA 92651
(714) 497-6049

Sand County Ventures
1010 El Camino Real
Suite 300
Menlo Park, CA 94025
(415) 324-4414

5

A Guide to Socially Responsible Investments— Stocks

Whether you are a first-time or seasoned investor, you've turned to the right page. First-time investors, be assured—stocks are *not* too complicated for you to understand! They are probably not too risky for at least part of your portfolio, even if you are in the most conservative of circumstances. And if you can't afford the substantial cash it takes to invest in a diverse stock portfolio, you very likely can afford shares of stock mutual funds.

For you stock market veterans, this chapter will show how to blend your ethical considerations with the conventional stock fundamentals. And in our review of those fundamentals, you may discover a few things you didn't know. Certainly you will want to check out the lists of SRI stocks at the chapter's end.

We want all our readers to know that stocks, along with mutual funds, are the most thoroughly researched SRI vehicle. They are the one major category of individual investment that we didn't cover in the previous chapter, simply because there is so much information about their ethical and financial virtues that they merit a chapter to themselves.

As with our other investment advice, the information below will guide you in managing your own portfolio if you choose to do so. You will find it just as edifying if your goal is to participate intelligently with a professional adviser in drafting your financial strategies.

THE "SO WHAT" SLUMPS OF 1929 AND 1987—A CRASH COURSE IN THE STOCK MARKET

As you may have gathered by now, we feel that—from a financial standpoint, anyway—stock ownership is probably the best possible investment you could make for the growth portion of your portfolios. Historically stocks have proven to be the small investors' best chance of outpacing inflation. No other investment vehicle offers so attractive a combination of appreciation potential and liquidity. And there is an impressive array of socially screened companies whose financial performance merits your consideration.

However, if you are new to investing, you may be looking to put your money toward anything *but* stocks, because the stock market suffers from a terrible image problem. It's much like the difficulty that a rehabilitated criminal faces when applying for a job. The parolee may now be a sweetheart and the job's most qualified applicant, but the stigma of past sins is impossible to erase. In a similar way, if you have no experience with the stock market, you may have trouble overlooking the rather indelible blotches in *its* past—specifically, the crash of 1929 and, far more recently, the crash of 1987.

Ironically, both crashes actually illustrate the tremendous value of stocks as investments. But before we prove our point, let's do a little time travel to better understand the dynamics of those two events.

The Crash of 1929

The 1929 crash was far from the first time the stock market had taken a tumble. The market had collapsed in 1785, 1791, 1819, 1857, 1869, 1873, and 1907. But in the giddily prosperous postwar decade, with the market hitting record highs, most Americans came to believe that good times were here to stay. Reflecting as well as feeding the mood of the populace, newly elected President Herbert Hoover declared in 1928 that "we shall soon, with the help of God, be in sight of the day when poverty shall be banished from the nation."

Americans were also enchanted by a new type of tycoon—men such as Jesse Livermore and Charles E. Mitchell, who, unlike the great industrialists Henry Ford and Andrew Carnegie before them, made their fortunes simply by trading paper. It seemed so easy to do what Mitchell and Livermore and their like had done, or at least to share in some of the bounty, because the market had been going up for eight straight years. So for the first time in history, working-class Americans—secretaries, shoe clerks, janitors—began investing in stocks.

The 1920s had seen another financial first—consumer credit for Americans of ordinary means. In keeping with the runaway optimism of the decade, buying on credit became a way of life that extended all the way to Wall Street. Using a speculative ploy known as *buying on margin,* an estimated six hundred thousand—or 40 percent of all shareholders—bought shares with only a small down payment (typically 20 percent, but sometimes with no money down at all). As the market continued to soar late in the decade, the margin buyers were the brokers' biggest customers, parlaying their paper profits into more and more stock.

So unshakable was the faith of the ordinary stockholders that not even the thinly disguised (and then legal) manipulation of the market by its biggest players could deter them from their frenzied buying. Wealthy investors commonly pooled their funds, bought huge blocks of stock to create an artificial demand, and then sold the shares at their now inflated price to an eager public. Of course, when the big boys took their profits and went home, the demand disappeared and the price of the stock plummeted. The public knew exactly what was going on, but they kept buying in the hope that they could time their own sales better and share in the plunder.

The esteemed members of the Federal Reserve Board were among the boomtime's few knowledgeable skeptics. The board had been established in 1913 and empowered by Congress to play a number of key roles in regulating the economy. One of those roles was to monitor and, if necessary, restrict margin buying. Fearful that this wild prosperity the market was generating was based largely on borrowed money, the board met in March of 1929 to decide if they should levy margin restrictions. However, they knew going into the sessions that they were faced with a classic damned-if-you-do, damned-if-you-don't dilemma, for they realized that eliminating the credit buys could itself destroy the market. After days of deliberation, they remained unwilling to act or even to comment, and their hesitancy shattered the public's confidence. A wave of panicked selling followed. Charles Mitchell, from his position as president of the National City Bank, stopped the plunge with a huge injection of cheap consumer credit, but the die had been cast.

Margin buying did in fact finally do in the market in October 1929. Not only had the majority of shares traded been purchased with only a small proportion of cash, but those shares had then been pyramided into more shares. It was just as the Federal Reserve Board had feared— the boom market was a house of cards with a deceptively small amount of real wealth propping up the structure. At the same time, the nation's margin buyers were a mass panic waiting to happen because the only "money" most of them had to pay off their brokerage loans was wrapped up in the artificially high prices of their stock. Once those prices began to erode in late 1929, a nation of terrified margin buyers sold their holdings to salvage what little cash value was left. In their minds, the end of the financial world was nigh.

The crash of 1929 was so traumatic for the country that the federal government took a number of strong steps to insure that such a serious slide never occur again. In 1933 Congress passed what came to be known as the Truth in Securities Act, proposed by President Franklin Roosevelt shortly after beginning his first term. Designed to eliminate the bogus and near bogus stock issues that proliferated in the boom years prior to the crash, the measure required companies to disclose all pertinent financial data before offering its stock to the public. It also enumerated new and broader remedies for defrauded investors.

The following year saw the passage of the Securities Exchange Act. Whereas the Truth in Securities Act had required only an initial dis- closure, the new act ordered companies to disclose and register facts on a continuing basis. It also included new standards for margin credit, prohibitions against market manipulations, and a design for self- regulation of stock exchanges and broker/dealers in an atmosphere of government oversight. The act provided for that oversight by creating the Securities and Exchange Commission (SEC) to enforce both pieces of legislation and to punish violators. That same year the Federal Reserve Board finally began to set binding limits on margin buying, exercising the authority that Congress had granted it from its inception. As of this writing, the minimum down payment is 50 percent, and there have been times since 1934 when it has been 100 percent.

The reforms worked. No stock market slump since has been nearly so severe. By the time the slide that started in 1929 reached its lowest point in 1932, the average stock had lost 90 percent of its value. In 1937–38 the next most serious slide occurred, with stocks dropping an average 50 percent in value over an eight-month period. Meanwhile the measures had begun to transform the nature of the market itself— more long-term investors, fewer speculators. The "crash" of 1987 that we remember so frightfully simply dropped the market back to about the level it had reached the year before—a level, by the way, that investors were ecstatic about in 1986.

The Crash of 1987

By comparison with 1929, the crash of 1987 was a little stumble on the sidewalk. The fall seemed so precipitous simply because the market had shot upward in the previous year at a rate rarely equaled in history. Nevertheless, the market did drop more than five hundred points in a matter of hours on October 19, with blood pressures all over the world rising in inverse relationship.

Perhaps with a finer sense of déjà-vu, more experts would have seen this coming. Despite the federal government's effective post-1929 market regulations, there remained no way to regulate investors' emotions. And many of the same emotional dynamics that helped undo the market in 1929 were at work prior to the 1987 crash. Much like Herbert Hoover's speeches in the late 1920s, Ronald Reagan's "power of positive thinking" political rhetoric—plus some radical economic surgery on the swollen inflation rate left over from the late 1970s—made most Americans more confident about their country's economic future than they had felt in a long time. That was especially reflected on Wall Street. A bull (or surging) market that started in 1982 gathered tremendous momentum in the several months prior to October 1987. The same giddiness that helped bring about the crash of 1929 had taken hold of the population again—the belief that prosperous times were here forever. Riding that belief for all it was worth and thus inflating stock prices were the inevitable speculators, limited now in credit, but never in greed or foolishness.

Historians still debate over what shook the faith that was holding this raging bull market together. The mass realization that Reaganomics was a sham, the international community's fears about the gargantuan federal budget deficit run up during the decade—these are only two of many theories. Whatever the specific reasons for the fall, however, the larger, constant truth about the stock market is that what goes up must come down (and vice versa). Every boom in the market's history has been followed by a crash, and every crash has been followed by a boom. The eventual crash of the 1980s market surge—although not its timing—was entirely predictable.

And just as predictable was the market's rapid recovery, this time with relatively little blood spilled because a much wiser breed of investor now populated Wall Street. Patient investors who bought early in the 1980s boom and held tight during the crash still had healthy profits to show for the decade, even at the crash's lowest point. Smart bargain hunters and dollar cost averagers, already attuned to the market's ups and downs, scored some incredible buys during the crash (because, after all, the crash didn't mean that once solid companies were failing—it just meant that the paper value of their stock had had a hellacious hiccup). The big losers, of course, were the speculators who were "playing" the market from day to day as if it were a big casino, and the poor planners who found they suddenly needed cash and had only their undervalued stock with which to produce it.

At the start of this section, we made the rather brash statement that the crashes of 1929 and 1987 actually proved the worth of stocks as investments. It's time to back up our claim, which actually comes down to two main points:

First, investors have always done exceedingly well by the stock market, even in bad times. Despite the market's peaks and valleys, the overall long-term trend is continually upward. So stocks held for the long term—the investor's strategy—will almost always weather the storm of any market downturn, recover their value, and eventually increase in worth. As you will discover below, many socially responsible investors have done significantly better even than the market's averages.

The "Wealth Indexes of Investments" graph (see page 181) demonstrates another outstanding feature of stocks. As you can see, the rate of growth in the market almost always at least keeps pace with inflation and usually does much better than that. The compound rate of inflation over the last sixty years has been about 3 percent. During that same period, the average share of common stock traded on the New York Stock Exchange—including the stocks of all companies that have failed in that time—has appreciated in value at a compound annual rate of approximately 10 percent. (This figure assumes that dividends are reinvested.) That's a real-interest return of about 7 percent annually—outstanding!

The reasons stocks generally do this well are straightforward. When you own stock, you own *things* (in this case, a portion of the assets and products of a company). Inflation means the price of things goes up, so the price of your company's things will tend to follow that trend.

Also, American business tends continually to create wealth, because in our economic system businesses tend to expand. This is such a con-

Wealth Indexes of Investments in U.S. Stocks, Bonds, Bills, and Inflation: 1926–1990

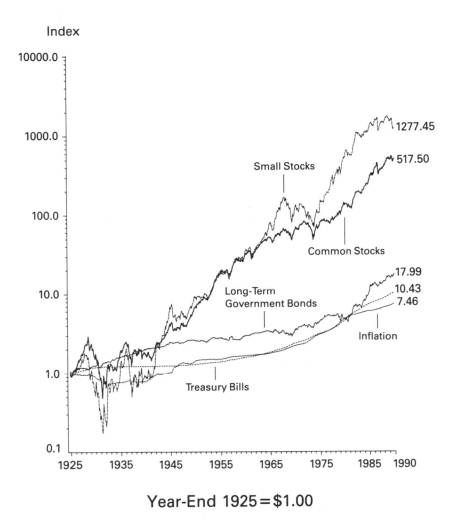

Year-End 1925 = $1.00

stant that even in times of recession the gross national product usually increases (even though the rate of growth may be slowed significantly). In fact, only in a few odd years has the GNP actually declined. If you associate your money with this wealth-creating momentum by investing in a diverse portfolio of stocks, it should be obvious to you that your chances of sharing in that growing wealth are excellent.

The second major point is that the admittedly horrendous crash of 1929 will never happen again. Yes, we know that "never" is a very long time, but no crash since 1929 has ever approached that level of severity—predictably. You see, the 1929 crash and the ensuing Great Depression made a heck of an *im*pression on the people running this country. For at least once in our history, we actually did learn from our mistakes. Government controls on market speculation and manipulation not only were well conceived, but they have also been fairly rigorously enforced. Ask Ivan Boesky—the once prescient-seeming deal maker was jailed for trading on the basis of illegal insider information. Although cooperation with the SEC shortened his sentence considerably (he got three years and served one), his testimony implicated significant others, including Michael Milken.* Ask Wes Groshans. The former chief executive officer of SFT, Inc.—which operated a small family of mutual funds—sold a great deal of his own Genoa Corporation stock to four of the funds, an obvious conflict of interest that also artificially inflated the stock's price. The SEC noticed and tore into action. Within thirty days order was restored: Groshans was banned from the business pending full investigation, SFT was sold to the very reputable William Penn Funds, restitution was promised frightened shareholders, who took a loss to bail out, and, most important, a meltdown of the funds' value was averted. And of course, ask Milken and his pals at Drexel Burnham. If only the government would get as tough on our environmental defilers as it's been on Wall Street pirates.

This is not to say that there will never be another major stock market slide. In fact, there almost certainly will be. The eminent economist John Kenneth Galbraith is simply reading history when he predicts that a wave of overly optimistic speculation will bring on a serious market crash every twenty or thirty years "because that is the length of the financial memory." And when the market goes down the tubes, the speculators will go down with it. The investor who patiently holds his

---

* *Los Angeles Times*, 7/27/90, D1.

or her stocks will turn profits eventually, because just as inevitably the market will recover and then grow.

SOCIALLY RESPONSIBLE STOCK PERFORMANCE

As well as stocks perform as investments overall, many socially screened stocks and stock mutual funds do even better. The Dow Jones Industrial Average, an index compiled from the average stock prices of thirty leading manufacturers and distributors, is the best-known barometer of stock market behavior. For the period between 1976 and end of 1990 (a period of both bad and good times), the DJIA gained over 160 percent. During that same period, the Good Money Industrial Average—*Good Money* publication's index of the stocks of thirty leading socially screened companies—gained over 500 percent (see page 184). The Good Money Utility Average—an index of ethical utility stocks—similarly outperformed the Dow Jones Utility Average (see page 185).

After October 19, 1987 ("Bloody Monday" of the 1987 crash), the Dow Jones Industrial Average had lost all of its gain for the year, which had been a whopping 43 percent on August 25. It was now up only 73 percent for the decade. The Good Money Industrial Average was still showing a slight gain (5 percent) for the year (it had peaked October 1, up 46 percent for the year) and was up 421 percent for the decade.

In its first 5½ years, the Clean Yield Model Portfolio (established March 1985 and comprising socially screened emerging growth companies) appreciated 311 percent compared with about a 119 percent rise for the market overall as measured by the Standard & Poor's 500 Index (another well-known market barometer). Although the portfolio's performance understandably dropped some during the onslaught of economic bad news in 1990 and early 1991, it continued to outperform the Standard & Poor's 500 by about the same substantial margin. As of January 1991, Clean Yield's individually managed socially screened accounts were also outperforming the Standard & Poor's 500 (see page 186).

Another way to measure socially screened stock performance is to examine the performance of mutual funds invested in socially screened stocks. Calvert-Ariel Growth Fund—which buys stocks of undervalued or underfollowed companies that are not involved in South Africa,

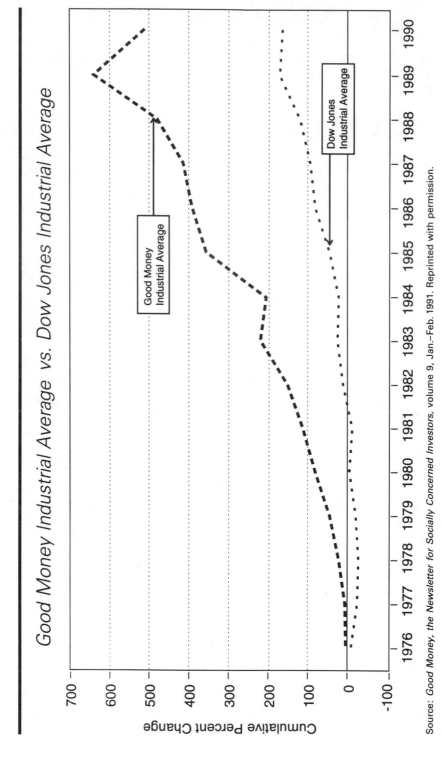

Good Money Industrial Average vs. Dow Jones Industrial Average

Good Money
Industrial Average

Dow Jones
Industrial Average

Cumulative Percent Change

1976 1977 1978 1979 1980 1981 1982 1983 1984 1985 1986 1987 1988 1989 1990

700
600
500
400
300
200
100
0
-100

Source: *Good Money, the Newsletter for Socially Concerned Investors,* volume 9, Jan.–Feb. 1991. Reprinted with permission.

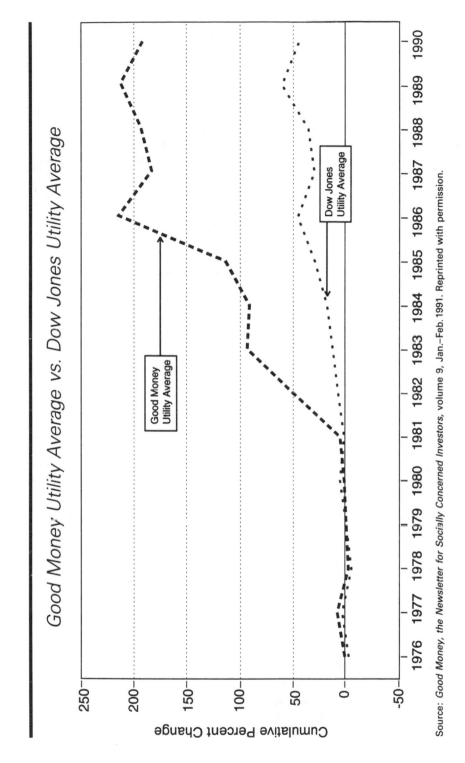

Good Money Utility Average vs. Dow Jones Utility Average

Source: *Good Money, the Newsletter for Socially Concerned Investors*, volume 9, Jan.–Feb. 1991. Reprinted with permission.

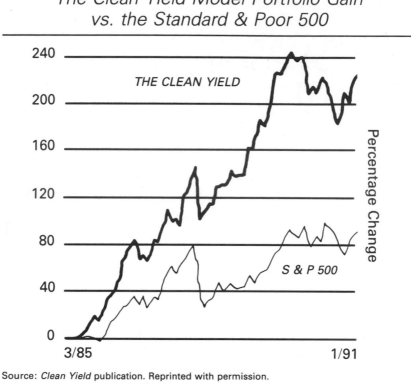

The Clean Yield Model Portfolio Gain vs. the Standard & Poor 500

Source: *Clean Yield* publication. Reprinted with permission.

weapons, or nuclear energy—earned 69.5 percent for the three years ending May 31, 1990. The Standard & Poor's 500 gained 37.9 percent in the same period, the average long-term growth fund 28 percent.

Many mainstream (non-SRI) investment professionals sniff at the suggestion that socially responsible stock investing could ever pay big returns, saying that the universe of screened stocks is too limited, or the criteria too limiting, to assemble a top-flight portfolio. Such statements are difficult to support with facts, but some of those same professionals defend their ignorance by announcing glibly, "I'm too busy making money for my clients." Forgive us if we gloat a bit, but we think the performance statistics and stock lists provided in this book, and particularly in this chapter, prove that socially screened stocks are a prudent choice even for the most cautious of professionals. And we dare anyone to state that the field is too limited after examining our list of socially screened stocks on page 221, the top ten holdings of the socially screened mutual funds listed in chapter 6, and the socially screened stock indexes detailed in appendix B.

STOCK BASICS

Now that we have you all pumped up about stocks, let's go over a few details about them. To review from earlier in this book, each share of stock represents a percentage of ownership in a company. Beyond this general meaning, the following distinctions apply:

Common stock includes certain ownership rights:

- The right to exercise one vote per share on corporate issues and in the election of corporate directors. (You may exercise these rights either in person at annual shareholder meetings or by mail through proxy statements.)
- The right to be informed about corporate financial standing through quarterly income statements and balance sheets and annual audited financial reports.
- The right to receive a portion of company profits in the form of stock *dividends*. (Different companies pay out different portions of their profits in dividends—see *"Income vs. Growth"* on page 192. A few companies pay dividends in the form of additional stock instead of cash.) In good times a company may increase its dividends; in bad times it may decrease them.
- For stockholders owning $1,000 or more of stock, the right—under certain conditions—to propose resolutions regarding company policy that will be put before all the stockholders. As we mentioned in chapter 1, some ethical investors with an activist bent use this right to influence the social behavior of a company.

Preferred stock differs from common stock in the following ways:

- Should the company enter financial straits, preferred stockholders would be paid before common stockholders (but after the company's debt holders and secured creditors—meaning after corporate bond-holders, among others).
- The dividends of preferred stock are fixed, like any other fixed income investment.
- The typical preferred stock does not carry voting rights.

Convertible preferred stock can be converted into common stock at the holder's discretion, although the rate of exchange is fixed and conversion rights may be limited to a defined period of time.

Listed stocks are marketed on either national or regional *stock exchanges*. The New York Stock Exchange (the largest) and the American Stock Exchange are the two national stock exchanges. The Pacific Stock Exchange and Midwest Stock Exchange are examples of major regional stock exchanges. Stock exchanges are the physical places where the buy and sell orders for stock are processed. The purpose of stock exchanges and the over-the-counter market (see immediately below) is to provide an orderly marketplace where 1) stockholders can instantly find buyers (and buyers can find sellers) and 2) transactions can be recorded such that new owners of stock will be recognized as bona fide stockholders. Each exchange trades only those stocks that are listed on it. Many stocks are listed on more than one exchange.

Over-the-counter stocks are available for public purchase but are usually not listed on national or regional stock exchanges. The over-the-counter market is not an actual place like the stock exchanges—instead it is a network of broker-dealers who conduct trades according to specific guidelines established by the SEC. Many stocks of small, newer companies are traded this way; some stocks of major corporations and public utilities are sold only over the counter as well. Over-the-counter stocks are also known as *unlisted stocks*. This is an inexact term, however, because listed stocks are sometimes dealt over the counter at a negotiated price, as opposed to the public auction conducted by the exchanges.

Privately held stock is stock held by the owners of a corporation that has not *gone public*—stock, in other words, that is not for sale on the public markets. Few major corporations today are privately held. Obviously our remaining discussion of stock will pertain to publicly traded stock.

So why do corporations sell stock in the first place? Normally corporations go public—that is, publicly sell part or all of their ownership—to raise capital for future expansion. The larger a company grows, the more it costs to expand its operations. Selling ownership raises cash, usually at a large profit, for the corporation—an attractive alternative to borrowing the money and paying interest on it. The profit is made only on the initial stock offering, however; the stock that the corporation has sold will be traded in the future between stockholders with no direct benefit to the corporation itself.

No company can issue stock (or bonds, for that matter) without first registering all pertinent facts—assets, liabilities, profit-and-loss history, operation details, and so on—with the SEC. Publicly owned companies sometimes issue new stock (with the approval of a majority of their shareholders) as a way of raising further cash or investment capital. These actions, too, are closely regulated by the SEC.

WHAT MAKES STOCK LIQUID?

Despite our assurance that stocks are liquid, it may have already occurred to you to ask a troubling question: What happens when a stock has more sellers than buyers? Luckily the same question occurred long ago to the operators of the stock exchanges. The answer they arrived at—*the specialist system*—ensures that for every seller there will be a buyer.

Specialists are brokers who have chosen to deal only in certain stocks. (A highly active blue chip stock like General Motors may have two or three specialists competing against one another. Specialists in less active stocks may handle a few dozen. The 1,200 stocks listed on the New York Stock Exchange are represented by about 350 specialists.) Specialists "make markets" for the stocks they represent—that is, when there is a gap between the supply and demand of one of their stocks, they are obligated to buy or sell shares for their own account to stabilize the situation. So if sell orders for a stock deluge the exchange on a day when there are nowhere near the buyers to match, the specialists in that stock will buy shares until a demand reappears.

Risky business on the face of it, isn't it—although most readers of this book would trade their net worth for the average specialist's any day of the week. Specialists have one other duty, by the way. When you place an order to a broker to buy or sell a stock, he/she is obligated to get you the best possible price. The broker accomplishes this by passing on the order to the specialist in that stock, who never leaves the spot at the exchange where that stock—and all the other stocks he/she handles—is sold. The specialists' immobility, so to speak, also helps guarantee that in a chaotic or crisis market they will immediately be available to stabilize prices.

Specialists earn their living off the fees they charge for processing orders in their stocks. Since they are obviously in a position to manipulate prices, they are closely supervised by both the SEC and the exchange's governors. There has been relatively little flagrant abuse of the system, and those few cases have been swiftly prosecuted. The system does have its critics—some feel that many specialists were rather slow to buy plummeting stocks in the crash of 1987, for example, causing the market to drop much farther than it might have had they fully lived up to their responsibility. But the system has held up remarkably well for over sixty years.

The point simply is that—theoretical criticism notwithstanding—the specialist system does work. American stock exchanges are the envy of the world for their orderliness and stability, and the specialists are a big part of the reason so many of the world's major investors prefer to invest in American stocks. American stocks are dependable in terms of their long-term appreciation value and liquidity.

By the way, the over-the-counter market also has its "market makers." These are not specialists in the strict sense, because the OTC market is a way of doing business between dealers, not a physical place. But they serve much the same function, buying or selling shares for their own inventory to ensure an orderly market.*

THE MECHANICS OF BUYING YOUR STOCK

Technically you don't need a stockbroker to buy or sell stock for you, nor do you need the services of a stock exchange. You can find your own buyer or seller, negotiate the price, and get the new ownership of the shares properly registered. In actual fact, of course, very few stock transactions take place without brokers. Not many investors are knowledgeable enough or have enough time to transact stocks on their own, and most of those who could would still rather pay a commission to a broker to select stocks and handle transactions for them. Commissions on stock transactions are modest compared to most in the investment world. And if you know exactly what stocks you want and need a broker only to do the actual transacting, you can save considerably more by using a discount broker. If you trade in amounts between $2,500 and $20,000, full-service brokers will probably charge you between 1½ and 2½ percent commission. Discount brokers will probably charge you 30 to 70 percent less but usually only transact—they generally do not advise. (The differences between a discount and full-service broker are covered more fully in chapter 6.)

---

* For a fuller understanding of the intricacies of "market making" and the inner workings of stock transactions in general, read Louis Engel's classic, easy-to-understand book, *How to Buy Stocks,* listed in our bibliography.

Whether you go with a full-service or discount broker, you will start the business of stock investment by filling out a new account card. The card will ask for basic information, including your name, address, phone numbers, occupation, employer, age, spouse's name, Social Security number, citizenship, bank reference, and how you want the ownership of your stock registered.

With your account opened, you will be able to buy and sell stock simply by calling in your order to your broker. You can make your purchases in either *round lots* or *odd lots*. An odd lot means a purchase of less than one hundred shares in a corporation. A round lot means one hundred shares or more. You will pay a small premium on your purchase of odd lots—usually around one-eighth of a percentage point—so it is financially advisable to purchase in round lots when you can.

A couple more wrinkles:

A substantial odd lot purchase may make some sense if an otherwise promising stock is so high-priced that you can afford only fifty shares or so. But don't buy little Janie two shares of Ben & Jerry's Homemade as a cute gift to go with her Ben & Jerry's birthday ice-cream cake. The typical minimum stock transaction fee is $45 no matter what the price of the shares. Janie won't think that's a very cute price to pay when she grows a little older and decides to cash in her "investment."

Companies sometimes administer a *stock split* to make round lot purchases more affordable for the average investor. A stock split increases the number of shares outstanding by giving stockholders a proportional number of new shares. Here's a (fictitious) example: Monkey Wards, the nation's fastest-growing builder of simian health care facilities, has seen its stock rise to $100 a share. Wanting to ensure that demand for its shares stays strong, the company administers a two-for-one stock split—investors are given new shares equal in number to the number they already hold. This action doubles the number of shares outstanding and halves the price of the stock. (A split never in and of itself changes the value of a stock investment.) A Monkey Wards stockholder who before the split owned two hundred shares of $100 stock (total value $20,000) now holds four hundred shares (two-for-one) of $50 stock (still a total of $20,000). New investors, however, can now buy a round lot of one hundred shares for $5,000 as opposed to $10,000 before the split. This is a significant difference.

CONSIDERATIONS IN CHOOSING STOCKS

Common vs. Preferred

The name *preferred stock* is misleading in a number of ways. First of all, few individual investors would "prefer" it over common stock, for reasons that will soon be clear. Second, although it is stock in the sense that it represents shares of ownership in a corporation, preferred stock behaves financially more like a fixed income investment. As with other fixed income vehicles, the dividend rate does not change and the stock has little ability to appreciate in value. Common stock will skyrocket if the company's fortunes do, too. Preferred stock increases in value only when the dividends it pays look good relative to prevailing interest rates. As is the case with bonds, preferred stock increases in value when interest rates fall and decreases when rates rise. However, the potential increase in value is almost inconsiderable when compared to a growth common stock.

Something else to keep in mind is that although preferred stock pays higher dividends than the company's common stock when it is first issued, the dividends paid on the company's common stock—which, again, are not fixed—may grow to outstrip those of the preferred.

Here's strike three (or four, but who's counting?) against preferred stock as a personal investment. Robert Gardiner, former CEO of Dean Witter, points out that "corporations enjoy certain tax benefits from owning the preferred stock of other corporations. Corporations therefore bid up the price of preferred stock to reflect those benefits. This makes preferred stock in most cases a noncompetitive investment for individuals."*

Although there may be cases where certain preferred stocks would be suitable investments for investors who wanted some income flow, those cases are fairly isolated. From here on in this book, when we say "stock" we mean common stock exclusively.

Income vs. Growth

Stocks are thought of generally as a growth investment because they almost always do a better job of outpacing the inflation rate than do conventional income-oriented investments, and as a group they are

---

* Gardiner, *The Dean Witter Guide to Personal Investing*, p. 81.

considered riskier than income-oriented investments (the inevitable risk/reward trade-off).

However, within the broad range of stocks available, some will produce considerable income flow in the form of high dividends. Essentially these are stocks of the more fully developed companies. With their fundamental growth now behind them, these companies can afford to pay out most of their earnings in dividends and still expand as needed. The other side of their maturity, of course, is the fact that their stocks are not likely to fluctuate in price—down *or* up—as much as stock in companies whose potential is still in front of them. Thus the higher dividends and stability are a financial trade-off for lessened possibility of appreciation.

Growth stocks—mainly the stocks of young, still developing companies plus stocks in volatile industries such as energy exploration—are the opposite in just about every way. Growth companies pay small dividends or sometimes no dividends at all. Instead they plow back their earnings into their own expansion or into ventures that otherwise promise a high rate of return. Investors can ultimately make fortunes in growth companies because of their tremendous ''up side.'' In 1981, for instance, the stock of San Diego–based Price Co.—which operates the Price Club cash-and-carry warehouses in several states and Canada—was selling for $25 a share shortly after its initial offering. Since then the stock has split four times—three times at two to one and once at three to one, for a total of twenty-four to one. On March 25, 1991, Price Co. was selling for $45 a share. That means each of those early $25 shares had grown in value to $1,080!

You can lose a lot of money investing in developing companies as well. The small, younger companies do not have the financial resources to withstand much adversity the way the majors can, so the risk of failure is always there. If you buy growth stocks, the risk to your principal can, of course, be offset by diversifying your investments, including buying stock in the more mature companies.

Company Attributes

As a socially responsible investor, you will be as concerned with what a company produces and how responsibly it conducts its business in society as you will with its particular financial attributes. Therefore you may rule out investments in certain sectors of the economy—such as the weapons industry, for example—right off the top. But you will find that within the economic sectors you do consider, many of the same things that make for a socially responsible business make for good business in general. That is one solid reason the stocks of many socially responsible businesses perform so well relative to the market averages.

The performance of Nordstrom department store stock is a perfect recent example of how socially responsible business practices and financial behavior dovetail. Nordstrom maintained a reputation for years as an exemplary business. Long noted for its high-quality merchandise and exceptional customer service, Nordstrom appeared to have a fine internal organization as well. Seemingly Nordstrom was the best of both worlds—a place where customers liked to shop and people liked to work. The company was growing rapidly, and its stock was favored by many socially responsible as well as conventional investors.

The first sign that all was not well at Nordstrom occurred on November 28, 1989, when the press reported that the company's customer service record had been exacted at a heavy price—a price paid by its employees.* Soon afterward Nordstrom sales staffers began appearing on national and local news shows, testifying as to how they had been pressured to provide customer services (such as writing thank-you notes and making home deliveries of merchandise in their off hours) without compensation. Along with this sad story ran another: The National Labor Relations Board was citing the Washington-based firm for not bargaining in good faith with the United Food & Commercial Workers Union in its home state.† Nordstrom stock, which had been a steady gainer, began to drop. Then, just as that little controversy seemed to be subsiding, federal labor officials charged that the company had unlawfully attempted to eliminate unionization, including paying employees to attend antiunion rallies.‡ Labor controversies were still swirling about the chain in November 1990 as its stock dropped all the way to 17⅛, down from its previous fifty-two-week high of 39¼. (Of course, with an absence of further damaging news, the stock had recovered to a price of 35 by March 14, 1991.)

Whether you are investing for income or growth, you want the company whose stock you buy to be a good one. Beyond even its social implications, a benevolent attitude toward its own employees is

* *Los Angeles Times*, 11/28/89, D2.
† *Ibid.*
‡ *Los Angeles Times*, 8/2/90, D1.

one sign of a well-managed business. Employees who feel supported and respected by management will not only work harder for their company, but will stay with the job longer, saving the company training costs and increasing its productive capacity. If given some say in company decisions, "line" employees can also be a valuable source of innovative ideas and other wisdom.

Other criteria that bode well for the financial future of a company include

- an experienced management team. Veteran leadership, vital in a big company, also lessens the risk of investing in a younger growth company. When we say "experienced," of course, we do not mean doddering, which can be the opposite sort of indication. Management must be "with it" and insightful enough to keep the company in step with inevitable changes in the market and in technology.
- products that are needed and wanted in society.
- the ability to deliver those products to their appropriate market efficiently and at a price people are willing to pay.
- a well-known, and respected, name in its field.
- a track record of success and steady growth. Obviously you have to be less demanding in this area if you are looking for younger growth companies.
- adequate emphasis on research and development. Stocks in companies that have allocated sufficient resources for their long-term development make more sense for the long-term investor.
- financial strength and stability. We will go into the specifics of what we mean here—and how to determine them—later. For now, though, consider that a company that is "highly leveraged"—that is, heavily in debt—will be unable to devote much money or energy toward future development. It is also not well girded for a downturn in the economy or in its market.

Note that we have not mentioned social attributes that have no obvious financial implications because those are personal criteria that you will establish for your investments. Not every reader of this book will have the same set of criteria. A worksheet in the final chapter will help you define your social priorities so you (and/or your broker) can select appropriate investments with those in mind. Refer also to the "good" and "bad" company lists in appendix A. You can cross-reference stocks that appeal to you from these lists with the lists of socially screened stocks that appear at the end of this chapter.

Identifying Attributes

Now that you've identified *what* to look for in a company, how do you identify specific companies that have those attributes? It's a big universe out there—almost three thousand companies listed on the New York and American exchanges, not to mention the thousands of unlisted possibilities. That's why most investors use brokers, who make it their business to pluck potential winners out of that almost endless pack.

Of course, not even the most thorough broker can be on top of every promising company in every field. Even if you plan to leave most of the rough sorting of stock investment prospects to your broker, don't sell your own hunches short. Perhaps you have special knowledge of promising companies within your own field or a friend's or relative's field. Perhaps you know of an impressive company because you use their products. Maybe your reading has tipped you off to a lead that your broker can follow up.

Beyond your particular insights, the following sources of information will help you discern those company attributes that we identified previously as being significant.

- The financial press. Prime organs include *The Wall Street Journal, Barron's, Forbes, Investor's Daily, Business Week,* and *Fortune.* The business section of *The New York Times* reports from the center of where the action is. Don't forget the business section of your local paper, either.
- People who work in the same field as the company in which you are interested. Have you ever noticed how inaccurate a newspaper story seems when it covers a subject you know intimately? Insiders in a field are likely to have far more in-depth knowledge of a company in that field than will the financial press.
- Trade journals. These are another excellent source of inside dope beyond what is covered by the financial press.
- SRI periodicals. We list a number of these in appendix D on page 379.
- Issue-oriented publications. Whatever your pet issues, you will find not only general periodicals and other writings on it, but also newsletters monitoring how that issue plays in specific fields. For example, Co-op America's *Building Economic Alternatives* covers "emerging, hopeful alternatives to 'business-as-usual' "; the AFL-CIO's *Label Letter* publishes a list of nationally sanctioned

AFL-CIO boycotts plus labor-related anecdotal reports; People for the Ethical Treatment of Animal's *PETA* publication reports investigations of companies' animal-testing and animal-oriented policies and practices. From this type of coverage, you can certainly pull together preliminary "good," "bad," and "maybe" lists of companies.

Statistical Indicators of Good Investments

Warning: We are about to get a little technical now—even use some basic (dare we say the word?) *math*. Do-it-yourself investors will want to know this material before venturing into the market. Even those of you who plan to work with a broker may want at least a passing knowledge of things like price-earning ratios, but in the main this is the kind of stuff your broker gets paid to be obsessed with. So if you have a broker, don't feel as though you need to pass a test on this material before you make your first stock investments.

We have organized the following information according to the source in which you would most easily find it. We begin with the most accessible source, the business page of your daily newspaper.

Stock Reports in Newspapers

The business pages of almost any major newspaper will include daily stock reports. These reports are useful not only in tracking stocks you already own, but also in locating new stocks to purchase. If your paper does not carry these reports, you should subscribe to a major daily like *The New York Times* or a financial periodical like *The Wall Street Journal* or *Investor's Daily* to stay informed.

Buying stocks is like buying eggs in the sense that you shop for the best value. That is, *the stock of a solid company is not necessarily a good buy if it is overpriced.* Some of the information in the newspaper stock reports will help you determine good values in the stock market. First examine our chart, "How to Read Stock Listings in the Daily Newspaper," to see just what all this stock report gobbledygook means. Here's an explanation of some of the terms:

52-Week High is the highest price the stock has reached during the last fifty-two weeks.

52-Week Low is the lowest price of the last fifty-two weeks.

Stock Name/Symbol is an abbreviation of the company's name, stan-

How to Read Stock Listings

PAST 52-WEEK PRICE		STOCK NAME	STOCK SYMBOL	ANNUAL DIVIDEND	
HIGH	LOW				
34	12¾	Wellman	WLM	.12	.6

Yield—
Annual Dividend
Divioed by
Closing Share Price

dardized for the stock reports; for example, Ben & Jerry's Homemade is BJICA on the NASDAQ National Market System, and Wellman is WLM on the New York Stock Exchange.

Dividend is the annual dividend paid on a share of this stock.

Yield is the dividend expressed as a percentage of the closing price of the stock that day.

P/E Ratio is the price-earnings ratio, a primary tool for evaluating the value of a stock. We will have more to say about the P/E ratio and yield columns later.

Sales is the number of shares of this stock that were traded that day, expressed in hundreds (add two zeros).

High is the highest price this stock reached that trading day.

Low is the lowest price of the day for this particular stock.

Closing Price is the price of this stock at the last trade of the day, or when the market closed.

Change is the net change in the stock's price from the previous trading day, expressed as a fraction.

Not all newspapers use the exact same column headings, although the information given should be comparable.

Of all these categories of information, the two that financial professionals pay most attention to are "yield" and "P/E ratio."

In the Daily Newspaper

PRICE TO EARNINGS RATIO (P/E)	NUMBER OF SHARES TRADED X 100	PRICE PER SHARE FOR THE DAY			
		High	Low	Closing	
11	748	21½	20⁵⁄₁₆	21	+ ¼

Share
Price Change from
Previous Day's
Closing Price

Note that stock listings may contain additional symbols and letter codes that will be explained in the explanatory notes published with the stock listings.

Source: *The Wall Street Journal*

The *yield* of a stock is actually significant only to investors who are looking for income from their stock investments. Yield tells you almost nothing about a growth stock and should not be a factor in your selection. Growth companies pay small dividends—and thus show small yields—because they are reinvesting most of their earnings in their own development instead of paying them out to investors. You invest in growth stocks for their long-term profit potential, not their ability to give you a little extra "mad money" each quarter.

Even when hunting for income stock, however, you should not take a stock's yield as an absolute indicator of anything. A company that is paying a large portion out of its earnings in dividends may not be retaining enough money for research and development (R&D). You hope that the yield is high because the company is so profitable that even the small percentage of earnings retained for R&D is sufficient to maintain the company's position in its industry. But you will need more information to determine that. The lesson here is that there are no surefire formulas in the stock market and no simple, reveal-all statistics, either.

The *P/E*, or *price-earnings, ratio* is determined by dividing the price of the stock by the company's earnings over the last twelve-monthperiod. The P/E ratio begins to tell you something about the value of a stock, but—as with the yield—what it tells you is far from absolute. Eggs at $1.39 a dozen are clearly a better value than eggs at $2.12 a dozen, everything else being equal. A stock with a P/E ratio of 8 is not necessarily a better value than a stock with a P/E of 20 because—unlike eggs—"everything else" is virtually never equal between two stocks.

Nevertheless, the P/E ratio is a good starting place for evaluating a stock's potential as an investment. Here's what the P/E ratio *does* tell you:

■ The P/E is an indication of investor confidence in a stock. And confidence breeds popularity. Fads drive Wall Street as surely as they do the fashion industry. You pay more for the latest fashion in blue jeans than you do for the classic Levi's 501s because consumer demand drives up the price. You will pay more for stocks that are popular with investors for the same reason. The companies with the highest P/Es are undoubtedly fast growers—that's why so many investors are flocking to them. Thus the short-term prospects for those stocks are excellent. But that is not the same thing as saying that the stock is a good value at its current price—especially for long-term investors. Plus there is no way of knowing in advance how short that short-term prospect is. It could be up the day you buy the stock.

A stock's real, as opposed to apparent, value can be known only in retrospect. If the stock continues to rise at its current rate, at least for some reasonable length of time, it obviously was a good buy even if you paid a fad follower's premium for it. If investor demand has driven up its price beyond its real worth, enough investors will eventually come to their senses to drop the price, perhaps precipitously.

■ The P/E ratio is an indicator of investor recognition of a stock. Investors in search of quick profits cluster around the growth leaders and thus pay a bid-up price to be part of the trend. That means there are some hidden gems out there, companies with low P/E ratios because they have not yet been "discovered" by the investing masses. A company may be overlooked simply because it is not in the "sexy" industry of the day.

For an example, let's take a time trip to the recent past. Computer companies are hot. So are athletic-shoe manufacturers. How is Tubular Yawn, a boring little pipe-making outfit, supposed to compete with that sort of hoopla? It probably can't—during the height of the fad, anyway. But sharp investors may notice that Tubular Yawn has been quietly growing at the same rate as the middle-of-the-pack computer and athletic-shoe outfits. And at a P/E ratio of 7, it is a far better buy than the high double-digit P/E ratios for those sexier stocks. You may have to hold on to your Tubular Yawn stock for a while, until the investing masses begin to notice the same things about it that you do. But since you *invest* in the market instead of speculate, that shouldn't be a problem, should it?

- A low P/E ratio increases the chance of a buy-out by another company. This is the tricky part of bargain hunting for the ethical investor. Corporate raiders are out poking through the underbrush with you as they search for stocks that are low-priced in relation to their real promise. If the socially responsible company you own stock in is bought out, you may see your stock's price go up considerably. But the new parent company may not be nearly the corporate citizen that the last one was.

By the way, consider as something of an exception to the above the fact that the stocks of older, developed companies also tend to have lower P/E ratios. Although the values of this class of stocks are more stable than the growth stocks, there will always be those within the class who ride a tide of popularity while others fall out of favor for a time. You can find some good values in the latter group with little risk of business failure. Remember that the growth potential of such stocks is likely to be of the steady, unspectacular variety.

- Expect stocks with a high P/E—say, over 20—to be quite volatile. These stocks are riding a wave of optimism in an environment where emotions are mercurial, to say the least. Growth stocks in general are more volatile than income stocks, because the stocks attract speculators and because growth companies have not yet amassed the resources to ensure stability and investor confidence.

C. Colburn Hardy, author of *Dun & Bradstreet's Guide to Your Investments,* advises avoiding (or selling) any stock whose P/E ratio reaches triple the market average, because statistically, he feels, it may be ripe for a fall. Is he right? Who's to say? Some analysts feel the best way to invest in the market is to associate with the hot stocks—ride the horse in the direction it's going, so to speak. Others avoid hot stocks as if they were hot potatoes and search for overlooked value stocks instead. Both schools of thought can offer proof of their point, but only by ignoring times when the formula hasn't worked.

Annual Reports

Annual reports to the shareholders are usually impressive-looking publications. They are often printed on coated paper and decorated with full-color photos and nifty graphics. Accompanying that flash and dash is an even slicker sales pitch to shareholders about the company's glowing prospects. Although companies are forbidden by law from saying fraudulent things about themselves, there are of course all kinds of legal ways to tweak words and numbers. Few investors who are not also trained accountants will be able to extract the entire reality from those pages. But even without that background, you can penetrate the mumbo-jumbo deeply enough to get a pretty good sense of the company's financial condition and an emotional *feel* for the company through what it says about itself. The latter may be especially important to your evaluation of its ethics.

The numbers (and, equally important, the explanatory footnotes) are in the back of the report. You will want to turn there first if you rough-sort investments by their financial promise before ruling out those that violate your ethics. Here's a few useful indicators to look for:

- **Ratio of assets to liabilities.** According to the common rule of thumb, financial health requires a ratio of assets to liabilities of at least two to one. If the company you're investigating checks out in that respect, check further by examining its listed assets. What proportion are intangibles like "goodwill,"and does that two-to-one ratio depend on their inclusion? If so, be skeptical. Do note, though, that the two-to-one figure does not hold for all industries. Financial analysts will accept lower ratios in railroads, for example. Conversely, they like to see ratios of three to one or four to one in such industries as chemicals and tobacco. Yes, these latter industries interest few ethical investors, but you get the point.
- **Earnings per share.** Of course, you want to see these rise continually over the years. (You can see the trend by checking previous years' annual reports.) By the same token, a recent drop in earnings should alarm you but not necessarily turn you off. There could be an acceptable reason for a drop—perhaps the company decided to retire (in other words, pay off) some debt that had been bleeding its resources and limiting development. Or perhaps the company has invested in a new expansion program that promises to return large profits in the future. It is also possible that business in general was

bad that year or that the company's industry suffered overall. So compare the company's performance both to others in its industry and to general business trends. The company may be doing poorly, but not as poorly as others.

The company will explain earnings slumps in the text of the report. Though you can expect them to put the best possible face on the facts, you should trust the gist of their explanation—it has to pass the muster of professional analysts at brokerages and institutional investors, too.

If the past year's earnings do reflect the growth you hoped for, make sure the earnings are based primarily on product sales. (A company might conceivably sell off a pricey asset, like one of its plants, to gussy up its earning picture.) A healthy, growing company will sell more and more products each year, and that should be the basis of its increased earnings.

- **Net sales, operating costs, and profit margin.** These three figures together and separately can reveal a great deal about a company's future prospects. Are sales increasing every year, as they should in a well-managed company? Are operating costs rising at a faster rate than sales? If so, those costs will erode profits. Are other companies in that industry showing the same cost increases? You determine profit margin by dividing operating costs by net sales. Is that figure increasing or declining compared with previous years, and how does it compare with analogous companies?

- **Inventory of unsold goods.** You want to see this figure small in comparison with annual sales. If it's not, it could mean that consumer demand has dropped for the company's product or that the company is having troubles getting its product to the marketplace. Either way, the company is in a dangerous position if prices drop.

  A high inventory of unsold goods should also send you scurrying back to reevaluate the company's assets-to-liabilities ratio. Given the above possibilities, the value of that inventory as an asset may be unrealistically assessed.

- **Working capital.** You know from the experience of managing your personal life that you can't raise your standard of living without increased capital. If you run a business, you know how vital increased capital is to business expansion, not to mention meeting rising costs. So if a company you are evaluating shows a decrease in net working capital (assets minus liabilities) from year to year, try to find out why. Temporary setbacks are a fact of life, but in general, organizations—like organisms—either grow or die. Although death may be a protracted, even decades-long, process, you clearly have nothing to gain by associating with a company in its winter years.

There are certainly other tidbits of information that you could extract from the numbers in the annual report, but these factors ought to do you for starters. So let's turn our attention now to what the annual report can tell you about a company's social consciousness. The SRI periodical literature is still your best source for comprehensive social evaluations of a company's operations, but the annual report text sections will give you at least good *inferential* information in the following areas:

- Management's attitudes toward its employees.
- Composition of the board of directors. Are there any progressive opinion leaders among its outside members? Are minorities and women represented on the board? Members are usually pictured.
- Description of the company's products and services. Are they environmentally sound and aimed at socially productive, or at least benign, uses?
- Social programs, progressive hiring and labor practices. If the company has any, they will likely boast about them in their annual report.

Read everything a company says about itself with a skeptic's eye. Particularly when the company talks about its environmental record, remember that everybody is "for" the environment.

Moody's, Standard & Poor's

As useful as annual reports are for providing basic information about a company, the information gathered must be placed in a broader context. All the indicators just discussed are best regarded both in comparison with previous (ideally ten) years' performances and performances by the company's competitors in the same industry. Since an annual report is only a one-year look at one company, you would have to gather dozens of reports to do a comparative, industrywide performance analysis over a ten-year period. You probably have already noted that much of what we suggest you answer for yourself about a company could not be answered without this additional material.

Fortunately there is an easily accessible and efficient approach to this problem. Moody's Investors Service and Standard & Poor's Corporation—the same financial research firms mentioned in chapter 4 for their quality ratings of bond issuers—publish financial data on virtually every publicly owned company in the United States. The most fundamental of these publications are Standard & Poor's *Corporation Register* and Moody's *Manuals*. These volumes, tens of thousands of pages in length, collect data pulled from annual reports but include the ten-year stock and business comparative data that annual reports lack. You can usually find either the Moody's or Standard & Poor's reports in your local library.

Both Standard & Poor's and Moody's offer other useful publications, including advisory publications, for mainstream investors. Another highly regarded advisory publication is *Value Line Investment Survey*. None of these publications take social criteria into account, however. For advisory purposes as well as social screening, socially responsible investors will be better served by the SRI periodicals. Publications such as *Insight* and *Clean Yield* give excellent investment advice. You can cross-reference their financial conclusions with the Standard & Poor's, Moody's, and Value Line publications to make your decisions with maximum confidence.

A listing of SRI advisory publications is provided in appendix D.

"FUNDAMENTAL" VS. "TECHNICAL" APPROACHES TO STOCK SELECTION

If you become a serious student of the market, you'll soon encounter these two terms. The *fundamental* approach is the one taken by old, uh, fogies like us in this book. It refers to analyzing a stock's value in terms of its fundamental economics: the company's business soundness, its growth potential, its place in the industry, the industry's promise in the economy, and so forth. *Technical* theorists attempt to predict a stock's performance independent of business factors. Instead they examine stock price trends, stock market volume, patterns of industry groups, and, primarily, the dynamics of supply and demand. Given that buying and selling on Wall Street is driven largely by emotion, there is some validity to this approach—except for one important factor: How do you predict what we crazy Earthlings are going to do at any particular time?

Every technician has his or her theory and formula to match. Some technicians meticulously track graphs of short-term market behavior, trying to anticipate the rise and fall of stock prices as if they obeyed the physics of some larger wave function. There are theories that link market movements with sociological patterns, political patterns, psychological patterns, you name it. (We've heard of one theory that associates annual market trends with the conference—AFC or NFC—of that year's Super Bowl winner.)

Many technicians market their advice to eager investors through letter services. Some of these technical theories hold up so well in practice that they predict just as well as fundamental analyses—in other words, there is little to choose statistically between the two. Advocates of each trumpet their successes and ignore their failures (although many fundamentalists use technical approaches to time their market actions).

With neither the fundamentalists nor the technicians able to demonstrate a clear predictive edge over the other, a third strain, the *contrarian* approach, has grown popular in recent years—as popular as something contrarian can be, anyway. Contrarians feel that both the fundamental and technical approaches lead to mass behavior, and since the masses are stupid, contrarians bet against the popular trends. Their methodology is straightforward—they simply note the recommendations of the top fundamentalists and technicians and stake their money in the opposite direction. How do they do? Sometimes well, sometimes not, just like everyone else.

Despite the fact that the fundamental approach enjoys no clear statistical advantage over any other approach, it does make more sense for socially responsible investors as well as long-term investors in general. Socially responsible investors care about what companies produce and how they behave in the society. These factors are both part of a fundamental analysis and reflected in it. Even if you are attracted to a particular technical approach, you will need to factor in these fundamentals to make your investing socially responsible.

For long-term investors, the strategy of investing in well-managed, financially solid and/or promising companies is more relevant than trying to predict short-term market behavior. As for the contrarian approach, realize that this is a formula, too. Believe us, if there were a surefire formula that would identify future stock superstars, we'd be the first to tell you.

TECHNICIANS VS. FUNDAMENTALISTS, PART II—HOW MARKET TRENDS RELATE TO BUSINESS TRENDS

The technicians are absolutely right in declaring that the short-term behavior of the market is more obedient to emotional "rules" than real developments in business. The crash of 1987 occurred during generally prosperous times, to cite just one noteworthy example. However, if you ignore the market's heart palpitations and examine the long-term patterns, you will see that the market behavior corresponds very closely to the performance of the economy. In bad business times—such as the recessional 1990—the market suffers, too. In good times it thrives. And overall it goes up, just like the GNP.

A CLOSER LOOK AT SPECULATION

Why would anyone take the risks of stock speculation when long-term investing is such a dependable winner? There is no good reason—if you are in the market to help secure your financial future, that is. But in spite of the odds against them and the government controls limiting the range of their activities, there will always be those who, driven by visions of untold wealth, feel there must be a way to beat the system. Such types seem to suffer a form of brain lock that causes them to keep throwing good money after bad, because far more speculators lose than win—short- or long-term.

We will now describe some of the methods of this madness in the hopes that you will be further dissuaded from ever depending upon them.

Buying on Margin

Buying on margin means buying stock with a down payment, with the stockbroker financing the remainder—for a price. (Current finance charges run 10 to 11.75 percent, depending on the balance in the margin account—the smaller the balance, the higher the rate.) This ploy is most often used by speculators to increase their profits on what they think is a hot stock, since they can buy more shares on margin than they could if they paid for their shares in full.

In broader investment terms, this approach is called *leveraging* a purchase. Leveraging works very well in real estate. Suppose you put 20 percent down ($20,000) on a $100,000 property, borrow the remaining 80 percent ($80,000) from a bank, and then sell the property three years later for $120,000. After you pay back the bank its $80,000 from the sale proceeds, you have $40,000—a 100 percent return on your original $20,000 investment (minus, of course, transaction costs and the interest paid on the loan). If you had bought the building for $100,000 cash and sold it for the same $120,000, your $20,000 profit would represent only a 20% return, not very impressive for a three-year investment.

Leveraging a stock purchase is a much riskier proposition for several reasons: the rules of the game, the volatility and unpredictability of the stock market, and the difference between real estate and stock values.

First rule—current federal regulations require a minimum 50 percent down payment on a stock purchase. The stockbroker puts up the balance, holding your stock as collateral. Fair enough. Now, the second rule—federal regulations require that the *equity* (the value of the stock minus the amount borrowed) in your margin account equal at least 25 percent of the current price of the stock. Should the equity drop below that figure, you must repay part of the borrowed amount in cash or collateral. If you are unable to make up that difference with other resources, the broker will make it up for you by selling part of your stock.

And that's the rub. Obviously you buy on margin because you think the stock will increase in price. If it should go down (and there is no way accurately to predict that it won't), your equity drops and you face the possibility of being forced to sell your stock at a loss. In simple terms, margin buyers make money faster than cash buyers when the market goes up and lose it faster when it goes down. Given that cash stock investments make such solid returns anyway when managed correctly, we think margin buying is a rather foolish, and greedy, risk.

Leveraging is wise in real estate investing because real estate does not behave with nearly as much volatility as stock. Only in rare circumstances—a serious recession, say, or a poorly considered piece of property—does real estate ever decline in value. It generally rises at least as fast as the rate of inflation, so its value is stable with the promise of increase. Over the short (and for margin buyers, most crucial) term, individual stocks rise and fall like waves in the ocean, drowning a lot of margin buyers in the process.

Selling Short

No speculative technique better demonstrates the difference between investing on Wall Street and playing it like a casino than selling short. Investors rely on the historical upward growth of the market to make their profit. In a casino, though, you can bet on any possibility. Short sellers gamble that the market will drop in the short term, playing the game with shares they don't own. The technique is best described through an example.

Suppose you think the stock of High Flyer, Inc., has peaked in value and is headed south. You arrange with your broker to borrow one hundred shares of its stock from a current stockholder and then sell those shares at their current price of $100 a share. Now you have $10,000 (minus the transaction cost) and owe somebody one hundred shares of High Flyer. If the stock drops in price to, for instance, 80, you can buy one hundred shares for $8,000 (minus your second transaction cost), repay the lending stockholder, and pocket your approximately $2,000 profit.

Great stuff, huh? But suppose High Flyer flies—to 120, say. What to do? You could conclude that you were wrong about High Flyer's prospects, grab the hundred shares now at $12,000 before it goes any higher, and accept your $2,000 loss (plus those two transaction costs). Or you could hang tough and wait for the drop you predicted.

It may be an expensive wait. First of all, you are obligated to pay out of pocket all dividends declared on the borrowed stock to the lending stockholder. Second, federal regulations require that short sellers maintain enough cash in their brokerage accounts to buy the borrowed number of shares at their current price. That sensibly ensures that the lending stockholder can always be repaid, but it means that you will have to feed your account with real money while waiting for High Flyer to crash. If you run out of feeder money before the market turns, your broker will use the money in your account to buy High Flyer at its current price. If that price is $120 a share, you will have lost $2,000 plus transaction costs plus dividends paid.

Margin buyers and short sellers alike are done in by the speculator's fallacy—that there is some reliable method by which you can foresee short-term market behavior. It's a crapshoot at best, and federal regulations raise the stakes even higher. Now, on to table number three at the Wall Street Golden Nugget.

Stock Options

Unlike margin buyers, who bet only when they think a stock's price is going up, or short sellers, who bet only when they think a stock will drop in price, those who purchase stock options can bet on either direction. Options—the right to buy or sell a specified number of shares of a specific stock at a specific price during a specific time period—are also called *puts* and *calls*. A call is an option to buy the stock from the options dealer at the current price during a specified period in the future—in effect, a bet that the stock price will go up during that time. A put is an option to sell to the dealer in the future at the current price—a bet that the price will drop. As with short selling, the mechanics of puts and calls are best illustrated through examples.

In a typical call scenario, you become convinced (Lord knows how) that the stock of the high-tech snack food company Micro Chips 'N Dips is undervalued and about to catch on big with stock purchasers. It sells now at $50. Having little cash to work with but wanting to score big on the action, you purchase a call to buy one hundred shares at $50 any time in the next six months. The premium—in other words, the cost of the option contract—is 5%, or $250. (Premiums, which can run as high as 15 percent for volatile stock, vary according to such factors as the duration of the contract, the price of the stock, the recent activity of the market, and the experts' sense of the stock's prospects. Smaller premiums are charged for shorter-term contracts and lower-priced and stable stocks.)

Now, let's say you are right. Micro Chips 'N Dips goes up to $60 in five months, and you decide to bite. For about $5,250 (minus brokerage commission), you now have $6,000 worth of stock. If you sell the stock right away to realize your profit, you will have made about $750 on your original $250 investment—or 300 percent in five months. Nice "work."

Let's look at the opposite scenario. Suppose Micro drops to $40 and stays there for six months. You let the option expire, and you are out your $250. End of episode. An option buyer's loss is always limited to the price of the contract.

For comparison's sake, imagine playing the same game with cash. You purchase one hundred shares of Micro at $50 a share for a total of $5,000. If it goes to $60 a share and you sell, you have made $1,000 on your original $5,000 (minus two transaction costs), about 20 percent before commissions. Good, but nothing like the 300 percent possible with the option.

On the down side, if Micro had dropped to $40 a share and you decided to cash out, you would have lost $1,000 plus commissions. Of course, when you play with cash, your hand is never forced. If you remain confident about Micro's future, you can hold the stock and see if your confidence is repaid.

Now, puts. Puts are the exact opposite of calls. Suppose you think Micro Chips 'N Dips is a loser and consumers will soon tire of snack foods shaped like tiny electronic components. You pay a premium for the option to buy the stock at a price lower than today's $50. If you are right and the stock drops to $40 during the contract's terms, you can exercise the put and the dealer must pay you $50 a share for stock you are buying at $40 a share. If you bet wrong, and the stock doesn't drop, you're out the price of the option.

The above presentation is admittedly simplistic. Suppose you purchased an option on a stock only to realize just before it expired that you had bet wrong. If you still like the stock's long-term promise, you might exercise the option anyway just to salvage part of its cost. The effect is like paying a surcharge on the stock—better than taking a total loss on the option, but worse than if you had bought the stock without messing with an option first. Some stockholders buy calls or puts to protect "paper" profits or as an insurance policy against losses. Some buy puts and calls on the same stock for the option of scoring in either direction. Another popular option game involves betting which direction the major stock indexes will head over the short term.

Clearly, a stock has to move significantly for the option buyer to profit. Going back to our example, remember that you paid a $250 premium for your six-month call on Micro Chips 'N Dips. That's 5 percent, so the stock will have to rise more than 5 percent— past 52½—sometime in those six months before you make any money.

By the way, there is a safer way to play the options game—by selling options instead of buying them. If you buy one hundred shares of Micro at $50 a share, you might then turn around and sell a six-month call on it at $50 for a premium of $250. The buyer, of course, would be a speculator convinced that Micro would go above 52½. Now that you've sold the option, you've effectively ensured that you won't lose money on your stock unless it drops below 47½. As long as Micro stays below 52½ in those six months, you'll have your $250 and your stock. If it goes above 52½, your option buyer will exercise the call and you will have to sell at the now below market price of $50. But you've got your principal back plus the $250 and any dividends earned in the meantime. That in itself represents a respectable return for those six months.

In fact, you do not really have to sell if your stock skyrockets. You can buy the call back (calls are sold on the secondary market just like other investments), keep the stock, and sell a new call on it at its current price to recoup part of the cost of the buy-back. There's much more to say about the economics and refinements of this strategy. But this is somewhat more complicated than we want to get on this topic. An excellent source on the subtleties of options is *Dun & Bradstreet's Guide to Your Investments* (updated annually).

Our main point here is that option buying is risky business, because like other forms of speculation, it presupposes that you know which way a stock will go over the short term. No matter how much research and technical analysis you do to back your hunch, statistics and history show that you're still flying nearly blind. The stakes aren't horribly high in option buying, but when you figure what that money could be doing for you compounding in relatively safe socially responsible investments, why would you want to take such a high risk of losing it?

Ethical investors should also note that options are sold through (four) special option exchanges and are offered only on a few hundred stocks at any one time. Thus the number of SRI stocks handled by these exchanges may be few and far between.

Penny Stocks

So-called penny stocks are the stocks of new companies with no track record yet of doing business. Often such a company has little more going for it at the moment than what it thinks is a hot idea. The product might not even be ready to market. Penny stocks are to be found on the over-the-counter market, and they actually sell for a few dollars a share (not pennies). They are considered a speculative investment because the chances of company failure are tremendously high. Yes, if you pick the right penny stock and the company hits it big, you will make big, big money. But . . . As you know by now, stock speculations have big "buts."

IF YOU WANT TO "BEAT THE SYSTEM," TRY DOLLAR COST AVERAGING

This methodology, in contrast with speculative systems, makes perfect sense because it incorporates the one predictable truth of the market—its long-term upward trend. We highly recommend dollar cost averaging. In case you are skipping around the book, or reading backward, we covered it in detail in chapter 3.

ANOTHER "EDGE" ON THE STOCK MARKET—DIVIDEND REINVESTMENT PLANS

Commissions paid to purchase stock shares do modestly reduce your returns, particularly over the short term. By the same token, any time you can reduce or eliminate commissions—without sacrificing investment quality, that is—you have gained an edge that will fatten your wallet. Dividend reinvestment plans, sponsored by many publicly traded corporations, allow you to increase your stock investment at costs far below those a broker would charge you to buy additional shares. If you have stock in a company offering such a plan, here's how to register for one and what will happen when you do.

Your initial purchase of stock in a company sponsoring a dividend reinvestment plan will be through your broker. Direct your broker to register the stock certificate in your name and send it to you. (Shares held by a broker and registered in "street name" are not eligible for the plans.) Once you have received the certificate, contact the company's shareholder relations department (the address can be found in the annual report or in a publication like *Value Line Investment Survey*) and request a prospectus and application for the plan.

Once you have filed your application, you will receive (instead of dividend checks) regular statements showing how many shares your dividends have purchased. Most companies pay all administrative and brokerage costs, so 100 percent of your dividend will be reinvested. Therefore your statement may show fractional shares, another advantage over a regular stock purchase since you don't have to worry about lot size.

Another benefit available through most plans is the ability to invest additional money beyond the dividend, again without commission and at a low minimum—typically $25. Some companies will even discount stock sold through a plan—up to 10 percent in some cases. Whether your stock is discounted or not, however, you still have a regular investment going with dollar cost averaging built in as well. Pretty nice deal if you've picked your stock well.

Our composite list of socially screened stocks on page 221 of this chapter and the Domini 400 socially screened stock index on page 369 have both been annotated for companies offering dividend reinvestment plans. If a stock that interests you is not so noted on our lists, check with the company to see if a plan is offered.

THE STOCK MARKET TODAY—MAGIC MOUNTAIN COMES TO WALL STREET

If anything, the market today fluctuates more wildly on a day-to-day basis than ever before. The fluctuations will have no long-term financial consequences for you if you are a long-term investor—other than the money you pay out for sedatives and antacids to get you through the wilder days. The purpose of the following section is to help you better understand the reasons behind the volatility and save on the pharmaceuticals.

Much of the force behind the fluctuations has to do with who is now doing most of the buying and selling. Until recent years the overwhelming majority of stocks traded daily were held by individual investors, with the balance traded by institutions such as mutual funds, banks, insurance companies, and pension funds. Today the ratio is reversed. Institutions do the overwhelming majority of the trading. And because they hold and trade huge blocks of stock, the effect of their activity can overwhelm the market, causing much sharper dips and climbs than used to result from the cumulative trades of masses of individuals.

Exacerbating this is what might be referred to as "the herd effect." Consider that those who manage the stock portfolios for large institutions are in essence betting their jobs on the performance of the portfolios. How else would their job performance be measured other than by how the portfolios do? In a rising market they will be expected to equal or beat the market's average gain. In a falling market they know they can't afford to fall much farther than the market average—at least not very often—if they expect to stay employed.

This puts each institutional portfolio manager in the position of trying to guess how other institutional managers will respond to market conditions, because the institutional buyers are the biggest influence on the market's rise and fall. So they talk to each other and often use the same analytical services. Then, as they bid against each other for the same stocks, the price shoots up. When they take profits at the same professionally advisable levels, the price nosedives. Both movements are swollen by the masses of individual investors who notice the activity and jump into the fray.

Technology also plays a big part in today's exaggerated market trends. Information is disseminated much more quickly and widely than ever before. Add that to a market that has always been quick to respond to the latest socioemotional ripple and you get some pretty hefty swings. Those swings are pushed farther in both directions by the prevalence of computerized buying and selling. Many big-time investors program into computers buy and sell parameters that track the market's every move. The result? Surgically sharp peaks and valleys on the graphs that track the market's day-to-day progress.

A terrific example of the market's responsiveness to minute-by-minute current events occurred on January 10, 1991. On that day, five days before the UN deadline demanding Iraq's withdrawal from Kuwait, U.S. Secretary of State James Baker and Iraqi Foreign Minister Tariq 'Aziz met for over six hours in a last-ditch effort to avert war in the Persian Gulf. Although there seemed little chance of success, there is always hope when two enemies talk, and Wall Street reflected that. The market, which had closed at 2509.41 the previous day, opened after the talks had begun and within minutes had climbed to over 2530. As hour after hour passed with Baker and Aziz still behind closed doors, the market crept up over 2550. It was still at about 2543 when Baker stepped outside for his press conference at 1:45 P.M. Within minutes the Dow Jones Industrial Average plunged about forty points. Jay Goldfinger of Beverly Hills brokerage Capital Insight, quoted in the *Los Angeles Times,* was one of the big players jumping out: "I noticed that Baker wasn't smiling when he walked in [to the conference], and I sold. . . . It was the only winning trade I've made all day."

What all this should mean to you, the long-term investor, is that it's crazier now to speculate than ever. One-day drops like we saw in October of 1987 are more likely to recur, followed by rapid recoveries like the one that succeeded that crash. So relax—in fact, look for sudden bargains—and otherwise apply the usual guidelines as to when to buy and sell. The long-term upward trend of the market is the same as it's always been.*

DIVERSIFICATION—PROTECTING YOURSELF AGAINST THE OCCASIONAL FAILURE

Is there a down side to all this good cheer that we've been spreading about the long-term view of the stock market? Of course there is. If the stock market were 100 percent certain, nobody would ever invest in anything else. For that matter, few would even keep money in a bank.

Any once solid company can fall on its face—management can get old, times can pass the company's product by, an environmental or other ethical gaffe by the company can lead to a consumer boycott and send the stock price plummeting. Just as harmful (if more innocent), should two or three institutions decide to take their profit on a particular stock on the same day with relatively few shares outstanding, its price could drop precipitously and take a long time to recover. Some stocks that take a battering in the market never recuperate even if the company stays in business.

Such occurrences are rare, and you certainly shouldn't let them discourage you from investing in the market. Nevertheless, you are taking a foolish risk if you've got too many eggs in one stock's basket. The risk/reward balance that we and most investment advisers find so favorable with stocks operates only if you invest in a diverse stock portfolio to lessen the impact of the collapse of any one stock. For analogous reasons, you should also consider diversifying in terms of economic sectors and industry groups.

Research shows that an array of five to eight stocks—from different industry groups and economic sectors—should give you adequate protection. In other words, this is not a case of more is better. A portfolio of fifty stocks does not buy you ten times more protection than a portfolio of five. In fact, we strongly advise against holding more than about twelve stocks at any one time. As we will explain in detail, it is important that you track the progress of your stocks even if your overall strategy is to hold your stocks for years. If you can track more than twelve stocks and take appropriate actions when called for, you must have an extra hemisphere in your brain.

* By the way, when the reporter on your car radio tells you that the stock market dropped thirty points today, that's no reason to drive into a bridge abutment. Thirty points off a Dow Jones Average of 2950 to 3000 barely qualifies as a bruise.

It does take a pretty good chunk of change to get into five stocks. For example, figuring a modest $40 per share as an average share price, buying a portfolio of five stocks in round lots of one hundred shares each would cost a total of $20,000. If that kind of cash is beyond your means, make your first foray into the market through stock mutual funds, which are by definition diverse stock portfolios. We review all available SRI mutual funds in the next chapter.

MANAGING YOUR STOCK PORTFOLIO—TRACKING STOCKS AND DETERMINING WHEN TO SELL

When we advise you to consider your stocks a long-term investment of your money, we don't mean that you should stow away your stock certificates in a safe and forget about them. Stocks are not passive investments, for one simple reason: The past does not guarantee the future. Sometimes stocks just fall out of favor with the buying public for reasons that have little to do with actual business conditions—when the demand drops, so does the price of the stock. Even the stocks of large, stable "name" companies can go south. IBM, at $80 a share in 1979, peaked at $176 just before the 1987 crash. The crash cut it off at the knees, and, as of this writing, the stock has not yet fully recovered. (On August 9, 1991, shares were priced at about $96.)

Yes, diversification of your stock portfolio will soften the effect of the odd stock that goes on a bummer. But better that you nip a potential bummer in the bud by selling the stock in time. If you track your stocks, you may see trouble in the making. At the same time, you can be weeding out slowpokes and replacing them with movers to maximize your stock portfolio's performance. It is not true that you never take a loss until you sell stock that is slumping. You are losing money if you could replace a stock with one that will do better.

We suggest that you check in on your stocks' prices weekly, or at worst monthly. And at least every six months to a year, examine the financial condition of the companies in which you own stock. You do this in-depth research with the same tools you use in evaluating stocks to buy—annual reports, the business press, Standard & Poor's, and Moody's. (Remember that Standard & Poor's and Moody's are indispensable for their historical data on a company and comparative data on the company's competitors in its industry.)

If you use a full-service brokerage for your transactions, another source of information available to you as a client will be the broker-age's reports on recommended stocks. Brokerage reports run the gamut from brief summaries to in-depth analyses running tens of pages. Keep in mind when reading these that brokerages are in the business of transacting stocks. They earn commissions on each transaction they mediate. Thus their recommendations tend to be biased toward more frequent trading than is healthy for your long-term strategy. You will probably do better, once you get the hang of stock investing, by doing your own research with the tools we mention or by discussing your stock with your broker individually as an *informed investor*.

So when is it appropriate to sell a stock? For the most part, you can figure it out by going back to the section on selecting stocks to buy. If, for example, an assets-to-liabilities ratio of less than two to one is a red flag when considering a stock to buy, it's also bad news when the ratio of a company in which you own stock has dropped to that level.

The following list of red flags are not absolute signals to sell. They are, however, signals to take a closer look at the company's total picture. Note also that in most cases we have refrained from giving you absolute formulas. Again, the total picture is the key. So analyze your stock in detail if you see

- **a cut in dividends.** Remember from our earlier discussion that there can be many reasons for a cut, and not all of them are bad.
- **a drop in earnings.** If others in the industry are suffering similar drops—for example, if it's a bad year for appliance sales in general—then this indicator by itself says little about your company. Check the company's annual earnings pattern, also—maybe it always does poorly in the third quarter, for instance, even though its annual gain is respectable. Some businesses are seasonal—toy companies make or break their year at Christmastime, to cite an obvious case.
- **a drop in net sales.** Again, compare with the industry. The same goes for the following two factors.
- **a significant increase in operating costs.**
- **a drop in profit margin.**
- **a drop in P/E ratio.** This may indicate a loss in investor confidence in your stock, or at least a waning of a fad, neither of which bodes well for future stock price gains. On the other hand, it could simply reflect a general market downturn, which is of little concern to a long-term investor.
- **a sharp increase in P/E ratio.** Is the company really doing this well, or is a buying frenzy pushing the stock price past its economic

value? If the latter, consider taking your profit now before the fad collapses.

- **a drop in the company's cash reserves.**
- **an increase in the company's inventories of unsold goods.**
- **an increase in accounts receivable.** Customers aren't paying their bills. Why?
- **an increase in accounts payable.** Your company doesn't seem to be paying *its* bills. Why?

Remember, there are two equally valid reasons for selling stock: first, *to cut losses,* and second, *to maximize profits.* Sometimes, regardless of your long-term strategy, it's just time to sell a stock—and not always because a company's fortunes change or business conditions change. Sometimes you sell because you change—say, from an aggressive growth strategy to a conservative, income-oriented strategy. Be forewarned: investors tend to get attached to their stocks the same way they get attached to their cars. It's not economic to keep driving your old clunker once the annual repair bills mount past a certain point. Some bright, shiny stocks become clunkers, too.

If all this tracking business seems like too much bother to you, now you understand why so many investors rely on the advice of full-service brokers or buy mutual funds or both. A full-service broker will review your portfolio for you at your request and make any needed recommendations. Portfolio managers—essentially full-service brokers plus luxury options—offer yet another level of scrutiny, research, and investment management for well-heeled investors with large, complex portfolios. The Clean Yield Group and other SRI firms manage individual, socially screened portfolios for their ethical investors. (See appendix D for a list of SRI portfolio managers.) Mutual funds, of course, do all the stock buying and selling for you, eliminating your need to track. You should check in on your funds' performances every so often, however.

Do note that there is a simplified method for locking in profits and cutting losses on your stocks. For each stock you own, figure out profit and loss limits. For instance, you may be holding a growth stock on which an 18 percent profit would be more than satisfactory to you and a loss of greater than 10 percent would ruin the sunniest day. You can give "good until canceled" sell orders to your broker at those levels and then leave on a long vacation worry-free. If your stock shoots up 20 percent and then drops out of sight overnight, you'll have earned your 18 percent and, better yet, saved yourself the agony of heavy losses. The levels you set are always up to you and can be revised anytime—upward in a surging market, downward for the bears.

THE SHORTCUT TO STOCK SELECTIONS—"OUR" RECOMMENDED STOCKS

You may very well be asking yourself at this point if there isn't an easier way to pick stocks. Believe us, we understand. It's a lot of work to manage your own stock portfolio. So you'll be pleased to know that the answer to your question is yes—quite simply, buy what the professionals are buying. The stocks purchased by the managers of mutual funds must usually meet strict standards of financial viability. In addition, the companies whose stocks are held by an SRI mutual fund will have been screened for various social issues, according to criteria published in the fund's literature.

However, before running to the phone to order a mutual fund's favorite stocks from your broker, make the following simple determinations:

Does the fund primarily seek long-term investments? If the answer is no, you will still want to check out the financial attributes of their stock more closely. The holding that interests you may have risen in price so much since they purchased it that there is little profit left for you to make. And for that or other reasons, they may be ready to dump those shares. If others in the herd follow, that stock's prices will plummet. By the way, even if the fund does invest mainly for the longer term, you are always better off confirming your selection's values with a little research. But your chances of finding stable values on the list of a fund with a long-term strategy are good.

Do the funds holding the stock have social goals that match yours? You want to be sure that the funds screen for the issues that concern you.

Are the funds' financial goals the same as yours? Obviously you should not be buying the stocks that only aggressive growth funds are buying if your interest is in high-yielding, conservative stocks.

Our Composite Stock Listings

Here we have listed the stocks held (as of December 31, 1990) by the various SRI stock mutual funds or reviewed by the prominent SRI advisory newsletters. Next to each stock are the names of the funds holding it. Cross-check this list with the various SRI mutual funds' top ten holdings (chapter 6, page 249) for a further indication of investment desirability. The fact that a stock would be held by more than one mutual fund is generally a good indication of its soundness ethically and financially, as is its presence on a top ten list.

Detailed information on the financial goals and social screens of each SRI mutual fund are contained in the next chapter.

Explanation of Codes

AA = Calvert-Ariel Appreciation Fund

AG = Calvert-Ariel Growth Fund

CB = Calvert Social Investment Fund: Bond Portfolio

CE = Calvert Social Investment Fund: Equity Portfolio

CM = Calvert Social Investment Fund: Managed Growth Portfolio

CY-MP = Clean Yield—Model Portfolio

CY-PC = Clean Yield—Profiled Companies

DR = Dreyfus Third Century Fund

GMIA = Good Money Industrial Average

GMUA = Good Money Utility Average

FI = Franklin Insight

NA = New Alternatives Fund

PR = Parnassus Fund

PX = Pax World Fund

RT = Rightime Social Awareness Fund

SP = Shield Progressive Environmental Fund

WA = Working Assets Money Fund

ADC Telecommunications—FI

AMP—CE,CM

AMR—CE

Acme Steel—PR

Acuson—CY-PC,DR

Adaptec—PR

Advanced Micro Devices—PR

Advatex Assoc.—SP

* Aetna Life and Casualty—FI,PR

Affiliated Publication—CM,CY-PC

A. G. Edwards—FI

Aircure—SP

* Air Products—FI,NA

Air & Water Tech, CL.A—DR

A. L. Labs—CY-PC

Alaska Airlines—FI,PR

Albertson's—CE,CM,DR,RT

Allwaste—SP

* Amerada Hess—RT

Amer. Capt'l & Research—SP

American Stores—RT

American Waste Services—SP

American Water Works—NA

American West Airlines—FI

Ameritech—CE,CM,RT

Ametek—GMIA,NA

* Amoco—FI,RT

Anchor Savings Bank—CB

Angelica—AG

* Apache—FI

Apogee Enterprises—FI

Apple Computer—CY-PC,FI, PR,RT

Archer Daniels Midland—CY-PC,NA

ARCO Chemical—FI

Argonaut Group—DR

* Arkla—RT

Armor All—AA,AG

Ask Computer—PR

Astra A Free—DR

* Atlanta Gas & Light—CY-PC,NA

* Atlantic Richfield—FI,RT

Automated Data Process—CE,DR

* Avon Products—RT

BHA Group—CY-PC,DR

Badger Meter—FI

Baldor Electric—NA,PR

Baker (Michael)—FI

Banta (George)—CY-PC

* Bausch & Lomb—AA

* Baxter International—CY-PC, FI,PR,PX,RT

Bay Banks—CM

Bay State Gas—NA,PX

Becton, Dickinson—CE,CM,RT

* Bell Atlantic—CE,CM,RT

* BellSouth—CE,CM,RT

Bell South Savings—AG

Ben & Jerry's—CY-PC,FI

* Beneficial—AG

Bergen Brunswig—AA

Betz Laboratories—CE,CM, CY-MP,DR,FI,NA

Black Hills—PX

* Bob Evans Farms—AG,FI,PR

Bonneville Pacific—FI,NA

Boston Acoustics—CY-PC

* Bristol-Meyers Squibb—RT

* Brooklyn Union Gas—PX

Burlington Resources—NA

CML Group—CY-PC,FI

* CPC International—RT

CRSS—SP

* CSX—RT

Caesar's World—AG

Calgon Carbon—FI,NA

Campbell Soup—RT

Capital Cities/ABC—DR

Capital Holding—CY-PC

Care Plus—DR

Carnival Cruise Lines—AA

Carolina Freight Group—PR

Carter Wallace—AA

C-Cor Electronics—FI

Celgene—SP

Central Newspaper—AG

Central Sprikler—AG,CY-PC

Chambers Development—FI

* Champion International—FI

* Chase Manhattan—DR

Chemfix Technologies—SP

Chemical Banking Group—PR

Chesapeake—NA

Church & Dwight—AA,CY-PC, FI,NA

Citizens & Southern Bank—CB

Citizens Utilities—GMUA,CY-PC

Clarcor—CY-PC

* Clorox Company—AG,CE,CM, RT

* Coca-Cola—DR,RT

Coca-Cola Enterprises—DR

* Columbia Gas Systems—RT

Comair—FI

Commerce Clearing House—CY-PC

Community Psychiatric—CY-PC, RT

Compaq Computer—RT
Computer Associates—AA
Connecticut Water—CY-PC
* Consolidated Natural
Gas—FI,GMUA,NA,RT
Consolidated Papers—CE,CM,
CY-PC,FI
* Consolidated Rail—RT
Consumers Water—CY-PC
Continental Health—CY-PC
Control Resources—CY-PC
Cooper Tire—FI
Corestates Capital Corp—CB
* Corning—FI,NA
Costco Wholesale—DR
Crawford & Co.—CY-PC,DR
Crompton & Knowles—DR
Cross, A. T.—GMIA
C-Tec—CY-PC
Cummins Engines—FI,GMIA,PR
* DPL—PX
DSC Communications—RT
Datascope—CY-PC
Davis Water & Waste—CY-PC
* Dayton Hudson—DR,FI,GMIA
DEKALB Enterprises—DR
* Delta Airlines—FI
* Delux Check Printers—CM,FI
Digital Equipment—CE,CM,
FI,GMIA,PR,RT
Dillard Dept. Stores—DR
Dime Savings Bank of NY—PR
Disney, Walt—GMIA,PX
* Donnelley, R. R.—AA,CE,
CM,FI
Donnelly—FI
Dun & Bradstreet—CE,CM
Duplex Products—AG
Durr-Fillauer—CY-MP
Eastern Environmental—SP
Echo Bay Mines—FI
* Eco Lab—AG
Ecology & Environmental—SP
Elan A.D.S.—DR

Electro Scientific Ind.—PR
* Energen—NA
Energy Convers. Devices—NA
Ennis Business Forms—CY-PC
* Enron—RT
* Enserch—RT
* Equitable Resources—CE,CM,
CY-MP,NA,PX
Everest & Jennings—PR
Farr—SP
Federal Express—CY-PC,FI
Federal Home Loan
Mortgage—PR
* Federal National
Mortgage—CY-PC,DR,FI,PX
First Brands—AG
First Federal of Michigan—PR
First Union Real Estate Investment
Trust—CM
* First Virginia Banks—GMIA
* Fleming—AA,AG
FlightSafety
International—DR,GMIA
* Fuller, H. B.—CY-PC,
FI,GMIA,PR
* GTE—DR
GWC—CY-PC
* Gannett—CE,CM,FI
Gehl—CY-PC,FI
Genelabs—DR
General Binding—AG
General Building—AA
* General Mills Inc.—PX,RT
* General Motors Cl.E.—DR
General RE—CE,CM
* Giant Food Inc., Class A—RT
Gidding & Lewis—CE,CM
Good Guys—CY-PC
Gould Pumps—FI
Graco—FI
Grainer, W. W.—CE,CM
Great Atlantic & Pacific Tea
Co.—AA,RT
* Great Western Financial—CB

Greenery Rehabilitation—CY-PC
Greiner Engineering—CY-PC
Grist Mill—CY-PC
Groundwater
Technology—CE,CM,CY-PC
Gundle Environmental
Systems—CY-MP,SP
* Handleman—AA,AG,CY-MP
* Hannaford Brothers—CE,CM,
CY-PC
Harding Associates—SP
Harleysville Group—CY-MP
* Hartmarx—GMIA
Harvey Hubbel—CE,CM
Hasbro—AA,AG
* Hawaiian Electric—FI,GMUA,
NA
Hechinger, Class A—CY-PC,PX
Hechinger, Class B—CY-PC,PX
* Heinz, H. J.—CE,CM,
CY-PC,PX
Helmerich & Payne—RT
* Hershey Foods—GMIA
Hewlett-Packard—PR
Hillenbrand Industries—DR
Homasote—CY-PC
Home Depot—DR
* Huffy—FI,NA
Humana Inc.—CE,RT
Hunt Manufacturing—AG,CY-PC
Hydrolic Company—CY-PC
IGI—NA
IMCO Recycling—CY-PC,FI
ISCO—CY-PC
Idaho Power—FI,GMUA,NA
Illinois Tool Works—CE,CM
Inland Steel—FI,PR
Intel—DR
Interface Floor—AG
Ionics—DR
John Harland—CY-PC
* Johnson & Johnson—GMIA,RT
Johnson Worldwide—AG,FI
* Jostens—CE,CM,FI

Juno Lighting—AA
* Kansas Power &
Light—CY-PC,GMUA
Kelly Services—CE,CM
Keystone International—CE
K mart Group—RT
Kimmins Environmental—SP
Laclede Steel—CY-PC
Laidlaw Transportation—CY-PC
Lands End—AA
Lawson Products—CE,CM
* La-Z-Boy—CM
Leggett & Platt—AG,CE,CM
Lifelines—FI
Lillian Vernon—CY-PC,FI
Liz Claiborne—CY-PC,FI
Longs Drug Stores—CE,CM,RT
Lotus Development—CY-PC,FI
* Louisville Gas & Electric—FI
* Lowes—FI
Luby's Cafeterias—CY-MP
Lydall—FI
MASCO—CB
MCI Communications—GMIA
Magma Power—CE,CM,DR,
GMUA,NA
Magna International—FI,PR
Manor Care—RT
Margaux—PR
Martech USA—SP
Matrix Services—SP
May Department Stores—CE,CM
* Maytag—CB,CY-PC,FI,
GMIA,PX
McCaw Cellular—DR
McCormick & Co.—AG,CY-PC
McDonald's—GMIA
* McGraw-Hill—CB,DR
* McKesson—FI
Medco Containment Services—DR
Medtronic—FI,RT
Melville Corporation—CE,CM,
GMIA
Mercantile Stores—CM

* Merck—FI,PX,RT
Meredith Corporation—GMIA
Merry Go Round Ent.—CY-PC
Metcalf & Eddy—SP
Met Pro—CY-PC,NA
Michael Foods—CY-PC
Michigan Bell Telephone—CB
Microsoft—DR,FI
Midwesco Filter—SP
Miller (Herman)—AG,CY-PC,
FI,GMIA,PR
Millipore—FI
* Minnesota Mining &
Manufacturing—GMIA
* Montana Power—GMIA
* Morrison Knudson—RT
Morrison—AA,AG
Mountain State Bell—AG
Mycogen—FI,NA
Mylan Labs—DR
Natec—NA
National Medical
Enterprise—AA,RT
Nature's Sunshine—CY-MP
NE Critical Care—FI
* New Jersey Resources—NA,PX
* New York Times—PX
Newell—DR
Nicor—RT
Nordson—CE,CM
Nordstrom—FI
Norfolk Southern—FI,RT
Northern Telecom Ltd.—RT
Northwest Natural Gas—FI
* Nucor—FI
Ogden Projects—NA
Oklahoma Gas &
Electric—GMUA
Omnicom Group—AA
Oneok—RT
* Orange & Rockland
Utilities—GMUA
Oregon Steel—CE,CM
Oryx Energy—RT

Oshkosh B'Gosh—AA,AG,CE,
CM
OtterTail Power Utilities—GMUA
Owens-Corning—NA
Pacific Enterprises—CE,CM,RT
* Pacific Telesis Group—DR,RT
PaineWebber CMO Trust—CB
* Pall—DR
Paramount Communications—DR
* Penney, J. C.—DR,PX,RT
* Peoples Energy—CY-MP,PX,
RT
Piedmont Aviation—CB
* Pitney Bowes—AA,CB,CY-PC,
FI,GMIA
Pienam Publishing—CY-PC
* Polaroid—FI,GMIA
* Premier Industrial—FI
* Procter & Gamble Co.—FI,PX,
RT
Progressive—CY-PC
Proler International—NA
Puritan Bennett—AA
* Quaker Oats—CE,CM,FI,PX,
RT
Quality Food Centers—FI
Raymond Corp.—FI,PR
Reebok International—PR
Regional Bancorp—CY-MP
Resource Recycling Tech.—NA
Reuters Holdings A.D.S.—DR
Riedel Environmental—CY-PC
Rouse Company—FI,GMIA
Rowan Companies Inc.—RT
* Rubbermaid—CE,CM,DR,FI
* Russell—AA,AG
* Ryder System—FI
SEI—AG
St. Jude Medical—DR,RT
Safeguard Business—CY-PC
* Safety Kleen—CB,CY-PC,FI,
NA
Salick Health—CY-PC
Sanford—AG

Sanifill—SP
Santa Fe Energy Resources—RT
* Santa Fe Pacific—RT
* Sara Lee—RT
Scantron—CY-PC
* Sears—RT
Seaman's Capital—CB
* Security Pacific—DR,FI
Sequent Computer Systems—DR
Service Master—FI
Shared Medical—CY-PC
Shorewood Packaging—AA,AG
Sierra Pacific Resources—PX
Sigma-Aldrich—CE,CM,PR
Smucker (J. M.)—CE,CM,
CY-PC,FI
Snap-On Tools—GMIA
Sonoco Products—CE,CM
Southern New England
Telecom—FI
Southwest Airlines—
CY-PC,FI,PR
* Southwest Gas—GMUA
* Southwest Public
Service—GMUA
Southwest Water—CY-PC
* Southwestern Bell—RT
Spec's Music—CY-PC
Spiegel—CY-PC
Stanhome—AA
Standard Register—CY-PC
* Stanley Works—AA,FI
Sterling Chemicals—FI
Stewart & Stevenson—NA
* Stride Rite—CY-PC,FI,GMIA
Stryker—CY-PC
Student Loan Marketing
Association—CE,CM,CY-PC,FI
Summit Health—CY-PC
Sunrise Medical—FI
Super Valu Stores—RT
Superior Surgical—AG
Syntex—PX
Sysco—CE,CM,CY-PC,RT

TECO Energy—CY-PC,FI,GMUA
TJ International—FI,PR
TRC Companies—CY-PC
Tandem Computers—DR,PR,RT
Tecogen—NA
Tektronix—PR
* Tennant—FI
Thermo Electron—DR,FI
Thermo Instrument—NA
Tootsie Roll—CY-PC
Topps—AG
Toys "R" Us—CY-MP,DR
Tribune Company—AA
Tri-Com—PR
T. Rowe Price—AG
UNUM—FI
United Coasts—SP
United Medical—CY-PC
United Stationers—AG
U.S. Air Equipment Trust—CB
* U.S. West—RT
U.S. Wind Power—CB
* United Water Res.—FI, GMUA
* Universal Foods—AA
VWR Corp.—FI
Vallen—DR
Venture Stores—CE,CM
VIVRA—AA
Volvo—FI,GMIA
Wahlco Environmental Sys.—SP
Walgreen—AA,RT
Wallace Computer—CY-PC
Wal-Mart—DR,FI,PX,RT
Wang Laboratories—GMIA,RT
Washington Energy—NA
Washington Post—CE,CM,GMIA
Watts Industries—AA,NA
Weirton Steel—FI
Wellman—AA,CE,CM,CY-PC,
FI,NA,PR,PX,SP
Western Publishing—AG
Wetterau—RT
* Weyerhauser—FI

Wheelabrator—NA
Wolohan Lumber—CY-PC
Worthington Industries—FI,GMIA

* Xerox—FI,PR
Zenith Electronics—PR
* Zurn Industries—DR,GMIA,NA

Here's a little bonus for your perusal—the SRI investment advisory publication *Clean Yield*'s favorite socially screened stocks:

Clean Yield Model Portfolio

Betz Laboratories
Durr-Fillauer Medical
Equitable Resources
Gundle Environmental
Handleman
Harleysville Group

Luby's Cafeteria
Nature's Sunshine
Peoples Energy
Regional Bancorp
Toys "R" Us

Source: *The Clean Yield,* volume 7, number 1, March 1991.

---

* Company offers dividend reinvestment plan (see page 213).

6

A Guide to
Socially Responsible
Investments—
Mutual Funds

You will probably feel more "at home" shopping for socially screened mutual funds than you will anywhere else in the investment universe. After all, SRI mutual funds were developed specifically for people like you—in most cases, by people who deeply share your values and concerns. Twelve mutual funds are broadly socially screened as of this writing. Many of these outperform the majority of their non–socially responsible competition. There is no more impressive testimony that socially responsible investing, both as an exercise of personal values and as a viable financial strategy, has come of age.

The universe of mutual funds that directly serve the needs of socially responsible investors actually extends far beyond the screened funds, because many, many funds are invested in socially benign or proactive vehicles like Ginnie Maes and municipal bonds. Believe us, you could put together a perfectly satisfying financial plan by doing all your investing through the mutual funds listed in this chapter. That said, some of you can do even better than that with investments covered in the previous two chapters, depending on the level of your resources and ability to monitor them. So before turning right to our mutual fund shopping list, consider our comments on the pros and cons of mutual fund investing. Then you can decide if mutual funds make sense for you.

WHAT IS A MUTUAL FUND?

In general terms, a mutual fund is a company that pools investors' money to assemble a portfolio of securities (usually common stocks and bonds) and/or other investment vehicles and then manages that portfolio for the investors. Mutual funds best serve the needs of those investors who do not have the resources to invest in a diverse portfolio of individual investments, and those who do not have (any combination of) the time, interest, or expertise to research and track the performance of individual investments. For socially responsible investors, SRI mutual funds provide an important additional benefit—an investment that is automatically socially screened.

Are you worried about letting some distant mutual fund company mess around with your money? Don't be. As with the stock market in general, most of the flagrant abuses in the mutual fund business occurred in the financial free-for-all of the 1920s. The crash of 1929 obliterated the old order, and the congressional reforms that followed—culminating in the Investment Companies Act of 1940—began the new on much more secure footing.

Under this law a mutual fund company must disclose its full intentions to the SEC before offering shares to the public. It must declare the type of vehicles (growth stocks, mature stocks, bonds, and so forth) in which it intends to invest; its investment philosophy (such as high income and safety through investment in U.S. government agency obligations); its social screens if any; and so on. The investor is by law due all this information in a prospectus before he or she invests. The SEC will monitor the fund on an ongoing basis to ensure that it remains true to its goals and to make sure it is not taking advantage of investors by running questionable securities through the fund. The rules prohibit the fund from speculative practices like short selling and margin buying and regulate a fund's diversity by requiring it to invest no more than 5 percent of its assets in any one company and no more than 25 percent in any one industry. There are also rules limiting to 10 percent the amount of a company's outstanding shares that any one institution can hold. Finally, the SEC takes steps to assure the public that a fund's managers are of "good moral character."

The following terminology will help you understand the different categories of mutual funds. By the way, mutual fund terminology varies slightly depending upon which source you consult. Our system combines a few popular classification schemes. We have divided fund types into four main categories: equity (primarily stock holdings although some funds include a small proportion of bonds), fixed income (primarily fixed income securities such as bonds and money market instruments), hybrid (portfolios combining attributes of equity and fixed income funds), and specialized funds. (Except for the specialized funds, funds in each category are listed in increasing order of risk and reward potential.)

Equity Funds

Growth and Income Funds

These funds are designed to serve investors looking for both long-term appreciation and some meaningful income for current living. To meet these goals growth and income fund managers invest primarily in the common stocks and convertible corporate bonds of mature, well-established corporations with a steady record of paying high dividends. Public utility stocks and convertible bonds also show up in force in many such funds' portfolios. These funds offer more overall diversification, less variable returns, and more income than other funds in the equity group. And they also offer less growth potential—but such is the price of all that stability. The greater income does increase the tax consequences. If you are in a high tax bracket, these funds might make most sense as part of a tax-advantaged retirement plan (see page 104).

Index Funds

The way to get your fund to match performance with a stock average like the Standard & Poor's 500, according to this concept, is to portfolio those same five hundred stocks. (Some funds try to beat the system by picking, say, the top two hundred of the S & P 500.) Considering that the S & P 500 has averaged a total return (with dividends reinvested) of 10.3 percent per year over the last sixty-five years, the concept would seem to have something going for it. These funds also tend to have lower management fees and pay fewer transaction fees because little or no research is required and there is little turnover in the holdings. The investor's costs will still be high, however, if the load (the sales charge) is on the high end.

The Domini Social Index Trust (see appendix B for a list of holdings) is a new SRI index fund comprising four hundred socially screened stocks. Its founders conducted a retrospective study that demonstrated that the Domini 400 index performed comparably to the S & P 500, so they expect their fund to perform comparably as well. Early returns, in fact, slightly beat the S & P.

Growth Funds

Growth fund investors seek long-term appreciation that outstrips that of the major stock indexes, but they are not so eager for hefty returns that they will accept a nerve-wracking degree of volatility in exchange. Growth fund managers attempt to satisfy their investors' needs by investing mainly in the stocks of older, established growth firms. Such firms also tend to pay moderate cash dividends.

Historically, growth funds have done well during surging, or "bull," markets and suffered during declining, or "bear," markets. Given the historical upward trend of the market, however, a well-chosen growth fund's up-and-down swings should average out to solid gains over the long term.

Aggressive Growth Funds

Designed for investors who can tolerate the high degree of short-term risk that usually accompanies high double-digit returns, aggressive growth funds concentrate in the stocks of younger companies still in their development phase and other smaller companies. Such stocks will pay low or no dividends and perform with more volatility than the stocks of mature companies—for reasons you will understand if you have read the previous chapter. The low-dividend income will keep your tax liability to a minimum while you wait patiently for your investment to appreciate and compound.

Since these funds do tend to swing wildly in the short term—shooting past the pack in a bull market and trailing most in a bear market—they are not for the faint-hearted or for the average retired investor. Nor should you consider this investment a source of emergency money even though (like all mutual funds) it is liquid. If you had to cash out during one of the downturns, you could lose a chunk of principal. When evaluating such a fund for possible purchase, take a several-year perspective. Periodic slumps go with the territory.

Fixed Income Funds

Government Funds, Taxable

These conservative, stable investments have a place in almost every financial plan, including socially responsible plans. Ethical investors must choose their government funds carefully, however. Taxable government issues include U.S. Treasury issues and U.S. agency issues. As explained previously, ethical investors usually stay clear of funds dealing in treasury issues because these vehicles support government undertakings indiscriminately. Few, though, will find anything objectionable in funds holding agency issues such as Ginnie Maes and Fannie Maes, which support single-family and low-income housing. Such funds as the Franklin U.S. Government Securities and Franklin Adjustable U.S. Government Securities are not designed as SRI funds, but because they deal in mortgage-backed securities exclusively, they are a popular income vehicle for ethical and mainstream investors alike. (Mortgage-backed securities also return about 1 to 1½ percent more than treasuries— a nice trade-off for an increased risk that is more technical than real.)

Municipal Bond Funds

Municipal bond funds, also known as *tax-free government funds,* combine positive social applications and tax advantages in a package almost built to order for well-heeled ethical investors seeking a conservative, income investment. If you can take full advantage of the tax savings, these funds are a good deal. You will trade a couple of percentage points in return for the tax savings but come out ahead when you figure in the money you've kept out of the IRS's clutches. (For example, a municipal bond fund paying 8 percent is equivalent to a taxable mutual fund paying 11.11 percent if you are in the 28 percent federal tax bracket; you won't find investments of comparable quality paying 11 percent in today's market.) If you are considering such a fund, be sure to examine a candidate fund's holdings to ensure that the bonds invested in pass your ethical tests. Note also that state-specific municipal bond funds will be double tax-free to residents of that state.

Corporate and Other Bond Funds

Corporate bond funds—like other fixed income funds—are designed for those who want a conservative, income-oriented investment but have neither the capital to invest in individual issues nor the time to research and track them. As with individual bonds, the price of the shares of the corporate bond fund will fluctuate in inverse relationship to the direction of interest rates. That is, when rates go up, the price of the shares will drop, and vice versa.

Corporate bond funds do not always hold corporate bonds exclusively. Some may include a smaller proportion of U.S. government bonds and other lending-type investments such as commercial paper. (An SRI bond fund like the Calvert Social Investment Fund Bond Portfolio contains relatively few corporate bonds in its blend because few of the companies that meet Calvert's social criteria issue bonds. Most of its holdings are in U.S. government agency bonds.) Non-SRI corporate and other bond funds often include U.S. Treasury issues too, verboten to most ethical investors.

High-yield bond funds—which come in two basic flavors, junk corporate bonds and junk municipal bonds—are an exception to the conservative philosophy of other fixed income investments. They are currently out of favor along with their principal instruments. For good reason, too—the failure rate of junk corporate bonds has been at near-record highs in the early 1990s, and the immediate future for junk munis does not look especially rosy.

Hybrid Funds

Convertible Bond Funds

Convertible bond funds, like the individual issues they pool, combine the income potential of bonds with the growth potential of common stocks. (Recall that convertible bondholders retain the option of converting the investment to common stock—at a specified ratio—if the stock appreciates sufficiently to justify the move.) Like balanced funds, convertible bond funds are also conservative investments based on hedged bets. Convertible bonds usually pay a lesser rate than nonconvertible bonds as a trade-off for their growth potential. The down side is protected, however, because at worst convertible bonds will still pay the guaranteed rate of the bond. As with other bond funds, though, shares in convertible bond mutual funds fluctuate in value inversely with the direction of interest rates.

Income Funds

The emphasis in these funds is on dividends and interest. The stocks they invest in have a history of paying solid dividends and the promise of continuing to pay, and ideally increasing, those dividends in the foreseeable future. Since these funds are designed to satisfy conservative investors, stock stability is another goal. Income funds normally include bonds and money market instruments as well as stocks. Because of their stock component, income funds are riskier than a fixed income investment—including a fixed income mutual fund. However, they have a far better chance of outpacing inflation because unlike fixed income investments, they can be expected to appreciate.

Balanced Funds

Like income funds balanced funds have some of the attributes of a diverse, planned investment portfolio. In other words, they include large proportions of both income vehicles (usually government and corporate bonds and preferred stock) and growth vehicles (common stock). These mutual funds attract conservative investors because the fund managers target high-quality bonds and stable, income-yielding stocks. In a surging stock market, they will probably not perform as well as a pure stock fund because the bond component will hold them back. However, in a deteriorating market they will probably not drop as far as stock funds, as long as the market decline is not caused by high interest rates. The mix of stocks to bonds and other income vehicles is usually about 60 percent to 40 percent in most balanced funds.

Specialized Funds

Sector and Precious Metals Funds

Sector funds concentrate on stocks of companies in particular economic sectors or industries. Examples include utility funds, high-tech funds, health care funds, and environmental (that is, environmental industry) funds.

A technically different but otherwise analogous investment is a precious metals fund. Precious metals funds (for instance, a fund invested in gold mining stocks) were frequently the most outstanding performers during the Persian Gulf crisis, because panicky investors—and cooler heads who bet on the panic—run for gold in times of international instability. They also dove as steeply as they climbed at other times during the tensions. Such behavior is typical of precious metals investments. They are as volatile an issue as is available in the investment spectrum.

Sector funds are an exception to the SEC edict that mutual funds be diverse by industry. They get around that regulation by clearing their intentions with the SEC and then disclosing their goals to the public in their prospectus. With utility funds a notable exception, sector funds tend to be volatile precisely because they are not diverse in terms of industries. (Utility stocks tend to be stable, conservative performers even in bad times because consumers who can delay a car or appliance purchase still need gas and electricity.)

Sector funds effectively expand the ethical investor's mutual fund universe. Even though it is not compiled by an SRI company per se, a health care sector fund, for instance, may meet your ethical requirements if you have no overriding objection to the conventional medical model.

Some sector funds bow a little in the SRI direction by applying limited screens to their acquisitions—such as precious metals funds screened for South African involvement included in our listings on page 245. Ethical investors will, however, want to be wary of funds advertising themselves as "environmental," a term that is becoming as meaningless as "natural" foods. This topic is addressed in more detail in the listings section.

International Funds

International funds are permitted by charter to invest some or all of their assets outside the United States. Because of the internationalization of major economic markets, this is becoming an increasingly popular approach to investing. Although international funds will fit into any of the above categories (growth, bond, balanced, and so on) depending upon the type of assets held, potential returns and risks will be on the upper end of the spectrum for that category, particularly since few if any international investment markets have the stability of the

American market. Some funds attempt to reduce that risk by diversifying by country. Other funds do their diversifying within single countries. One notable drawback to international investing is the difficulty of comprehending ethical and financial information obtained at such great geographical, and often cultural, distance.

The above scheme is general. There are lots of variations within the categories mentioned, and other fund philosophies and objectives that do not fit neatly into any scheme. Also, although we already covered money market funds in chapter 3, these too are in fact mutual funds. Please refer to page 88 for a detailed discussion of their attributes.

All currently available socially screened mutual funds fall into the category known as *open-end funds*. Open-end funds are permitted by the SEC to issue unlimited numbers of shares. A new investor is issued new shares created by the fund; the fund managers in turn expand the fund's holdings with the new investor's contribution.

The price of an open-end fund is based on the *net asset value (NAV)* plus (when applicable) the *load*. The NAV represents the total value of the investments held by the fund at yesterday's closing price divided by the number of shares outstanding that day. The *load* is the commission, or sales charge, that the investor pays to buy into the fund. (The majority of the load goes to the selling broker; a lesser amount goes to the firm with which the broker is associated.)

In an effort to undercut the competition, some funds do not charge a commission—these are called *no-load funds*. (SRI examples include the Pax World Fund, Dreyfus Third Century, and the Domini Social Index Trust.) Most open-end funds are *front-loading*—in other words, they charge a commission at the time of purchase. (The maximum is 8.5 percent—most SRI funds charge from 3.5 to 4.75 percent.) *Rear-loading* funds charge at the time the investor sells his or her shares, usually according to a sliding scale from 5 percent for a brief turnaround to 0 percent if the shares are held at least five years. (If your newspaper's business section carries daily mutual fund quotes, you can usually determine a fund's load from the quote. See page 292 to learn how.)

Closed-end funds offer only a fixed number of shares. Once those are sold, the fund is closed. At that point a new investor can buy shares only by purchasing them from an existing investor offering shares for sale. Closed-end funds, after closing, are traded through a public market like stocks, and the price per share is determined by the market. Thus some will sell below NAV, some above.

Note that on occasion an open-end fund will voluntarily close to new investors. (This does not make it a closed-end fund, by the way.) Why close a growing business? Because sometimes a fund's management decides that the fund's size is getting too unwieldy to maintain the diversity that the SEC mandates—at least without compromising the fund's quality. In such cases managers prefer to open a new, similar fund and start from scratch. (A perfect SRI example is the Calvert-Ariel Growth Fund, which closed in April 1989. The Calvert Group immediately thereafter opened a new fund called the Calvert-Ariel Appreciation Fund with a similar approach and goals.) For the same reason, smaller funds generally have a better chance of outperforming the market averages than the larger funds, which often end up holding so many stocks that they perform like an index fund even if that is not their stated intent.

A DOZEN REASONS TO INVEST IN MUTUAL FUNDS IN GENERAL AND SRI MUTUAL FUNDS IN PARTICULAR

We've made brief mention of some of these reasons already. They are repeated here—in expanded form—with the others so you can take in the entire picture.

1) *Low minimum opening investment, subsequent investments that are lower still.* For investors of limited means, this quality alone makes some mutual funds the investment of choice—as a first investment, anyway. For example, Ginnie Mae investments cost a minimum of $25,000 per. As previously mentioned, you can open an account with the Franklin U.S. Government Securities Fund, which invests in Ginnie Maes and is popular with many ethical investors, for $100. You can make subsequent contributions of as little as $25. In fact, Franklin allows you to open with only $25 as long as you do so as part of a monthly check-debit plan (see point eleven); the monthly contributions would also be $25 minimum. The Calvert Social Investment Fund portfolios require a $1,000 opening investment and $250 minimum subsequent investments, but the subsequents drop all the way to $50 a month with a commitment to a check-debit plan.

2) *Minimum demands on your time and attention.* Your mutual fund researches investments, buys and sells them as it thinks appropriate, and does most of your bookkeeping as well. There is little for

you to do but occasionally check in on the fund's performance to make sure it still meets your goals. The time-and-hassle savings over managing your own stock portfolio, say, is monumental, Consider, for just one example, that if you own Apple Computer stock and Apple skyrockets, you'll soon want to sell so you can grab your profit. Then you'll have to do research to find another good stock in which to invest so you can keep your money working for you. If you've picked your mutual fund as well as you picked Apple (so to speak), its strong performance is only further incentive to hang on to it.

3) *Professional management, research, and analysis.* Your fund is managed by experienced professionals who themselves are usually backed by extensive research and analytical departments. With SRI funds, the expertise extends into the social arena as well.

4) *Social screening "muscle."* As an individual, you may have trouble getting a company to respond to your questions about its ethics with any degree of forthrightness. SRI mutual funds have the knowledge base expertise to compile revealing questionnaires and the financial clout to get companies to answer them meaningfully.

5) *Shareholder activism as part of the deal.* Some SRI mutual funds will fight your social battles for you, pressing the companies in which they hold stock to become better corporate citizens. Pax World Fund, U.S. Trust (which advises the Calvert Social Investment Fund), and Working Assets Money Fund are three SRI-oriented organizations whose managers regularly dialogue with corporate decision makers. Last year Pax World Fund's managers urged Procter & Gamble's management to reconsider its importation of coffee from death-squad-ridden El Salvador. Around the same time Working Assets was encouraging Nordstrom management to be more receptive to unionized workers. It is also not uncommon for an SRI mutual fund to issue an explanatory communication to a company or even to the public at large when the fund divests the company's stock for ethical reasons. This too is a powerful statement, considering the huge blocks of shares in which funds deal. After Calvert sent representatives to Alaska's North Slope in 1989 to investigate Atlantic Richfield's environmental performance there, it issued a public announcement of its decision to divest its ARCO holdings (78,000 shares worth about $7.7 million). The reasons given were ARCO's complicity in the *Valdez* disaster and its failure to respond adequately afterward.

You can assume that ARCO, and the rest of the business world, took note.

6) *Budget-priced diversification.* You have to diversify your holdings within an investment category if you want to get the risk component in your financial plan down to a reasonable level. But for small investors, the cost (not to mention the time and trouble) of compiling a diverse portfolio of individual investments is an impossible dream. As we mentioned in the previous chapter, a minimally diverse portfolio of modestly priced stocks might cost $20,000 if the stocks were purchased in round lots. Mutual funds are by definition—and law—diverse (within the confines of their investment type and strategy, that is). They are the most obvious solution to the small investor's dilemma—particularly considering the broad array of fund types and goals now available from the industry. Again, a check-debit plan can make the opening and subsequent investments in some funds affordable to almost anyone with regular employment.

7) *Liquidity.* Closed-end funds are sold like stocks. Open-end mutual funds will redeem the shares of any shareholder who wants to sell and maintain cash reserves for this purpose in the event that more shareholders want to bail than new investors buy. You can redeem your money within five working days upon request and without interest penalty. If your fund is part of a family (see point ten) that includes a money market fund (most fund families do), you can move your money there by phone and gain instant check-writing privileges.

8) *Range of products for socially responsible investors.* Not only can you choose SRI funds from a wide variety of financial vehicles, but you can find funds screened for almost every conceivable social issue. There are funds screened according to religious/cultural principles (such as the Star family of funds, [based on pro-Israel sentiment]; Amana Fund [Islamic moral principles]; and Pax World Fund [Christian moral principles]); funds screened for "sin" issues—alcohol, tobacco, and gambling—involvement (such as the Pioneer funds); true environmental funds (such as New Alternatives and the Schield Progressive Environmental Fund); funds screened for a broad range of progressive social issues (the Calvert Social Investment Fund and Pax World Fund); and funds screened idiosyncratically (Dreyfus Third Century, for example—screened to encourage environmental protection and

improvement, consumer and employee safety, and equal opportunity employment, but not screened for involvement in defense or liquor).

9) *Performance of SRI funds.* Not only do you not have to sacrifice returns to invest in SRI funds, but you can do much better than the industry average, depending on your choice. For instance, the Calvert Social Investment Fund Managed Growth portfolio was one of the best performers in its class in the down markets of 1987, 1989, and 1990. This is a key indicator of fund strength. New Alternatives Fund has been one of the outstanding performers in the environmental sector class since its inception in 1982 and held up better than most stock vehicles in recession- and Mideast-troubled 1990. And, of course, the Pax World Fund was 1990's leading balanced fund. In fact, the August 1990 issue of *Changing Times* (a non-SRI publication) concluded that "most 'socially conscious' funds have delivered better total returns over the past one and three years than the average funds with similar investment objectives." Obviously SRI mutual funds also deserve a serious look from investors who care only about the bottom line.

10) *Investment maneuverability.* Many of the leading funds are part of a "fund family." That is, the mutual fund company manages a number of funds, with various types of investments and goals. Calvert Social Investment Fund, for example, manages a common stock portfolio, a balanced portfolio, a corporate bond portfolio, and a money market portfolio; the Calvert Group operates the Calvert-Ariel Appreciation Fund, a socially screened aggressive growth fund, as well. Fund families frequently allow you to switch your investment—by phone, if you wish—from fund to fund without sales charges. (Some do charge a minimal transfer fee.) This feature allows you to maneuver your money without the high commission costs normally incurred in transactions. Move your money to play a hunch or to change your investment strategy for whatever reason; the company does not care.

11) *A regular investment and tax-sheltered retirement programs made easy. Dollar cost averaging, too.* Most funds allow you to set up IRA, Keogh plan, and other tax-sheltered retirement plan accounts. Most funds will also, upon your authorization, regularly debit your bank checking account and invest your money into whichever of their products you have directed them. This makes disciplined investing a cinch, with lucrative long-term payoffs

thanks to the power of compound interest. If your fund is a stock fund, a bank withdrawal plan will effectively build dollar cost averaging into your investment program as well.

12) *Low-cost pension alternative.* Mutual funds typically charge much lower custodial and administrative fees than do authorized pension custodial companies. Why not ask your company's pension fund managers to set up an SRI mutual fund as an alternative for socially responsible employees?

HALF A DOZEN REASONS WHY MUTUAL FUNDS AREN'T FOR EVERYONE

Not only are mutual funds not for everyone, but not every financial expert is a fan of the concept. Following are some commonly expressed counterpoints to the attributes listed in the previous section:

1) *High sales charges.* This is the most frequently voiced objection to mutual fund investment, and it has some merit, particularly for funds charging the highest loads. Considering that commissions on stock transactions run between 1½ and 2½ percent on the average, you will save considerably when you buy a portfolio of individual stocks as opposed to purchasing shares of an equal dollar amount in an 8.5 percent load mutual fund. Looking at it another way, if you pay an 8.5 percent load for a fund returning 9 percent annually, you are giving up most of your first year's interest just to get into the fund. On the other hand, at some point you have to sell your stock to realize a profit (or to get rid of a dud), at which point you will pay another 1½ to 2½ percent. That means your total commissions on stock ownership are 3 to 5 percent, or about the same as a moderately priced mutual fund.

Besides, not all mutual funds charge commissions. In fact, there is no statistical correlation, as we've said elsewhere, between the size of the load and the performance of the fund. No-load funds as a group perform just as well as funds with sales charges. The Pax World Fund has been a strong-performing, no-load SRI balanced fund (although it has the drawback of not being part of a fund family). The Dreyfus Third Century Fund, an SRI no-load, has also performed strongly, as has the no-load Domini Social Index Trust (although still in its infancy).

High sales charges do make the liquidity of a mutual fund less meaningful. Yes, you can cash in your mutual fund shares as easily as you can shares of stock. But if you've paid a high price to jump into the fund, you'll be taking a big bite if you have to jump out anytime soon.

2) *Aggressive sales practices.* When high loads are involved, you can expect open-end mutual funds to be marketed aggressively, both by the mutual fund company and by individual brokers. The tremendous growth in the industry (from about 500 funds in 1985 to some 3,100 funds currently) is certainly due partly to a drive for those commission earnings. It is therefore incumbent on you the individual investor to know when a mutual fund investment makes sense for you and when it doesn't—or at least to request that your broker demonstrate why the fund he or she recommends is better for you than a portfolio of individual stocks.

3) *"Hidden costs."* Obviously, if you review the fund's prospectus before investing, you can't complain about hidden costs. But the costs of buying into a mutual fund can extend a ways beyond the load. In addition to the management fee, which is part of the load, the prospectus may disclose "12B-1" fees. These cover things like advertising, commissions, and other expenses that may be listed separately from the management fee. Currently 12B-1 fees run from .1 to 1.25 percent—typically about .25 percent. Some funds also charge "reload" fees, which are charges to re-invest your dividends and capital gains; "exit" fees, a charge to cash in any of your shares (usually either a flat rate or 1 to 2 percent of the proceeds); and transfer fees, which we've mentioned, for switching your investment to another fund in the family. All of these plus the load must be taken into consideration when figuring your actual returns from the fund. And, yes, no-load funds do charge management and other fees.

4) *Professional management does not guarantee professional-level performance.* True. Some mutual funds, including some of the SRI funds, have compiled impressive records of beating the market averages. But others consistently do worse. Sometimes the mutual fund industry as a whole tends to lag behind the market trend. Other times it beats the trend. In the decade from 1976 to 1986, the industry significantly outperformed Standard & Poor's index of five hundred stocks, with the funds gaining 302 percent to the index's 266 percent.* From 1986 to 1990,

* Gardiner, Robert. *The Dean Witter Guide to Personal Investing*, p. 112.

however, less than one mutual fund in seven beat the Standard & Poor's 500.* What's the conclusion? Pick your fund carefully, track its performance, and hold a solid performer for the long term.

5) *Some funds are overdiversified.* Recall that in the previous chapter we recommended that an individual hold no more than a dozen stocks and preferably fewer. Some funds hold hundreds of stocks. Even with professional managers, computers, and large staffs, tracking that many stocks effectively is both a formidable task and an unnecessary one. More is not better once you obtain an adequate level of diversity. Arguably, it's worse. How select can your fund's portfolio of stocks be if it includes hundreds of companies? (Excepting, of course, index funds.)

6) *The rules of the game can compromise the quality of the fund.* Given that a fund by law can have no more than 5 percent of its holdings in any company and 25 percent in any industry, management may sometimes invest in lesser stocks to stay within those percentages. That can adversely affect the fund's performance.

HOW TO EVALUATE A SOCIALLY RESPONSIBLE MUTUAL FUND

The above caveats noted, we feel that mutual funds are an excellent investment for most small-to-moderate investors, particularly since so many of the SRI funds carry modest loads and perform well. We had earlier cited a statistic on the phenomenal growth in the industry as evidence of the aggressive marketing of mutual funds; that statistic is also a sign that more and more people of limited means are discovering that mutual funds are the best vehicle with which to begin securing their future.

The following guide to identifying the better funds will help you understand our SRI fund ratings, which immediately follow this section:

- **Make sure the fund's type and financial goals match your needs.** So what if an aggressive growth fund meets all your social criteria when you are looking for an income investment?

- **Examine the fund's prospectus and other written material to make sure its social criteria are compatible with yours.** Such

---

* *Consumer Reports*, May 1990.

words as "environmental" and "life-supportive" may mean different things to the fund's managers than they do to you. Are those differences you can live with or not?

- **Be wary of funds with high management and other fees.** A management fee of more than 1½ percent in a stock fund, or 1 percent in a bond fund, is what we would call high. Note any additional fees disclosed in the prospectus: 12B-1 fees, "reload" or reinvestment fees, "exit" fees, and transfer fees. Add these to the stated management fee to determine your actual costs (known as "total operating expenses"). Remember that even no-load funds charge management fees and sometimes 12B-1 and other operating expenses, too. There is no formula for determining when total fees are too high. Look instead to the fund's track record—the bottom line is whether or not the fund's performance level offsets the fee structure.
- **If you are considering a no-load fund, realize that there is a trade-off in convenience for what you will save in commission.** All communications regarding initial purchase, subsequent investment, sale or partial sale, change of address, and so on, will be between you and the fund. A broker receiving a commission for a fund would (or should) provide those services for you and track the performance of your fund as well.
- **Consider the fund's features.** Are the load break points—gradually reduced loads at higher levels of investment—relevant to you? Can you reinvest dividends without charge? Is the fund part of a family, and if so, is there a charge for transferring to another fund? What is the minimum investment—initial and subsequent? Particularly if one of your primary reasons for investing in the fund is convenience, examine its convenience features. Can you invest or transfer by phone? Can you establish tax-sheltered retirement plans through the fund?
- **How long do you plan to stay with this fund?** Because of their loads, mutual funds should generally be considered long-term investments, and you certainly should not place any money that you might need in the next three years or so in a moderate- or high-load fund. The money is liquid, but your overall financial plan should include other tiers of liquid money—such as money market funds—that you can tap first.
- **Finally, and most important, examine the fund's track record of performance.** A ten-year record would be preferable, but few of the SRI funds have been around that long, so try for a *five-year* history

at least. You should also look at recent records, *one* and *three* years back, to make sure that performance hasn't deteriorated and to identify signs of gathering strength. Answer the following questions for yourself in your review:

How does the fund perform in a bear market relative to other funds and to the market as a whole? Good returns in a surging market don't mean that much—going up in a down elevator or even dropping only one floor when everyone else is dropping three is a much neater trick.

If the fund does perform well in a bull market, does it do better than the average? That, of course, is what you're looking for. If the fund doesn't run with the bulls, forget it—as long as you're comparing apples to apples in terms of fund type and goals.

If the fund has a load, does the amount of the load justify itself compared with similar low- and no-load funds? You're looking for consistent performance here, not just a recent outstanding year.

SRI MUTUAL FUNDS AND OTHER FUNDS OF INTEREST TO ETHICAL INVESTORS

Mutual Fund Listings Key

Ratings

- **Social ratings (1–5 hearts).** Each fund employs its own system for evaluating the social attributes of investments. Our rating makes a qualitative evaluation of the thoroughness of each fund's screening process. We base this both on the criteria stated in the fund's prospectus and on the actual practices of the fund. (Some funds evaluate more factors than they are willing to commit to on paper in their prospectus; others "consider" only the stated criteria as part of a larger social/economic evaluation. Nor do a list of stated criteria indicate how those criteria are variously weighted.)
- **Growth rating (1–5 $'s).** An evaluation of the past financial performance of the fund, based upon *annual average return* over the past five years (or the past three years for younger funds). If the fund is less than three years old, the fund is labeled NR—no rating. (Remember the rule of thumb that the greater an investment's growth potential, the greater the risk to the investor's principal.)

$ = <5%
$$ = 5–7.5%
$$$ = 7.5–10%
$$$$ = 10–12.5%
$$$$$ = >12.5%

■ **Risk rating (1–5 R's).** For stock and balanced funds, this is based on the fund's beta coefficient. (The beta coefficient is a standard measure of market risk. Funds with a beta greater than 1 are riskier than the market average as defined by the S&P 500 stock market index. Funds with a beta of less than 1 are less risky than the market average. To illustrate, if a fund has a beta of 1.3, the fund is predicted to fluctuate 30 percent more than the index in both up and down markets. Conversely, a fund with a beta of .7 should fluctuate only 70 percent as much as the market.)

For money market and bond funds, one "R" is assigned for interest rate risk. All listed funds hold investment-grade bonds with emphasis on the highest-quality bonds.

Funds in business less than three years are labeled NR—no rating.

R = Money market and bond
        funds only
RR = < 0.5
RRR = 0.5–0.7
RRRR = 0.7–1.0
RRRRR = > 1.0

Social Policies

These are a summary of stated policies from the fund's prospectus. As mentioned above, some of the funds consider factors besides those stated in the prospectus. Those factors will not appear in this summary.

Performance

■ **Average Annual Total Return.** Does not include sales charges. (The first year is not really an average, just the total return for 1990; three and five years are averages.)
■ **Return on Initial Investment of $10,000.** Sales charge (front-end load) is deducted from first year's total. This gives an accurate comparison of the actual returns between load and no-load funds.

Other Data

- **Current Yield.** Calculated by totaling all income distributions paid during the preceding twelve-month period and dividing this figure by the net asset value on June 30, 1991.
- **Turnover.** The rate at which the fund's portfolio turns over each year. (This is a measure of volatility—funds with a high turnover may be more volatile.)
- **Price on June 30, 1991.**

Largest Holdings

A general indication of the types of stocks held by the funds. Before investing, we advise that you request a fund's annual report to examine the holdings in full.

Portfolio Breakdown

Shows how the fund has allocated its money among various categories of investments. Those with higher amounts in stocks are subject to more risk.

Services

- **Family.** Other investment options available within the fund family.
- **Telephone Transfer.** "Yes" means the ability to redeem funds and have them wired to the investor's designated bank account. If the fund is part of a family, "Yes" also means investment can be transferred by phone to other family funds.
- **Automatic Purchase.** "Yes" means the investor may arrange for the fund to automatically transfer money from the investor's local bank account to his/her mutual fund. We have also referred to these as "check-debit" plans. This feature promotes regular investing and dollar cost averaging.
- **Automatic Distribution.** "Yes" means investors may elect to receive a quarterly or monthly check if they maintain a sufficient minimum balance.
- **Retirement Plans.** "Yes" means that investment can be set up as an IRA. Most such funds can also be used in conjunction with other retirement plan options, such as profit-sharing, money purchase, 403(b), 401(k), and so on. Contact the fund for specifics.

Purchasing
- **Minimum.** Gives initial and additional investments.
- **Fees and Commissions.** Maximum Initial Sales Charge is the front-end load. Total Operating Expenses are net expenses divided by total fund assets during 1990, expressed as a percentage; includes such items as management fees, custodial fees, printing costs, and 12b-1 (extra marketing and distribution) charges.

*Broadly Screened Socially Responsible
Stock and Bond Mutual Funds*

On the following twenty-five pages are our reviews of
eleven socially responsible stock and bond mutual funds.

Calvert-Ariel Appreciation Fund

	Social	Growth	Risk
	♥♥♥	NR	NR

Type:	Aggressive Growth	**Address:**	1700 Pennsylvania Avenue, N.W.
Incorporated:	1989		Washington, D.C. 20006
Manager:	Eric McKissack	**Telephone:**	(800) 368-2748 National
Net Assets:	$39 million		(301) 951-4800 Local

SOCIAL POLICIES

Selects companies that take positive steps toward preserving the environment.

Avoids companies with poor environmental records, South Africa business activities, weapons producers, and nuclear energy equipment and producers.

Largest Stock Holdings
(As of 11/30/90)

1. Church & Dwight
2. Universal Foods
3. Watts Industries
4. Bergen Brunswig
5. Vivra
6. Fleming Companies
7. Bausch & Lomb
8. National Medical Enterprises
9. Great Atlantic/Pacific Tea
10. Carter-Wallace

PERFORMANCE (as of 6/30/91)

	Average Annual Total Return (%)	Return on Initial Investment of $10,000		Current Yield:	1.1%
				Turnover:	N/A
1 year	8.97	$10,379			

Price on 6/30/91:

NAV: $17.77
Offer: $18.66

PORTFOLIO BREAKDOWN

Stocks: 97%
Cash Equivalents: 3%

PURCHASING

MINIMUM

Initial Investment: $2,000
Additional Investment: $250

FEES AND COMMISSIONS

Maximum Initial Sales Charge: 4.75%
Total Operating Expenses: 0.95%

SERVICES

Family: See Calvert Social Investment: Managed Growth
Telephone Transfer: Yes
Automatic Purchase: Yes
Automatic Distribution: Yes, with $10,000 minimum account balance
Retirement Plans: Yes

COMMENTS

The Ariel Appreciation fund was formed when the Ariel Growth Fund closed to new investors. Using a similar investment strategy, it has added environmental concerns to its screening criteria.

251

Calvert-Ariel Growth Fund

(Note: As of this writing, this fund is closed to new investors.)

Social	Growth	Risk
♥♥	$$$	RRR

Type: Aggressive Growth
Incorporated: 1986
Manager: John Rogers
Net Assets: $249 million

Address: 1700 Pennsylvania Avenue, N.W.
Washington, D.C. 20006
Telephone: (800) 368-2748   National
(301) 951-4800   Local

SOCIAL POLICIES

Will not invest in companies in South Africa, weapons producers, nuclear energy equipment and producers.

Largest Stock Holdings
(As of 11/30/90)

1. McCormick
2. Clorox
3. Fleming Companies
4. Ecolab
5. Russell
6. Sanford
7. Herman Miller
8. Handleman
9. Topps
10. Angelica

PERFORMANCE (as of 6/30/91)

	Average Annual Total Return (%)	Return on Initial Investment of $10,000
1 year	0.53	$9,575
3 years	9.18	$14,028

Current Yield:	1.2%
Turnover:	22.0%

Price on 6/30/91:

NAV:	$27.61
Offer:	$28.99

PORTFOLIO BREAKDOWN

Stocks:	95%
Cash Equivalents:	5%

PURCHASING

MINIMUM

Initial Investment:	$2,000
Additional Investment:	$250

FEES AND COMMISSIONS

Maximum Initial Sales Charge:	4.75%
Total Operating Expenses:	1.73%

SERVICES

Family:	See Calvert Social Investment: Managed Growth
Telephone Transfer:	Yes
Automatic Purchase:	Yes
Automatic Distribution:	Yes, with $10,000 minimum account balance
Retirement Plans:	Yes

COMMENTS

Has recorded impressive gains during bull markets, including a #6 ranking of all funds during 1988. Looks for small, undervalued companies, which increases its profit potential but leaves it vulnerable to large price fluctuations. Ariel Growth uses fewer screens than the other Calvert socially screened funds.

Calvert Social Investment Fund:
Bond Portfolio

	Social	Growth	Risk
	♥♥♥♥♥	$$$	R

Type: Income

Incorporated: 1987

Manager: Calvert Asset Mgmt. Co.

Net Assets: $28 million

Address: 1700 Pennsylvania Avenue, N.W.
Washington, D.C. 20006 National

Telephone: (800) 368-2748
(301) 951-4800 Local

SOCIAL POLICIES

Invests in companies that make a significant contribution to society through their products and services and through the way they do business. Criteria evaluated include: product safety, environmental protection, participatory management, labor relations, equal employment opportunities, commitment to human goals. Will not invest in nuclear energy equipment or producers, companies in South Africa or other repressive regimes, weapons manufacturers, alcohol or tobacco producers or gambling casinos.

Largest Bond Holdings
(As of 3/31/91)

1. Fed. Home Loan Bank Board
2. Fed. National Mortgage Assn.
3. Gvt. National Mortgage Assn.
4. Federal Farm Credit Bank
5. Financial Assistance
6. Collat. Mortgage Oblig. Trust
7. Student Loan Mktg. Assn.
8. Mountain State Bell
9. Michigan Bell Telephone
10. McGraw Hill

PERFORMANCE (as of 6/30/91)

	Aerage Annual Total Return (%)	Return on Initial Investment of $10,000
1 year	10.17	$10,493
3 years	9.42	$12,476

Current Yield:	7.4%
Turnover:	50.0%

Price on 6/30/91:	
NAV:	$15.84
Offer:	$16.63

PORTFOLIO BREAKDOWN

U.S. Govt. Agencies:	80%
Corporate Bonds:	19%
Other:	1%

PURCHASING

MINIMUM
Initial Investment:	$1,000
Additional Investment	$250

FEES AND COMMISSIONS
Maximum Initial Sales Charge:	4.75%
Total Operating Expenses:	0.65%

SERVICES

Family:	See Calvert Social Investment: Managed Growth
Telephone Transfer:	Yes
Automatic Purchase:	Yes
Automatic Distribution:	Yes, with $10,000 minimum account balance
Retirement Plans:	Yes

COMMENTS

The only socially screened income fund with both corporate and government agency bonds. They do not carry U.S. Treasury notes in their portfolio.

Calvert Social Investment Fund: Equity Portfolio

	Social	Growth	Risk
	♥♥♥♥♥	$$$$	RRR

Type: Growth
Incorporated: 1987
Manager: Dominic Colasacco (1987)
Net Assets: $32 million

Address: 1700 Pennsylvania Avenue, N.W.
Washington, D.C. 20006
Telephone: (800) 368-2748   National
(301) 951-4800   Local

SOCIAL POLICIES

Same criteria as all Calvert Social Investment Fund portfolios. See Calvert Social Investment Fund: Bond Portfolio, page 254.

Largest Stock Holdings
(As of 3/31/91)

1. AMP
2. W. W. Grainger
3. Equitable Resources
4. May Department Stores
5. Gannett
6. H. J. Heinz
7. General Re Corp.
8. Becton Dickinson
9. R. R. Donnelley & Sons
10. Albertson's

PERFORMANCE (as of 6/30/91)

	Average Annual Total Return (%)	Return on Initial Investment of $10,000
1 year	5.15	$10,016
3 years	10.76	$12,942

Current Yield:	2.2%
Turnover:	8.0%

Price on 6/30/91:
NAV:	$18.66
Offer:	$19.59

PORTFOLIO BREAKDOWN

Stocks:	96%
Municipal Bonds:	2%
Cash Equivalents:	1%
Convertible Bonds:	1%

PURCHASING

MINIMUM
Initial Investment:	$1,000
Additional Investment	$250

FEES AND COMMISSIONS
Maximum Initial Sales Charge:	4.75%
Total Operating Expenses:	0.78%

SERVICES

Family:	See Calvert Social Investment: Managed Growth
Telephone Transfer:	Yes
Automatic Purchase:	Yes
Automatic Distribution:	Yes, with $10,000 minimum account balance
Retirement Plans:	Yes

COMMENTS

This portfolio offers a more aggressive (and riskier) opportunity with the same thorough screening as the Managed Growth portfolio. It has performed well.

Calvert Social Investment Fund:
Managed Growth Portfolio

	Social	Growth	Risk
	❤❤❤❤❤	$$	R

Type: Balanced
Incorporated: 1982
Manager: Dominic Colasacco
Net Assets: $287 million

Address: 1700 Pennsylvania Avenue, N.W.
Washington, D.C. 20006

Telephone: (800) 368-2748   National
(301) 951-4800   Local

SOCIAL POLICIES

Same criteria as all Calvert Social Investment Fund portfolios. See Calvert Social Investment Fund: Bond Portfolio, page 254.

Largest Holdings
(As of 3/31/91)

Stocks:

1. Albertson's
2. AMP
3. W. W. Grainger
4. Becton Dickinson
5. R. R. Donnelley & Sons
6. Quaker Oats
7. Sysco
8. May Department Stores
9. Bell Atlantic
10. Equitable Resources

U.S. Government Agencies:

1. Federal Home Loan Bank Board
2. Federal National Morgage Association
3. Government National Mortgage Association

PERFORMANCE (as of 6/30/91)

	Average Annual Total Return (%)	Return on Initial Investment of $10,000
1 year	7.08	$10,199
3 year	9.11	$12,372
5 years	7.38	$13,600

Current Yield:	4.7%
Turnover:	46.0%

Price on 6/30/91:

NAV:	$27.47
Offer:	$28.84

PORTFOLIO BREAKDOWN

Stocks:	43%
U.S. Govt. Agencies:	54%
Cash Equivalents:	1%
Other:	2%

PURCHASING

MINIMUM

Initial Investment:	$1,000
Additional Investment	$250

FEES AND COMMISSIONS

Maximum Initial Sales Charge:	4.75%
Total Operating Expenses:	1.52%

SERVICES

Family: The only family of socially screened funds. In addition to the Managed Growth portfolio, the Calvert Social Investment Fund offers money market, bond, and equity portfolios. Calvert also operates the two socially screened Ariel funds and various nonscreened funds.

Telephone Transfer:	Yes
Automatic Purchase:	Yes
Automatic Distribution:	Yes, with $10,000 minimum account balance
Retirement Plans:	Yes

COMMENTS

The most popular socially screened fund, the Managed Growth portfolio employs a conservative investment strategy that has protected its investors during bear markets while providing steady long-term earnings. Since its inception, the fund has never had a losing year.

Domini Social Index Trust

	Social	Growth	Risk
	♥♥♥	NR	NR

Type: Index
Incorporated: 1991
Manager: Amy Domini

Address: 6 St. James Avenue
Boston, MA 02116
Telephone: (800) 762-6814

SOCIAL POLICIES

Invests in companies that make up the Domini Social Index 400. Excluded are companies that derive more than 4 percent of their revenues from military weapons, tobacco products, alcohol, gambling, or nuclear power. Also eliminated are companies with operations in South Africa. Also evaluated are secondary criteria such as the quality of a company's products, environmental performance, corporate citizenship, and employee relations.

Largest Stock Holdings
(As of 8/12/91)

1. Wal-Mart
2. Merck
3. Coca Cola
4. Procter & Gamble
5. Bell Atlantic
6. Sears
7. PepsiCo
8. Amoco
9. Amer. Intern. Group
10. Southwestern Bell

PERFORMANCE

(Began Investing in May, 1991)

PURCHASING

MINIMUM
Initial Investment: $1,000
Additional Investment: None
FEES AND COMMISSIONS
Maximum Initial Sales Charge: 0%
Total Operating Expenses: 0.75%

SERVICES

Family: None
Telephone Transfer: Yes
Automatic Purchase: No
Automatic Distribution: No
Retirement Plans: Yes, IRA only

COMMENTS

This new fund offers investors the opportunity to invest in the broad range of socially responsible companies listed in the Domini 400. Performance should be comparable to other index funds that are based on the Standard & Poor's 500 Index.

Dreyfus Third Century Fund

	Social	Growth	Risk
	♥♥♥	$$$$	RRR

Type: Growth and Income
Incorporated: 1971
Manager: Howard Stein
Net Assets: $278 million

Address: 144 Glenn Curtiss Boulevard
Uniondale, NY 11556-0144
Telephone: (800) 782-6620 National
(516) 296-6958 Local

SOCIAL POLICIES

Companies which contribute to the enhancement of the quality of life in America. Criteria evaluated include: protection and improvement of the environment and proper use of natural resources, occupational health and safety, consumer protection and product purity, and equal employment opportunity. Avoids companies operating in South Africa.

Largest Stock Holdings
(As of 5/31/91)

1. Astra A Free
2. Coca-Cola
3. Medco Containment Service
4. Microsoft
5. General Motors
6. Wal-Mart Stores
7. Sigma-Aldrich
8. Humana
9. Sysco
10. Novell

PERFORMANCE (as of 6/30/91)

	Average Annual Total Return (%)	Return on Initial Investment of $10,000
1 year	10.95	$11,095
3 years	15.60	$15,450
5 years	12.10	$17,698
10 years	10.79	$27,864

Current Yield:	1.8%
Turnover:	53.0%

Price on 6/30/91:

NAV:	$7.48
Offer:	$7.48

PORTFOLIO BREAKDOWN

Stocks:	75%
US Treasury Bills:	24%
Cash Equivalents:	1%

PURCHASING

MINIMUM

Initial Investment:	$2,500
Additional Investment	$50

FEES AND COMMISSIONS

Maximum Initial Sales Charge:	0%
Total Operating Expenses:	1.05%

SERVICES

Family: Dreyfus is a large family of mostly no-load funds. None of its other portfolios are socially screened.

Telephone Transfer:	Yes
Automatic Purchase:	Yes
Automatic Distribution:	Yes, with $10,000 minimum account balance
Retirement Plans:	Yes

COMMENTS

A moderately conservative fund that has been a steady performer over its history. Does not screen out defense contractors, and does use U.S. Treasury notes for holding cash. The fund currently has its largest investments in the health care industry.

New Alternatives Fund

Social	Growth	Risk
♥♥♥♥	$$$$	RRR

Type: Sector-Natural Resources
Incorporated: 1982
Managers: David and Maurice Schoenwald

Net Assets: $20 million

Address: 295 Northern Boulevard
Great Neck, NY 11021

Telephone: (516) 466-0808
(collect okay)

SOCIAL POLICIES

Selects companies that have an interest in solar and alternative energy development, including solar photovoltaic cells, cogeneration, biomass, passive and active heating and cooling, and natural gas, hydroelectric, wind, and geothermal energy. Uses socially conscious, federally insured banks for holding cash. Avoids energy production from petroleum or nuclear sources, companies with business activities in South Africa, and nuclear weapons producers.

Largest Stock Holdings
(As of 6/28/91)

1. Stewart & Stevenson
2. Wellman
3. Corning
4. Thermo Instrument
5. Zurn Industries
6. Burlington Resources
7. Hawaiian Electric
8. Archer Daniels Midland
9. Baldor
10. Equitable Resources

PERFORMANCE (as of 6/30/91)

	Average Annual Total Return (%)	Return on Initial Investment of $10,000
1 year	5.83	$ 9,984
3 years	12.01	$13,256
5 years	10.10	$15,261

Current Yield:	1.6%
Turnover:	26.0%

Price on 6/30/91:

NAV:	$28.50
Offer:	$30.21

PORTFOLIO BREAKDOWN

Stocks:	84%
U.S. Treasury:	10%
Socially Responsible Bank Money Market:	3%
Cash Equivalents:	3%

PURCHASING

MINIMUM

Initial Investment:	$2,650
Additional Investment	$500

FEES AND COMMISSIONS

Maximum Initial Sales Charge:	5.66%
Total Operating Expenses:	1.24%

SERVICES

Family: None.

Telephone Transfer:	No
Automatic Purchase:	Yes
Automatic Distribution:	No
Retirement Plans:	No

COMMENTS

New Alternatives is the most proactive investment in the field of solar and alternative energy, and has been a solid performer. The fund has kept its risk moderate by diversifying into a wide range of energy related companies such as natural gas utilities. While strongly environmental, the fund's social screens do not cover some issues, and they do use U.S. Treasury bills for holding cash.

Parnassus Fund

	Social	Growth	Risk
	❤❤❤❤❤	$$	RRRR

Type: Growth
Incorporated: 1985
Manager: Jerome Dodson
Net Assets: $27 million

Address: 244 California Street
San Francisco, CA 94111

Telephone: (800) 999-3505   National
(415) 362-3505   Local

SOCIAL POLICIES

Selects companies that have an "enlightened and progressive" management by evaluating the quality of their products, customer service, positive contribution to the community, labor relations, environmental protection, equal employment opportunities, and a history of ethical business practices. Avoids alcohol and tobacco producers, gambling operations, weapons contractors, firms with operations in South Africa, nuclear power producers.

Largest Stock Holdings
(As of 6/30/91)

1. Hewlett-Packard
2. Ask Computer
3. Tandem Computers
4. Raymond Corporation
5. Cummins Engine
6. Herman Miller
7. Federal Home Loan Mortgage
8. Digital Equipment
9. Electro Scientific Industries
10. Advanced Micro Devices

PERFORMANCE (as of 6/30/91)

	Average Annual Total Return (%)	Return on Initial Investment of $10,000
1 year	3.83	$10,020
3 years	4.19	$10,914
5 years	5.99	$12,905

Current Yield:	1.0%
Turnover:	30.0%

Price on 6/30/91:

NAV:	$21.91
Offer:	$22.70

PORTFOLIO BREAKDOWN

Stocks:	84%
Cash Equivalents:	16%

PURCHASING

MINIMUM

Initial Investment:	$2,000
Additional Investment	$100

FEES AND COMMISSIONS

Maximum Initial Sales Charge:	3.5%
Total Operating Expenses:	1.86%

SERVICES

Family:	None.
Telephone Transfer:	No
Automatic Purchase:	Yes
Automatic Distribution:	Yes, with $10,000 minimum balance
Retirement Plans:	Yes

COMMENTS

This fund's contrarian policy of selecting out-of-favor, undervalued companies makes it one of the riskiest socially screened investments. In 1988 it was the fourth best performing mutual fund out of over 1,600 funds. It took a beating the following two years, then climbed 42 percent from 9/90 to 6/91.

Pax World Fund

	Social	Growth	Risk
	♥♥♥♥♥	$$$$	RR

Type: Balanced

Incorporated: 1970

Manager: Anthony Brown

Net Assets: $145 million

Address: 224 State Street
Portsmouth, NH 03801

Telephone: (800) 767-1729 National
(603) 431-8022 Local

SOCIAL POLICIES

Seeks investments in companies that produce life-supportive goods and services, preserve the environment, and have fair employment practices. Avoids defense and weapons contractors, liquor or tobacco producers, and gambling operations.

Largest Stock Holdings
(As of 7/19/91)

1. Hechinger
2. Peoples Energy
3. Equitable Resources
4. Bay State Gas
5. Brooklyn Union Gas
6. DPL
7. H. J. Heinz
8. Advanced Logic
9. Sierra Pacific Resources
10. New Jersey Resources

Government Agencies:

1. Federal National Mortgage Association
2. Federal Home Loan Bank System
3. International Bank for Reconstruction and Development

PERFORMANCE (as of 6/30/91)

	Average Annual Total Return (%)	Return on Initial Investment of $10,000
1 year	14.79	$11,479
3 years	15.97	$15,598
5 years	10.39	$16,393
10 years	13.81	$36,451

Current Yield: 4.1%
Turnover: 58.0%

Price on 6/30/91:
NAV: $15.35
Offer: $15.35

PORTFOLIO BREAKDOWN

U.S. Govt. Agencies:	52%
Stocks:	46%
Cash Equivalents:	2%

PURCHASING

MINIMUM
Initial Investment: $250
Additional Investment: $50

FEES AND COMMISSIONS
Maximum Initial Sales Charge: None
Total Operating Expenses: 1.43%

SERVICES

Family: None.
Telephone Transfer: N/A
Automatic Purchase: No
Automatic Distribution: Yes, with $10,000 minimum account balance
Retirement Plans: Yes

COMMENTS

The oldest socially screened fund. Its strong performance and low risk have earned Pax high ratings in the investment industry. The fund was one of the best performing balanced funds of 1990. With a low $250 minimum initial (and only $50 subsequent) investment, even investors of very modest means can begin a regular investment program.

Rightime Social Awareness Fund

	Social	Growth	Risk
	❤❤❤	NR	NR

Type:	Specialized Growth
Incorporated:	1990
Manager:	David Rights
Net Assets:	$6 million

Address:	The Forst Pavilion, Suite 3000
	Wyncote, PA 19095-1594
Telephone:	(800) 242-1421 National
	(215) 887-8111 Local

SOCIAL POLICIES

Invests in companies which, relative to other companies in the same industry, contribute to enhancing the quality of human life. Analyzes and ranks their relative social performance on criteria, such as: extent of South African operations and adherence to Sullivan principles, environmental preservation, fair employee relations, amount of energy production and its source (i.e. alternative energy), weapons production, and corporate citizenship.

Largest Stock Holdings
(As of 4/30/91)

1.	Coca-Cola	6.	Roadway Services
2.	Northern Telecom	7.	Walgreen
3.	Avon Products	8.	Albertson's
4.	Manor Care	9.	NIKE
5.	Rubbermaid	10.	Fleetwood Enterprises

PERFORMANCE (as of 6/30/91)

	Average Annual Total Return (%)	Return on Initial Investment of $10,000
1 year	6.50	$10,144

Current Yield:	1.7%	
Turnover:	N/A	
Price on 6/30/91:		
NAV:	$26.08	
Offer:	$27.38	

PORTFOLIO BREAKDOWN

Stocks:	96%
Repurchase Agreements:	4%

Note: Portfolio makeup varies widely over time.

PURCHASING

MINIMUM

Initial Investment:	$2,000
Additional Investment:	$100

FEES AND COMMISSIONS

Maximum Initial Sales Charge:	4.75%
Total Operating Expenses:	2.55%

SERVICES

Family:	Rightime has four other funds in its family, none of which are socially screened.
Telephone Transfer:	Yes
Automatic Purchase:	Yes
Automatic Distribution:	Yes, with $10,000 minimum account balance
Retirement Plans:	Yes

COMMENTS

Rightime uses a market-timing strategy, moving in or out of the market depending on its evaluation of economic forecasts. Its goal is to minimize risk during down markets while capitalizing on gains during up periods.

Schield Progressive Environmental Fund

Social	Growth	Risk
♥♥♥♥	NR	NR

Type: Sector—Environmental Services
Incorporated: 1990
Manager: Glenn Cutler
Net Assets: $4 million

Address: 390 Union Boulevard, Suite 410
Denver, CO 80228

Telephone: (800) 826-8154 National
(303) 985-9999 Local

SOCIAL POLICIES

Invests in companies that contribute to a cleaner and healthier environment. Companies are involved in areas such as: solid and hazardous waste management, pollution control, landfills, recycling, air filtration, leak detection, waterway cleanups, pollution reduction systems, incineration, nuclear waste reduction, waste to energy, and producers of environmentally safe products. Avoids polluters, unethical business practices, and other violations of socially responsible principles.

Largest Stock Holdings
(As of 6/28/91)

1. Sanifill
2. Wahlco Environmental
3. Harding Associates
4. Midwesco Filter Research
5. Martech USA
6. CRSS
7. ICF International
8. Ground Water Technology
9. American Waste Service
10. Plants for Tomorrow

PERFORMANCE (as of 6/30/91)

	Average Annual Total Return (%)	Return on Initial Investment of $10,000	Current Yield:	2.4%
			Turnover:	N/A
1 year	-11.42	$8,459		

Price on 6/30/91:

NAV: $5.61
Offer: $5.87

PORTFOLIO BREAKDOWN

Stocks: 86%
Cash Equivalents: 14%

PURCHASING

MINIMUM
Initial Investment: $1,000
Additional Investment: $100
FEES AND COMMISSIONS
Maximum Initial Sales Charge: 4.5%
Total Operating Expenses: 2.5%

SERVICES

Family: Schield manages four other nonscreened funds.
Telephone Transfer: Yes
Automatic Purchase: Yes
Automatic Distribution: Yes, with $5,000 minimum account balance
Retirement Plans: Yes, IRA only

COMMENTS

The only environmental sector fund with full social screening. The environmental industry is experiencing rapid growth, so environmental stocks should outperform the market over the coming decade. However, investors should be aware of the high risk associated with this sector. Schield holds many small company, low-capitalization stocks, making it highly volatile. During its first year the fund has been on a roller coaster, going up or down as much as 30 percent in a quarter.

Ten-Year Annual Performance Data for Socially Screened Mutual Funds and Stock Market Indices

	Jan.-June 1991	1990	1989	1988	1987	1986	1985	1984	1983	1982
Calvert-Ariel Appreciation	22.3%	-1.2% (started 1/15/90)								
Calvert-Ariel Growth	20.4%	-16.1%	25.1%	39.9%	11.4%					
Calvert Social: Bond	3.7%	8.3%	13.6%	8.0%						
Calvert Social: Equity	11.4%	-4.9%	27.5%	14.8%						
Calvert Social: Managed Growth	6.3%	1.8%	18.7%	10.7%	5.4%	18.1%	6.8%	11.3%	11.37%	4.8%*
Domini Social Index	(started in 1991)									
Dreyfus Third Century	18.2%	3.5%	17.3%	23.2%	2.6%	4.6%	29.8%	1.4%	19.9%	4.6%
New Alternatives	15.8%	-7.6%	26.1%	23.9%	-3.2%	22.5%	24.0%	-0.3%	13.4%	
Parnassus	36.2%	-21.2%	2.8%	42.4%	-8.0%	2.4%				
Pax World	9.8%	10.5%	24.8%	11.7%	2.5%	8.4%	25.9%	7.4%	24.2%	18.3%
Rightime Social Awareness	14.2%	-7.0% (started 3/1/90)								
Schield Progressive Environmental	10.4%	4.0% (started 2/5/90)								
Dow Jones Industrial Avg.	10.4%	-4.3%	27.0%	11.9%	2.3%	22.6%	27.7%	-3.7%	20.3%	19.6%
Standard & Poor's 500	12.4%	-6.6%	27.3%	12.4%	2.0%	14.6%	26.3%	1.4%	17.3%	14.8%

Note: Performance data does not include sales charges.

Source: Rugg & Steele Mutual Fund Selector, July 1991

*Started October 1982

*Broadly Screened Socially Responsible
Money Market Funds*

On the following four pages are our reviews of two
socially responsible money market funds.

Calvert Social Investment Fund: Money Market Portfolio

	Social	Growth	Risk
	♥♥♥♥♥	$$	R

		Address:	1700 Pennsylvania Avenue, N.W.
Type:	Money Market		Washington, DC 20006
Incorporated:	1982	Telephone:	(800) 368-2748 National
Manager:	Colleen Trosko		(301) 951-4820 Local
Net Assets:	$182 million		

SOCIAL POLICIES

Same criteria as all Calvert Social Investment Fund portfolios. See Calvert Social Investment Fund: Bond Portfolio, page 254.

PURCHASING AND PERFORMANCE

MINIMUM

Initial Investment:	$1,000
Additional Investment:	$250

FEES AND COMMISSIONS

Maximum Initial Sales Charge:	None
Total Operating Expenses:	0.85%

1990 ANNUAL YIELD: 7.7%

SERVICES

Family	See Calvert Social Investment: Managed Growth
Telephone Transfer:	Yes (to local bank)
Automatic Purchase:	Yes, $50 minimum
Automatic Distribution:	Yes, with $10,000 minimum balance
Retirement Plans:	Yes
Check Writing:	Unlimited, free; must be for $250 or over.

COMMENTS

As part of the Calvert family, this fund gives investors the flexibility to move in and out of the other portfolios. This has been the highest-yielding socially screened money market fund.

Working Assets Money Fund

	Social	Growth	Risk
	♥♥♥♥♥	$$	R

Type:	Money Market	**Address:**	230 California Street
Incorporated:	1983		San Francisco, CA 94111
Manager:	Ken Yeoh	**Telephone:**	(800) 533-3863 National
Net Assets:	$220 million		(415) 989-3200 Local

SOCIAL POLICIES

Seeks out companies with positive records on the environment, affirmative action, labor relations, charitable giving, alternative energy, housing, small business, family farming, and education. Avoids military contractors, U.S. Treasury securities, repressive regimes (including South Africa), nuclear power equipment or producers, polluters, and employers with poor records on discrimination, labor relations, or worker safety.

PURCHASING AND PERFORMANCE

MINIMUM

Initial Investment: $1,000

Additional Investment: $100

FEES AND COMMISSIONS

Maximum Initial Sales Charge: None

Total Operating Expenses: 1.05%

1990 ANNUAL YIELD: 7.4%

SERVICES

Family: None

Telephone Transfer: Yes (to local bank)

Automatic Purchase: Yes

Automatic Distribution: Yes, with $10,000 minimum account balance

Retirement Plans: Yes

Check Writing: Unlimited. Checks under $250 cost $.35 per check; $250 or over are free

COMMENTS

The largest socially screened money market fund, it uses one of the most comprehensive screening processes. Yields have been slightly lower than the Calvert fund. Small investors may appreciate its low minimum for additional purchase and automatic purchase.

Five-Year Annual Performance Data for Socially Screened Money Market Funds and Money Fund Index

	1990	1989	1988	1987	1986
Calvert Social Investment: Money Market	7.7%	8.8%	7.1%	6.0%	6.2%
Working Assets Money Fund	7.4%	8.4%	6.7%	5.9%	6.1%
Donoghue's All Taxable Money Fund	7.8%	8.9%	7.1%	6.1%	6.3%

Note: Effective annual yields include reinvestment of dividends

Source: Provided by the funds

Environmental Funds: Nonscreened

Along with the explosion of environmental awareness in the general population has come an explosion of industrial environmental regulations. There is a lot of money to be made cleaning up environmental disasters and helping industries otherwise comply with the new regulations, so environmental companies promise to be "hot" for decades to come. The investment field has responded to the opportunity with a number of new mutual funds that invest in the environmental sector. Although many of these funds are promoted as being socially responsible and making a positive impact on the environment, we urge our readers to regard them warily. These funds do not screen out companies cited for unethical conduct, nor do they avoid investments in "environmental" companies that have themselves been guilty of polluting. Following is a list of some of the better known funds, offered for information purposes only. See the preceding reviews of the Schield Progressive Environmental and New Alternatives funds (pages 272 and 264) for true SRI environmental funds.

Alliance Global Environment Fund
1345 Avenue of the Americas
New York, NY 10105
(800) 221-5672
(Note: This is a closed-end mutual fund.)

Fidelity Select—Environmental Services Portfolio
P.O. Box 193
Boston, MA 02101
(800) 544-8888

Freedom Environmental Fund
One Beacon Street
Boston, MA 02108
(800) 225-6258 National
(800) 392-6037 Massachusetts

Kemper Environmental Services Fund
120 South LaSalle Street
Chicago, IL 60603
(800) 343-6677

Oppenheimer Global Environment Fund
Two World Trade Center
New York, NY 10048-0669
(800) 525-7048

"Sin" Funds

There are three mutual fund companies that practice a limited form of social screening. These funds screen only to avoid companies involved with three issues: alcohol, tobacco, and gambling. However, these policies are not mentioned in their literature.

The Pioneer Group of Mutual Funds and the William Penn Funds use this policy on each of their portfolios. Pioneer operates six stock funds, three bond funds, and three money market funds. William Penn operates one stock fund, three bond funds, and a money market fund. The American Funds Group screens for these issues on only two of their funds: the American Mutual Fund and the Washington Mutual Investors Fund.

American Funds Distributors, Inc.
135 South State College Boulevard
Brea, CA 92621
(800) 421-9900, ext.11

The William Penn Funds
P.O. Box 1419
Reading, PA 19603
(800) 523-8440

The Pioneer Group of Mutual Funds
60 State Street
Boston, MA 02109
(800) 225-6292

Special Interest Funds

	AVG. ANNUAL TOTAL RETURN			SALES CHARGE	NET ASSETS (MILLIONS)	MIN. INVEST.	PHONE
	1990	3 YEAR	5 YEAR				
ENVIRONMENTAL—NOT SCREENED							
Fidelity Environmental Services Portfolio	-2.48%			3.00%	$94	$1,000	(800) 544-6666
Freedom Environmental Fund	-10.64%			4.50%	$50	$1,000	(800) 225-6258
Kemper Environmental Services Fund	(Started 5/18/90)			4.75%	$45	$1,000	(800) 621-1048
Oppenheimer Global Environment Fund	(Started 3/2/90)			4.75%	$45	$1,000	(800) 255-2755
ISLAMIC							
Amana Mutual Funds—Income	-3.41%	9.04%		0.00%	$5	$100	(800) 542-5496
PRO-ISRAEL							
StarTrade Fund (four portfolios)	(Started 7/23/90)			4.75%		$1,000	(800) 221-7827
"SIN" FUNDS							
American Funds Group							
American Mutual Fund	-1.62%	11.62%	11.47%	5.75%	$3,200	$250	(800) 421-9900
Washington Mutual Investors Fund	-3.85%	13.42%	12.62%	5.75%	$4,868	$250	(800) 421-9900
Penn Square Mutual Fund	-5.17%	10.81%	10.01%	4.75%	$184	$500	(800) 523-8440
Pioneer Mutual Funds							
Pioneer Fund	-10.52%	9.32%	8.96%	8.50%	$1,336	$50	(800) 225-6292
Pioneer II	-12.03%	9.39%	7.97%	8.50%	$3,592	$50	(800) 225-6292
Pioneer Three	-12.95%	10.88%	6.90%	8.50%	$563	$1,000	(800) 225-6292
Pioneer Bond Fund	7.31%	8.85%	7.98%	4.50%	$74	$1,000	(800) 225-6292

Total Return data does not include sales charges.

Source: Rugg & Steele Mutual Fund Selector, January 1991

Other Special Interest Funds

Pro-Israel: The StarTrade Fund

The StarTrade Fund is a newly organized family of mutual funds that invests in companies that have a presence in or maintain good trade relations with the state of Israel. According to the fund, a large number of U.S. corporations are unofficially honoring a commercial boycott of Israel promoted by certain foreign countries. This fund hopes to demonstrate to these corporations that it is in their economic interests to engage in trade with Israel.

StarTrade operates four portfolios: stock, bond, equity, and money market.

StarTrade Fund
747 Third Avenue
New York, NY 10017
(800) 221-7827

Islamic: The Amana Fund

Amana is a no-load mutual fund designed to meet the special needs of Muslims (but open to any investor) by investing in accordance with Islamic principles. These principles require that investors share in profit and loss, receive no usury or interest (which eliminates banks or loan associations), and do not invest in companies involved with liquor, gambling, or pornography. However, this fund would not be acceptable to most other socially responsible investors, as its investments include major defense contractors, industrial polluters, and producers of nuclear energy.

Amana Mutual Funds Trust
520 Herald Building
1155 North State Street
Bellingham, WA 98225
(800) 728-8762

Gold and Precious Metals Funds

For socially responsible investors who wish to invest in gold funds (some of these funds invest in other precious metals and natural resources), there are many choices that exclude South African companies. Survey of funds conducted in March 1991.

API Trust Precious Resources
P.O. Box 2529
Lynchburg, VA 24501
(800) 544-6060

Benham Gold Equities Index Fund
1665 Charleston Road
Mountain View, CA 94043
(800) 472-3389

**Colonial Advanced Strategies
Gold Trust**
One Financial Center
Boston, MA 02111
(800) 248-2828

Excel Midas Gold Shares
16955 Via del Campo
San Diego, CA 92127
(800) 783-3444

Fidelity Select—American Gold
Fidelity Investments
P.O. Box 31460
Salt Lake City, UT 84131-9968
(800) 544-6666

Financial Strategic—Gold
P.O. Box 2040
Denver, CO 80201
(800) 525-8085

Kemper Gold Fund
Kemper Financial Services
120 South La Salle Street
Chicago, IL 60603
(800) 621-1048

**MFS Lifetime Gold & Precious
Metals**
MFS Service Center
P.O. Box 2281
Boston, MA 02107
(800) 225-2606

Mainstay Gold & Precious Metals
51 Madison Avenue, Room 2402
New York, NY 10010
(800) 522-4202

Monitrend Gold Fund
P.O. Box 701
Milwaukee, WI 53201
(800) 251-1970

Pioneer Gold Shares
P.O. Box 9014
Boston, MA 02205-9014
(800) 225-6292

**Rushmore Fund—Precious
Metals**
4922 Fairmont Avenue
Bethesda, MD 20814
(800) 343-3355

Scudder Gold Fund
P.O. Box 2291
Boston, MA 02107-2291
(800) 225-2470

Strategic Gold & Minerals Fund
Strategic Silver Fund
Strategic Management
2030 Royal Lane
Dallas, TX 75229
(800) 527-5027

**Thomson Precious Metals and
Natural Resources**
One Station Place
Stamford, CT 06902
(800) 628-1237

**United Services Global
Resources Fund**
United Services World Gold Fund
United Services Funds
P.O. Box 29467
San Antonio, TX 78229-0467
(800) 873-8637

Van Eck Gold/Resources Fund
Van Eck Securities
122 East 42nd Street
New York, NY 10168

INVESTING FOR INCOME: ADDITIONAL FUNDS OF INTEREST TO SOCIALLY RESPONSIBLE INVESTORS

Government Mortgage-Backed Bond Funds

Government mortgage-backed bond funds—invested primarily in Ginnie Mae securities—expand the universe of taxable income funds appropriate for socially responsible investors. These funds provide both dependable income and a high degree of safety, as explained earlier. Ginnie Mae securities support affordable housing, a valuable social goal in the eyes of most in the SRI community. Do note that some Ginnie Mae funds also hold U.S. Treasury securities, which support the general activities of the government, including activities with which many social investors take issue. The bond funds chart (see page 286) compares the performances of funds holding only a small percentage of treasuries or none at all.

Municipal Bond Funds

Municipal bond funds are also a source of safe, dependable income generated by instruments approved of by most social investors. These funds are best suited to those who can take full advantage of their tax-free attributes. Our municipal bond chart (see page 287) lists the top-performing funds in both the high-quality (less risk, lower returns) and high-yield (the converse) categories. This chart is followed by a list of municipal bond fund sponsors by state, for all states in which such funds are available. If you invest in a fund sponsored in your state, you will benefit from *double* tax-free benefits.

Government Mortgage-Backed Bond Funds
With Holdings Containing Less Than 5 Percent U.S. Treasury Issues

NAME (LISTED ALPHABETICALLY)	AVG. ANNUAL TOTAL RETURN			YIELD	U.S. TREAS.	SALES CHARGE	ASSETS ($ MILLIONS)	MIN. INVEST.	PHONE
	1990	3-YEAR	5-YEAR						
Benham GNMA Income Fund	10.2%	10.8%	9.3%	8.8%	0.0%	NL	319	1,000	(800) 472-3389
Cardinal Government Obligations	10.1%	9.9%		9.0%	1.0%	4.75%	115	1,000	(800) 848-7734
Fidelity GNMA	10.5%	10.5%	9.0%	8.1%	0.0%	NL	649	2,500	(800) 544-8888
Fidelity Mortgage Securities	10.4%	10.2%	8.9%	8.0%	3.4%	NL	380	2,500	(800) 544-8888
Franklin Adjustable U.S. Gov't	9.5%	8.8%		9.8%	0.0%	4.00%	632	100	(800) 342-5236
Franklin U.S. Gov't Securities	10.3%	10.0%	8.8%	9.4%	2.0%	4.00%	11,137	100	(800) 342-5236
Liberty Advantage U.S. Gov't	10.3%	9.0%		8.7%	0.8%	4.50%	314	500	(800) 872-5426
Pacific Horizon GNMA Extra	10.3%			8.5%	0.0%	4.50%	5	1,000	(800) 332-3863
Princor Government Securities	9.5%	11.1%	9.1%	7.6%	0.0%	5.00%	71	1,000	(800) 247-4123
Prudential-Bache GNMA Fund—"B"	8.1%	8.6%	7.4%	8.1%	4.9%	NL	219	1,000	(800) 648-7637
SAFECO U.S. Government Securities	8.7%	9.8%		8.2%	0.0%	NL	29	1,000	(800) 426-6730
Scudder GNMA Fund	10.1%	9.9%	8.6%	8.4%	3.5%	NL	245	1,000	(800) 225-2470
Security Income U.S. Gov't	9.8%	9.2%	8.1%	8.0%	0.0%	4.75%	5	100	(800) 888-2461
Smith Barney—Monthly Pay.Gov.	9.6%	11.0%		8.6%	0.0%	4.00%	21	10,000	(800) 544-7835
Smith Barney—U.S. Gov't	9.6%	10.8%	9.0%	8.6%	0.0%	4.00%	329	3,000	(800) 544-7835
Sunamerica Gov't Securities Port.	5.5%	7.3%		9.4%	0.0%	4.75%	38	1,000	(800) 821-5100
U.S. Government Securities Fund	8.6%	9.8%		8.4%	0.0%	NL	12	10,000	(800) 225-8778
Vanguard GNMA	10.3%	11.3%	9.5%	8.7%	0.0%	NL	2,330	3,000	(800) 662-7447

Note: Total return data not adjusted for sales charges. Percentage of U.S. Treasury issues in portfolio taken from fund's annual report.

Source: Rugg & Steele Mutual Fund Selector, January 1991

Federally Tax-Free Municipal Bond Funds
(Listed by Performance for 1990 Total Return)

Type High-Quality

NAME	AVG. ANNUAL TOTAL RETURN			YIELD	SALES CHARGE	ASSETS ($ MILLION)	MIN. INVEST.	PHONE
	1990	3-YEAR	5-YEAR					
First Prairie Tax Exempt—Insur.	7.8%			6.5%	4.50%	2	1,000	(800) 537-4938
General Municipal Bond Fund	7.6%	10.6%	8.4%	7.8%	0	216	2,500	(800) 782-6620
First Prairie Tax Exempt—Inter.	7.6%			6.4%	4.50%	6	1,000	(800) 537-4938
Steinroe Intermediate Municipal	7.5%	7.3%	7.4%	5.9%	0	100	1,000	(800) 338-2550
Mutual of Omaha Tax Free Income	7.4%	9.9%	9.9%	6.3%	4.75%	352	1,000	(800) 228-9596
Premier Municipal Bond	7.3%	10.4%		7.3%	4.50%	150	1,000	(800) 242-8671
Compass Capital Group Municipal	7.3%			5.9%	0	5	2,500	(800) 451-8371
SIT New Beginning Tax-Free	7.3%			8.2%	0	37	2,000	(800) 332-5580
Muni Bond Fund—Limited Term	7.3%			6.2%	2.00%	34	10,000	(800) 544-7835
Lord Abbett Tax Free—National	7.3%	9.8%	9.7%	6.4%	4.75%	320	1,000	(800) 874-3733
High-Yield								
Plymouth High Income Municipal	10.3%	11.7%		6.9%	4.75%	22	1,000	(800) 544-6666
Fidelity High Yield Tax-Free	8.5%	10.7%	9.4%	6.8%	0	1,683	2,500	(800) 544-6666
Steinroe High-Yield Municipals	7.7%	10.9%	10.5%	7.2%	0	314	1,000	(800) 338-2550
Fidelity Aggressive Tax Free	7.5%	10.1%	9.8%	7.7%	0	533	2,500	(800) 544-6666
Alliance Muni National	7.4%	10.7%		5.7%	4.50%	184	250	(800) 227-4618

Note: Total Returns do not include sales charges.
Earning from some funds may be partially exempt from state income taxes.

High-quality municipal bond funds invest in municipal bonds rated Baa, BBB, or better.
High-yield municipal bond funds invest in municipal bonds rated Baa, BBB, or lower.

Source: Rugg & Steele Mutual Fund Selector, January 1991

Major Sponsors of State and Federal Tax-Free Municipal Bond Funds

Alabama
Franklin (800) 632-2180
MFS (800) 225-2606

Arizona
Aquila (800) 228-4227
Flagship (800) 227-4648
Franklin (800) 632-2180
Prudential-Bache (800) 225-1852
SLH (212) 528-2744

California
Alliance (800) 221-5672
Benham (800) 321-8321
Colonial (800) 225-2365
Dean Witter (800) 869-3863
Dreyfus (800) 782-6620
Eaton Vance (800) 225-6265
Fidelity (800) 544-6666
First Investors (800) 423-4026
Franklin (800) 632-2180
IDS Financial Services
   (800) 328-8300
Lord Abbett (800) 223-4224
Kemper (800) 621-1048
Merrill Lynch (800) 637-3863
MFS (800) 225-2606
Muir Investment Trust
   (800) 648-3448*
Nuveen (800) 621-7227
Oppenheimer (800) 525-7048
Pacific Horizon (800) 332-3863
PaineWebber (201) 902-7341
T. Rowe Price (800) 638-5660
Prudential-Bache (800) 441-9490
Putnam (800) 225-1580

Safeco (800) 426-6730
Scudder (800) 225-2470
Seligman (800) 221-2450
Shearson Lehman Hutton
   (212) 528-2744
Smith Barney (800) 544-7835
Vanguard (800) 662-7447
Van Kampen Merritt
(800) 225-2222

Colorado
Aquila (800) 228-4227
Flagship (800) 227-4648
Franklin (800) 632-2180
Seligman (800) 221-2450

Connecticut
Dreyfus (800) 782-6620
Fidelity (800) 544-6666
Flagship (800) 227-4648
Franklin (800) 632-2180

Florida
Dreyfus (800) 782-6620
Franklin (800) 632-2180
Seligman (800) 221-2450

Georgia
Carnegie (800) 321-2322
Flagship (800) 227-4648
Franklin (800) 632-2180
MFS (800) 225-2606
Prudential-Bache (800) 225-1852
Seligman (800) 221-2450

Indiana
Franklin (800) 632-2180

*As of this writing, only socially screened municipal bond fund. Concentrates on education, affordable housing, and the environment.

Kentucky
Flagship (800) 227-4648

Louisiana
Flagship (800) 227-4648
Seligman (800) 221-2450

Maryland
Franklin (800) 632-2180
MFS (800) 225-2606
T. Rowe Price (800) 638-5660
Prudential-Bache (800) 225-1852
Seligman (800) 221-2450

Massachusetts
Colonial (800) 225-2365
Dreyfus (800) 782-6620
Fidelity (800) 544-6666
First Investors (800) 423-4026
Franklin (800) 632-2180
John Hancock (800) 225-5291
IDS Financial Services
    (800) 328-8300
MFS (800) 225-2606
Nuveen (800) 621-7227
Prudential-Bache (800) 225-1852
Putnam (800) 225-1581
Scudder (800) 225-2470
Seligman (800) 221-2450

Michigan
Colonial (800) 225-2365
Fidelity (800) 544-6666
First Investors (800) 227-4648
Franklin (800) 632-2180
IDS Financial Services
    (800) 328-8300
Prudential-Bache (800) 225-1852
Putnam (800) 225-1581
Seligman (800) 221-2450

Minnesota
Carnegie (800) 321-2322
Colonial (800) 225-2365

Fidelity (800) 544-6666
First Investors (800) 423-4026
Franklin (800) 632-2180
IDS Financial Services
    (800) 328-8300
Prudential-Bache (800) 225-1852
Putnam (800) 225-1581
Seligman (800) 221-2450

Missouri
Franklin (800) 632-2180
Flagship (800) 227-4648
Seligman (800) 221-2450

New Jersey
Dreyfus (800) 782-6620
Fidelity (800) 544-6666
First Investors (800) 423-4026
Franklin (800) 632-2180
Prudential-Bache (800) 441-9490
Putnam (800) 225-1581
Shearson Lehman Hutton
    (212) 528-2744
Seligman (800) 221-2450
Vanguard (800) 662-7447

New York
Alliance (800) 221-5672
Colonial (800) 225-2365
Dean Witter (800) 869-3863
Dreyfus (800) 782-6620
Fidelity (800) 544-6666
First Investors (800) 423-4026
Franklin (800) 632-2180
John Hancock (800) 225-5291
IDS Financial Services
    (800) 328-8300
Kemper (800) 621-1048
Kidder, Peabody (212) 510-5351
Lord Abbett (800) 223-4224
Merrill Lynch (800) 637-3863
MFS (800) 225-2606
Nuveen (800) 621-7227

Oppenheimer (800) 525-7048
PaineWebber (201) 902-7341
T. Rowe Price (800) 638-5660
Putnam (800) 225-1581
Scudder (800) 225-2470
Shearson Lehman Hutton
   (212) 528-2744
Seligman (800) 221-2450
Smith Barney (800) 544-7853
Value Line (800) 223-0818
Vanguard (800) 662-7447

North Carolina
Flagship (800) 227-4648
Franklin (800) 632-2180
MFS (800) 225-2606
Prudential-Bache (800) 225-1852

Ohio
Carnegie (800) 321-2322
Colonial (800) 225-2365
Fidelity (800) 544-6666
First Investors (800) 423-4026
Flagship (800) 227-4648
Franklin (800) 632-2180
IDS Financial Services
   (800) 328-8300
Nuveen (800) 621-7227
Prudential-Bache (800) 225-1852
Putnam (800) 225-1581
Scudder (800) 225-2470
Seligman (800) 221-2450

Oregon
Aquila (800) 228-4227
Franklin (800) 632-2180
Prudential-Bache (800) 225-1852
Seligman (800) 221-2450

Pennsylvania
Delaware (800) 523-4640
Dreyfus (800) 782-6620
Fidelity (800) 544-6666
Flagship (800) 227-4648
Franklin (800) 632-2180
Oppenheimer (800) 525-7048
Penn Square (800) 523-8440
Provident Mutual (800) 441-9490
Prudential-Bache (800) 225-1852
Putnam (800) 225-1581
Scudder (800) 225-2470
Seligman (800) 221-2450
Van Kampen Merritt
   (800) 225-2222
Vanguard (800) 662-7447

Puerto Rico
Franklin (800) 632-2180

South Carolina
MFS (800) 225-2606
Seligman (800) 221-2450

Tennessee
Flagship (800) 227-4648
MFS (800) 225-2606

Texas
Dreyfus (800) 782-6620
Fidelity (800) 544-6666
Franklin (800) 632-2180
Lord Abbett (800) 223-4224

Virginia
Flagship (800) 227-4648
Franklin (800) 632-2180
MFS (800) 225-2606

West Virginia
MFS (800) 225-2606

TRACKING YOUR MUTUAL FUND

One of the biggest reasons so many investors choose mutual funds is that mutual funds are largely self-managing. Why track a fund's performance when the fund's professional managers are paid to do just that—by you, as a matter of fact, through the load and management fee?

The fact is, you do not need to track a fund's performance nearly as closely as you would a portfolio of stocks. But you should check in once in a while—annually, at least. A fund's managers can change, grow old, or lose their touch as easily as any corporation's, and if that shows up as a downward trend in the fund's performance, you will probably want to get out—loss or no loss.

As of this writing, there is no comprehensive SRI source for long-term tracking of socially responsible mutual funds. However, you can track all SRI funds in mainstream sources. Among the mainstream publications, *Barron's* publishes the most comprehensive guide to mutual fund performance, using Lipper Analytical Service data. The guide comes out about six weeks after the close of the fiscal quarter. Daily publications like the *The Wall Street Journal* and newspaper business pages usually list only funds with more than 1,000 shareholders and $1 million in assets. Publications such as *Money* magazine, *Changing Times,* and *Consumer Reports* are more limited still. Of course, you can track the performance of any fund you own or are considering owning through its annual and semiannual reports. The former are available free from the fund. Funds in which you are invested will send you copies automatically.

For our computer-handy readers, the Rugg & Steele Mutual Fund Selector is a software program designed to simplify and expedite mutual funds research. It contains a data base on virtually the entire universe of equity and fixed income mutual funds, enabling the user to search and sort for many kinds of data. You can select individual fund summaries and comparisons of a range of funds, as well as print a variety of reports. The program is exceptionally user-friendly and has an on-screen help system that provides clear instructions for accessing the commands. Updates are available on a monthly or quarterly basis. Most of the mutual fund research for this book was done using this program. For more information, contact Rugg & Steele, Inc.: 6433 Topanga Canyon Boulevard, Suite 108, Canoga Park, CA 91303; (800) 678-3863 or (800) 237-8400, ext. 678.

The daily fluctuations in a mutual fund's NAV shouldn't trouble you too much—short of an over-the-cliff plunge, of course. Everything fluctuates, even the value of your house. You wouldn't sell your house just because average property values dipped in November, would you? Nevertheless you may want to look in from time to time, and major-city newspapers usually do carry mutual fund quotes. You can also calculate a fund's load from the quotes.

How to Read Mutual Fund Listings in the Daily Newspaper

Fund and Portfolio Name	NAV	Offer Price	NAV Change	Description
Calvert Group:				
Ariel	22.73	23.80	+.11	Ariel Growth, a front-end load fund.
PaxWld	13.96	NL	+.05	Pax World, a no-load fund.

KEY:

a. The NAV (net asset value) may also be called *bid*. This is the price you will pay for a no-load fund.

b. The second price listed may be called the *ask, public offering price* (POP), or *offer price*. This is the price you will pay for a front-loading fund. A no-load fund may show an "NL" or similar designation in this column.

c. Some newspapers may include other abbreviations after a fund's name, indicating other charges or features. These should be explained in a legend at the beginning or end of the listings. Example: "r" means redemptions charge may apply.

d. To figure a load from the quotes: Divide the first (NAV or bid) price by the second (ask or POP or offer price). Subtract that figure from 1. The remainder, expressed as a percentage, is the load. Example: The NAV is $7.01, the POP is $7.30. So $1 - (7.01/7.30) = 1 - 0.9602739$. Round 0.9602739 to the nearest hundredth, since you are concerned with percentages—in this case, that is 0.96. Thus $1 - 0.96 = 0.04$. The fund has a 4 percent load.

Note: This method identifies front-loading funds only. Rear-loading funds will probably appear in the quotes as no-load funds, and show an additional abbreviation—such as r—to signify redemption charges.

7

Building Your Financial Plan, Stage I—Balancing Wants and Needs

Congratulations. You have now devoured enough material about financial planning and SRI to impress your friends at a dinner party. What you are still missing, though, is some nuts-and-bolts information about applying this stuff in your daily life. In this chapter some of that information will be coming from your files, not ours, because our generalities will work for you only when they are merged with your unique circumstances and perspectives. We've been telling you all along that any person of average means can become wealthy without compromising ethics. When you complete the worksheets provided below and then follow the advice in these next two chapters, you'll discover that, true to our environmentalist pledge, we're not just blowing smoke.

WHERE DO YOU DRAW THE LINE? PRIORITIZING YOUR SOCIAL CONCERNS

As we have emphasized repeatedly, there are few, if any, purely ethical investments. The world—especially the financial world—is just not that cut-and-dried. Most of your investment decisions will involve at least some minor ''inorganic'' elements unless you stash your money in a 100 percent wool sock somewhere. Worksheet I will assist you in clarifying your social priorities and help your broker find investments for you that will meet your ethical criteria. (Worksheets I through IV have been adapted from models created by Co-op America from their *A Socially Responsible Financial Planning Guide.*)

Worksheet I: Social Priorities

Part A

Indicate your concern for each issue listed. (Rank as top priority [T], important [I], or not important [NI]):

Environmental protection	____	Community develop-ment	____
Affordable housing	____	Consumer protection and product purity	____
Infrastructure develop-ment	____	Producing jobs in low-income areas	____
Animal protection	____	Peace	____
Researching, selling, using renewable energy products	____	Equal opportunity for women, gays, and mi-norities	____
Elderly care	____	International develop-ment	____
Child care	____	Other _____	____
Recycling of resources	____		
Participatory or demo-cratic workplace	____		

For each issue listed, indicate how important it is that your money be diverted from it (same ranking system):

Involvement in South Africa	____	Other unfair labor prac-tices	____
Involvement in other repressive regimes: (list) _____	____	Environmental pollution or degradation	____
Nuclear weapons pro-duction	____	Tobacco	____
Other weapons/ firearms production	____	Alcohol	____
Nuclear power	____	Production of other un-healthy products	____
Racist, sexist, or anti-gay labor policies	____	Production of unsafe products	____
		Gambling	____
		Animal experimentation	____
		Other _____	____

Part B

Mark each statement that is true for you.

_____ I will consider any investment that does not directly contradict my social values.

_____ I will consider investments that may have incidental social negatives as long as they are primarily socially responsible.

_____ I (only/primarily) wish to invest in products that have social screens.

_____ I (only/primarily) wish to invest in products that actively contribute to the realization of my social goals.

_____ I will sacrifice market-level returns for investments that actively contribute to my social goals.

LIFE IN BALANCE—YOUR NET (BUT NOT YOUR TRUE) WORTH

We live in a culture that places a high value on acquiring material wealth. Thus you should forgive yourself if you tend to equate your personal value with your net financial worth. This is not a self-help psychology book, so we won't even pretend to unravel that little piece of cultural confusion for you. We will simply affirm that your net worth has about as much to do with your personal worth as does your hair color.

Changing your financial condition takes a lot more preparation than changing your hair color, however. You need to map your strategy with a financial plan. And you certainly can't start planning for the future until you know what you have now.

Determining your net worth is simple addition and subtraction—add up your assets, add up your liabilities, and subtract the latter from the former. Complete Worksheet II following to find out just where you stand.

Worksheet II: Determining My/Our Net Worth

Note: Your property assets should be valued at their current market value, not their purchase price. Also, your pension plans from work are an important part of your overall financial scheme. If you do not know your current balance, ask the benefits administrator at your job.

Assets (What I/We Own)

A. INVESTMENT ASSETS

Cash/Checking Accounts	$ _____
Savings Accounts	$ _____
Certificates of Deposit	$ _____
Stocks/Bonds	$ _____
Money Market Funds	$ _____
Mutual Funds	$ _____
TOTAL LIQUID ASSETS	$ _____
Life Insurance Cash Value(s)	$ _____
Community Loan Funds	$ _____
Real Estate	$ _____
Equity in Business	$ _____
Partnerships	$ _____
Loans Owed to Me	$ _____
Other	$ _____
TOTAL INVESTMENT ASSETS	$ _____

B. DEFERRED ASSETS
(TO BE REALIZED IN THE FUTURE)

Social Security*	$ _____
Annuities	$ _____
Pension Plans	$ _____
IRA	$ _____
SEP IRA	$ _____
TSA	$ _____
401(k)	$ _____
Money Purchase	$ _____

* The Social Security Administration will send you a form to help you project this figure. Call (800) 937-2000.

Profit-Sharing	$ _____
Other	$ _____
TOTAL DEFERRED ASSETS	$ _____

C. PROPERTY ASSETS

Residence (market value)	$ _____
Other Real Estate (market value)	$ _____
Vehicles	$ _____
Household Furnishings	$ _____
Other	$ _____
TOTAL PROPERTY ASSETS	$ _____
TOTAL ASSETS (Add A, B, and C)	$ _____

Liabilities (What I/We Owe)

Current Bills	$ _____
Home Mortgage (loan balance)	$ _____
Other Real Estate Expenses	$ _____
Auto Loans (balance)	$ _____
Furniture Loans (balance)	$ _____
Home Improvement Loans (balance)	$ _____
Education Loans (balance)	$ _____
Other Installment Credit	$ _____
Credit Cards	$ _____
Other Outstanding Debts	$ _____
TOTAL LIABILITIES	$ _____
NET WORTH (Assets Minus Liabilities)	$ _____
% DEBT	$ _____
(Total Liabilities Divided by Total Assets)	
% LIQUID ASSETS	$ _____
(Total Liquid Assets Divided by Total Assets)	

CASH FLOW—THE PLUMBING OF YOUR FINANCIAL STRUCTURE

Knowing your net worth is one-half of the picture upon which your financial plan will be based. The other half is your cash flow. You will want to know just how much money you have left, if any, each month after all your regular expenses are paid. Worksheet III will reveal that to you, plus give you an idea of where you might squeeze out some extra savings to feed your investments.

Now, if you should come up with a negative cash flow, do not despair. You still qualify for the financial planning game. It's just that your goals are a little more predetermined by your circumstances than if you had some surplus. Obviously you want first to get to the point where you are financially breaking even on your life. From there you can start looking for some surplus so you can begin investing. A financial planner, or even a fiscally wise friend or relative, should be able to help you right your ship.

So keep playing here along with everyone else. Examine your cash flow worksheet in detail and set goals on the next worksheet, even though you have no idea where the money will come from to achieve them.

Worksheet III: Determining Your Cash Flow

Monthly Income

Earnings (Take Home)	$ _____
Moonlighting Income	$ _____
Investment Income	$ _____
Other Income	$ _____
TOTAL INCOME	$ _____

Monthly Expenses

A. FIXED EXPENSES

Savings and Investments	$ _____
House/Rent Payments	$ _____
Auto/Transportation	$ _____
Consumer and Other Debt	$ _____
Food	$ _____
Utilities	$ _____
Child Care	$ _____
Retirement Plan	$ _____
Other	$ _____
SUBTOTAL A	$ _____

B. VARIABLE EXPENSES (Enter annual total
   divided by 12 [mos.])

    Tax Payments $ _____

    Insurance Payments $ _____

    Household Repairs $ _____

    Auto Maintenance $ _____

    Medical Expenses $ _____

    Clothing $ _____

    Education $ _____

    Entertainment $ _____

    Charitable Donations $ _____

    Other $ _____

SUBTOTAL B $ _____

TOTAL MONTHLY EXPENSES

(Add Subtotals A and B) $ _____

TOTAL DISCRETIONARY DOLLARS $ _____

INFLATION PLANNER CHART

Inflation in recent years has run about 5 percent per year, modest indeed. As recently as 1979, admittedly an aberrant year, inflation was 13 percent per year. So at what rate do you figure inflation into your planning for the future? Well, are you an optimist or a pessimist? The following chart will serve you in either case.

Estimated Average Annual Inflation Rate Until Retirement

Years Until Retirement	5%	8%	10%	12%	15%
10	1.63	2.16	2.59	3.11	4.05
11	1.71	2.33	2.85	3.48	4.65
12	1.80	2.52	3.14	3.90	5.35
13	1.89	2.72	3.45	4.36	6.15
14	1.98	2.94	3.80	4.89	7.08
15	2.08	3.17	4.18	5.47	8.14
16	2.18	3.43	4.60	6.13	9.36
17	2.29	3.70	5.05	6.87	10.77
18	2.41	4.00	5.56	7.69	12.38
19	2.53	4.32	6.12	8.61	14.23

Years Until Retirement	5%	8%	10%	12%	15%
20	2.65	4.66	6.73	9.65	16.37
21	2.79	5.03	7.40	10.80	18.82
22	2.93	5.44	8.14	12.10	21.64
23	3.07	5.87	8.95	13.55	24.89
24	3.23	6.34	9.85	15.18	28.63
25	3.39	6.85	10.83	17.00	32.92
26	3.56	7.40	11.92	19.04	37.86
27	3.73	7.99	13.11	21.32	43.54
28	3.92	8.63	14.42	23.88	50.07
29	4.12	9.32	15.86	26.75	57.58
30	4.32	10.06	17.45	29.96	66.22
31	4.54	10.87	19.19	33.56	76.14
32	4.76	11.74	21.11	37.58	87.57
33	5.00	12.68	23.23	42.09	100.70
34	5.25	13.69	25.55	47.14	115.80
35	5.52	14.79	28.10	52.80	133.18

Figuring Inflation into Your Retirement Planning

Step 1: How much monthly income would you want if you were retiring today?

$ _____

Step 2: How many years until you plan to retire? Find the number on the left-hand column of the table above. What do you think the average inflation rate will be between now and then? Select a percentage figure from those across the top of the table. Find the figure on the table where your "Years Until Retirement" intersects with your estimated average annual inflation rate until retirement. Enter that figure here:

$ _____

Step 3: Multiply your dollar figure from step 1 by the figure from step 2 to see what monthly retirement income you'll need in future dollars. Enter that figure here:

$ _____

Do you see why we say a comfortable retirement means planning *now?*

LIFE INSURANCE—DO YOU NEED IT?

There is no absolute formula for figuring your life insurance needs, if in fact you need to have life insurance at all. (Remember from chapter 3 that only those whose loss would create a financial hardship for others need consider life insurance and that there are several alternative ways to meet those needs.) Estimating your insurance needs is best done with an expert, preferably a comprehensive financial planner rather than one who has a vested interest in selling you a lot of insurance. Standard recommendations of minimal coverage to purchase will range from five to ten times your annual income.

We have provided the following list of factors that go into this calculation so you can be an informed consumer:

Costs to Consider After the Insured's Demise

- "Final costs" (primarily deceased's funeral and unpaid medical bills).
- The percentage of your income (probably 60 to 70) needed to maintain everyone in your family on their current standard of living.
- College costs for the kids.
- Child care and housekeeping costs created by your demise.
- Inflation.

Cost-Reducing Factors

- Kids eventually grow up and earn their own incomes.
- Surviving spouse has an income or potential income.
- Surviving spouse eventually retires and receives a pension and/or Social Security.
- Your Social Security (if your spouse or minor children qualify to receive it).
- At some point in time survivors may be able regularly to withdraw some principal from the insurance policy as well as live off the interest.
- Any estate you have accumulated that your dependents will inherit.

Because there is so much variability in these factors, it's a good idea to reassess your insurance needs periodically.

QUALIFYING YOUR INSURANCE COMPANY

This being the 1990s—that is, the decade after the disastrously unregulated eighties—some of our once rock-solid insurance companies have turned gelatinous. During the previous decade, many insurance companies snatched up the same sorts of dicey junk bond and commercial real estate investments that had enticed their banking buddies. We're talking about a small minority of the industry here that are now suffering for their lack of prudence. Nevertheless, given that your insurance with an insurance company is not itself insured, it's a good idea to check out the creditworthiness of a company before committing to any of its products. A. M. Best is the rating agency best known for rating insurance companies, although Standard & Poor's, Moody's, and an outfit called Duff & Phelps rate them, too. A. M. Best's top rating is A+, which about 270 companies enjoy currently. You can look up the rating in the agency's publication, *Best's Insurance Reports: Life-Health Edition,* available at many major libraries. You can also ask your insurance agent for the company's rating.

Since ratings are not foolproof, reassure yourself further by querying any potential insurer on what percentage of its investment portfolio is invested in junk bonds. Steer clear of companies holding more than 10 percent.

Of course, if you bring your social values into the evaluation process, your selection process will be a short one. There is currently only one socially responsible company—Consumers United, the only insurance company that invests according to broad social criteria. One other will soon be available, formed by the SRI financial advisers' group First Affirmative Financial Network in cooperation with American Bankers Insurance Company. Tentatively titled First Affirmative Insurance Company, the company expects to offer term, universal, and whole life insurance backed by government agency bond investments by October 1991.

COLLEGE COSTS PLANNING

Kids provide their parents pleasures and satisfaction that nobody could put a price on. It's a good thing, too, because those pleasures and satisfactions do not come cheaply. Fortunately the biggest expense—college—comes right at the end of their childhood, so you have plenty of time to let compounding interest help you foot the bill. College expenses have been increasing at a *faster* rate than inflation, so starting a college account for little Larry when he's still in diapers is not too early. Use the following table for some guidelines:

The Cost of a College Education			
		Four Years*	
If your child is now: [years of age]	He or she will enter college in:	at a PUBLIC COLLEGE will cost:	or at a PRIVATE COLLEGE will cost:
15	3 yrs.	$25,895	$70,567
14	4 yrs.	$27,449	$74,802
13	5 yrs.	$29,095	$79,290
12	6 yrs.	$30,841	$84,047
11	7 yrs.	$32,692	$89,090
10	8 yrs.	$34,653	$94,435
9	9 yrs.	$36,732	$100,101
8	10 yrs.	$38,936	$106,107
7	11 yrs.	$41,273	$112,474
6	12 yrs.	$43,749	$119,222
5	13 yrs.	$46,374	$126,376
4	14 yrs.	$49,156	$133,958
3	15 yrs.	$52,106	$141,996
2	16 yrs.	$55,232	$150,515
1	17 yrs.	$58,546	$159,546

* Projected cost upon child's entrance to college for four years at a public or private college. Figures are based upon a 6% average annual rate of inflation applied to average annual total expenses reported by the College Board for the 1990–91 school year of $4,970 for a four-year public college and $13,544 for a four-year private college.

Source: College Board, 1991

By the way, you very well may want to take advantage of the Uniform Gifts to Minors Act in your college education investments. The act is discussed on page 323.

GOALS—HAVING EXACTLY WHAT YOU WANT BY PREPARING FOR IT

Now that you know where you *are* financially, you can begin homing in on where you want to *go*. Establishing goals is part wish list and part game plan. Having read this far, you are now a financially sophisticated person, so you realize that financial goals of any substance, such as maintaining your standard of living in retirement, do require a strategy. You also know that depending on how far in the future you have set your retirement or other goal, your cost of living may double or triple or more with even modest inflation. You know that you not only have to plan for inflation, but need to implement your plan ASAP to allow the mathematics of compounding to work its magic for you.

A significant portion of your goals, though, may have nothing to do with any goals we have advised, because only you know what you want. Far be it from us to make up your wish list for you. However, the following suggestions will help you with the pragmatic aspects of your goal setting, which in turn will make it more likely that the fantasies on your wish list can come true.

- Try to save/invest at least 10 percent of your income. One excellent way to do that is to authorize a socially responsible mutual fund to invest that money for you automatically by making a monthly withdrawal from your checking account. This method also automatically introduces dollar cost averaging into your investments.
- Reduce your debt to no more than 30 percent of your total assets. To figure your current debt percentage, take the "Total Liabilities" figure that you calculated on Worksheet I and divide it by the "Total Assets" figure from the same worksheet. If the figure exceeds 30 percent, you should include a definite plan to reduce it on your goals worksheet. Your long-term goal should be to be debt-free, or at least nearly so, at retirement. If early retirement is in your game plan, obviously you should reduce your debt that much more aggressively.
- Make sure that an adequate portion—normally about 25 percent—of your assets are liquid. You should normally have at least six months' worth of expenses available to you in the form of liquid assets, to protect you against a loss of job or other financial emergency. Refer to Worksheet III; multiply your "Total Monthly Expenses" by six and compare that figure to the "Total Liquid Assets" figure from Worksheet I. If you come up short, emphasize liquid investments in your plan until you have reached the appropriate ratio.

You may also want to consider some of the following on your goals list:

- **Short-term goals.** Pay down credit cards, summer vacations, new car, education and personal growth, cultural activities, and reserve fund for change of job/career.
- **Midterm goals.** Create additional income to reduce the number of hours you need to work, finance an expansion of your business, facilitate a new home purchase, provide help to retired parents, plan college educations for your kids, offer aid to your children for home purchase or other needs in early adulthood, subsidize home improvements, pay for special celebrations such as weddings and graduations, or establish an emergency fund.
- **Long-term goals.** Comfortable retirement, frequent travel in your leisure years, increased medical coverage for your later years, establishing an estate for your children and/or favorite cause.

In our experience, two of the biggest reasons for financial troubles later in life are the failure to set goals at an early stage, followed by failure to implement those plans that are made. Therefore, the following worksheet is not just some "airy-fairy" exercise. It is the map by which you can navigate your family toward a secure and comfortable life:

Worksheet IV: Financial Goals Action Planner

My Short-term goals*   Target Date
1. _____   _____
2. _____   _____
3. _____   _____
4. _____   _____

Estimated Cost   Already Saved†   Savings Monthly
$ _____ $ _____ $ _____
$ _____ $ _____ $ _____
$ _____ $ _____ $ _____
$ _____ $ _____ $ _____

My Midterm goals*   Target Date
1. _____   _____
2. _____   _____
3. _____   _____
4. _____   _____

Estimated Cost	Already Saved†	Savings Monthly
$ _____	$ _____	$ _____
$ _____	$ _____	$ _____
$ _____	$ _____	$ _____
$ _____	$ _____	$ _____

My Long-term goals*   Target Date

1. _____   _____
2. _____   _____
3. _____   _____
4. _____   _____

Estimated Cost	Already Saved†	Savings Monthly
$ _____	$ _____	$ _____
$ _____	$ _____	$ _____
$ _____	$ _____	$ _____
$ _____	$ _____	$ _____

* You realize by now, of course, that goals that are farther in the future must be figured in future dollars.
† Remember that the amount you need to save will be offset somewhat by compounding interest.

8

Building Your Financial Plan, Stage II—Smart Money, the SRI Way

You probably know at least a few people in your life for whom money never seems to present a problem. They appear to be on top of every financial detail in their daily lives, they live in admirable comfort, and they have planned for every possible future contingency. They also seem to know the most financially advantageous way to do things in almost every instance. In short, they are smart about money.

It really doesn't take much effort to be one of those people. You don't have to be obsessed with money to be smart about it. In fact, you don't even have to be smart. Handling money in your life is simply a matter of applying yourself to the topic.

This chapter is about all the little financial subtleties that "those other people" seem to know. With an SRI spin, of course. But don't wait to invest until you've mastered these things. If you have some investment capital now (or think an expert might help you create some from your other resources), turn right to the section "Finding a Financial Adviser," find one, and get rolling. You can read the rest of the chapter at a leisurely pace while your money is earning you solid, ethical returns.

TAX-ADVANTAGED RETIREMENT PLANS

You should give careful consideration to including some form of tax-advantaged retirement plan in your overall financial strategy. Wait, let's put that more strongly: Any money you feel comfortable leaving invested for several years should be invested through a retirement plan up to the maximum the IRS allows. Tax-deferred compounding, featured in all these plans, produces the same spectacular results of regular compounding. In some cases, your employer will match part or all of your contributions to a plan—and that, friend, is free money!

All plans detailed below have some rules in common:

- If you withdraw your money before age 59½, you will pay a 10 percent penalty (in other words, 10 percent of the amount withdrawn) unless you have become disabled since opening the account. For example, if you withdraw $1,500, you will pay ordinary income tax on the withdrawal plus an additional $150 in federal penalty. (Additional state penalties may apply.) Certain exceptions to the rules allow you to withdraw before age 59½ without penalty.

- You must begin withdrawing your money by April 1 following your 70½ birthday so Uncle Sam can collect his cut as taxes. You do not have to withdraw the entire amount at once. You can withdraw according to an IRS payment schedule that figures in factors such as life expectancy, marital status, and so on.

- Money compounds tax-free under each of these plans. Therefore never, never, never put your retirement plan contributions into a tax-free investment such as a municipal bond mutual fund. First of all, since retirement plan money is taxed upon withdrawal, you will have transformed a tax-free investment into a taxable one. Second, tax-free investments pay lower rates as a trade-off for tax benefits, and your retirement plan is already tax-advantaged (although not in the same way).

- Retirement plan funds are generally transferable from one plan to another, typically within sixty calendar days after closing an account and receiving the check. Be sure to consult an adviser before maneuvering, because certain movements are not allowed. For example, you are not allowed to roll over an IRA into a money purchase plan, but the reverse is allowed.

Individual Retirement Accounts (IRAs)

Eligibility and Contribution Limits

Individuals who are not covered by an employer-sponsored plan, plus those eligible pension plan participants whose adjusted gross income is below a given amount (according to a graduated IRS scale), may contribute up to $2,000 per year to an IRA. Contributions can be excluded from taxable income (for example, if you earned $24,000 last year but contributed $2,000 to an IRA, you will pay income tax only on $22,000). In a one-income family, additional contributions may be made on behalf of the nonearning spouse.

Types of Investments

Almost any type of investment can be used as an IRA account: certificates of deposit, mutual funds, annuities, stocks, bonds, and real estate investment trusts, among others. Types of investments generally forbidden include life insurance, precious metals, collectibles, and securities bought on margin.

Note: You can have as many IRA accounts as you wish, as long as the total amount contributed among them does not exceed the annual allowed limit. Therefore you could maintain both growth- and income-oriented IRA accounts.

Withdrawal

When you retire after age 59½ or reach age 70½, you have two choices. You can withdraw your IRA in a lump sum, in which case the entire amount will be taxed as ordinary income; or you can withdraw in installments, in which case the minimum withdrawal will be governed by IRS tables on life expectancy and taxed as received.

Other Rules, Features, and Strategies

- Your IRA contributions can be made from borrowed money.
- You can transfer your money from one IRA account to another without tax penalty as often as you wish, as long as you don't take possession of the money.
- Exception to the above: You can use your IRA money once per year without tax penalty, as long as you replace it within sixty days.
- If you expect your IRA to provide a significant portion of your retirement income, you will want to invest the money conservatively. How conservatively is a function largely of your age. If you are younger, you may want to risk some principal in exchange for realizing some appreciation, in which case stocks or a stock mutual fund would be a wise choice. (Of course, there are a wide range of

risk/appreciation ratios available within the stock universe, and you will probably want to look toward the conservative side of that spectrum.) If you are near retirement age, you should be more concerned with the safety of your principal than with growth, in which case a fixed income investment, such as a Ginnie Mae mutual fund, would be the most prudent selection. (See chapters 4, 5, and 6 for more on the various categories of investments.)

- If you are over 59½ and still earning wages, set up your savings account as an IRA. Your money will compound tax-free until you are 70½, and you can make withdrawals without tax penalty as well.
- Even if you do not pass the above-mentioned IRS tests for full IRA eligibility, you are still allowed to open an IRA account and contribute up to $2,000 annually. You will not be able to exclude your contributions from your taxable income, but your IRA money will compound tax-deferred. Stripping an IRA of its tax exclusion benefit weakens its power as an investment, of course. Depending upon your personal circumstances, you may do better by investing in tax-free municipal bonds or even selected taxable investments. Consult a financial professional for the best fit to your situation.

Ethical Considerations

Most of the socially screened mutual funds and money market funds allow you to set up IRA accounts. Other socially responsible options would include Ginnie Mae mutual funds and unit investment trusts, certificates of deposits at socially responsible banking institutions, socially screened stocks, socially responsible REITs, and socially screened corporate bonds. (See chapters 4, 5, and 6 for more on these vehicles.)

Salary Reduction Plans

Salary reduction plans are offered by some employers to their employees. With these plans the IRS will allow you to set aside some of your salary and not pay taxes on it until the money is withdrawn. In some cases employers will match or otherwise make additional contributions to your own; in terms of taxes, this money is treated exactly like your personal contributions.

The employer will ultimately determine how the money is to be invested, although you may have your choice of employer-selected options. In some cases your money buys shares of the company stock; in other cases you may have your choice of several mutual funds. You are allowed to withdraw the money penalty-free if you should retire, become disabled, or suffer financial hardship (some companies will permit you to maintain an active investment in the company plan if you should leave for another reason). In any case, you always retain the option of rolling over into an IRA the entire amount of your plan investment.

There are two categories of salary reduction plans, 401(k)s and 403(b)s:

401(k) Plans

Eligibility and Contribution Limits

IRS Tax Code Section 401(k) regulations are designed for employees of for-profit companies. If your employer offers a 401(k), you may elect to have up to 25 percent of your salary set aside—up to a maximum of $8,475 annually (1991 limit, indexed for inflation for future years). Your contributions are deducted from each of your paychecks by your employer and paid into the investment.

Types of Investments

Determined by employer.

Withdrawal

When you leave employment, you are allowed to roll over the proceeds of your 401(k) into an IRA without penalty. Should you withdraw your 401(k) as a lump sum of cash, you will receive more favorable tax treatment than with a similar withdrawal from an IRA. A lump sum IRA withdrawal is taxed as ordinary income. A lump sum 401(k) withdrawal is eligible for five-year income averaging.

Other Rules, Features, and Strategies
■ You can participate in both an IRA and a salary reduction plan for a maximum total tax exclusion of $10,475 per year ($2,000 IRA plus $8,475 salary reduction).

- If you can't afford to contribute to both an IRA and a salary reduction plan, the latter has the following advantages:
  1) You are allowed a higher maximum contribution.
  2) You receive more favorable tax treatment of a lump-sum withdrawal.
  3) Your employer pays all plan administrative costs.
  4) Your employer withdraws the contribution directly from your paycheck, so you are not tempted to spend the money before it is invested.
  5) You may borrow against accumulated salary reduction funds but not against IRA funds.
  6) Perhaps most important, many employers match part or all of your contribution. Why turn down free money?
- Under certain conditions in which a great disparity exists between what the company's top executives and lowest-level employees are paid, the company may be required to set up and contribute to employee accounts even when employees have chosen not to reduce any of their own salary through the plan.

Ethical Considerations

Your company may not offer an SRI option through its 401(k). If it doesn't, ask the plan managers if they would consider adding one. (Calvert, the largest of the socially screened mutual fund companies, now offers a 401(k). So does Dreyfus Third Century, another SRI fund. Use this book to help demonstrate the financial viability of making such an option available.) You might also examine the existing options to see if they include an ethically acceptable alternative, such as a Ginnie Mae mutual fund. However, Ginnie Maes are very conservative investments that may not be a good fit if your retirement is a long way off.

403(b), AKA TSAs, Tax-Sheltered Annuities, or Tax-Sheltered Accounts

Eligibility and Contribution Limits

These plans are designed for individuals employed by nonprofit organizations such as schools, hospitals, and charitable foundations. The 403(b) plan permits salary reduction of about 20 percent of gross income (according to a formula) up to a limit typically of $9,500 per year. (There are exceptions to the maximum: for example, under certain conditions, you may make greater contributions if you are in your last years of working.)

Types of Investments

As with other salary reduction plans, 403(b) investments are determined by the employer. Traditionally, the overwhelming majority of 403(b) investments have been in annuities managed by insurance companies that have qualified for such plans with the government. Some nonprofit employers' plans may give you a choice of mutual funds as well.

Withdrawal

As with 401(k)s, when you leave employment you are allowed, without penalty, to roll over the proceeds of your plan into an IRA. However, you will not receive the same income-averaging tax treatment that 401(k) plan participants enjoy should you choose to withdraw your money in a lump sum.

Other Rules, Features, and Strategies
- You are allowed to participate in both an IRA and a salary reduction plan. Your total tax exclusion probably will be about $1,000 higher with a 403(b) than is possible with a 401(k) because of the greater tax exclusion allowed under 403(b)s.
- If you can't afford to contribute to both an IRA and a 403(b), the latter has the following advantages:
    1) You are allowed a higher maximum contribution.
    2) Your employer withdraws your contribution directly from your paycheck, so you are not tempted to spend the money before it is invested.
    3) The IRS allows you to borrow against accumulated salary reduction funds but not against IRA funds. If your 403(b) funds are in an annuity, most insurance companies will permit you to borrow against the annuity to purchase a residence or in the event of a personal emergency or hardship. Normally you will be charged an interest rate about 2 percent over the inter-

est the company is paying to you (for example, if your annuity is earning 8 percent, you will pay 10 percent interest if you borrow against it). The loan is not taxed, but it must be repaid—generally within five years, or within thirty years for a home loan. If your 403(b) money is in a mutual fund, you may not be able to borrow against it.

Ethical Considerations

The insurance industry is almost entirely devoid of companies that maintain socially screened investment portfolios. Therefore you may be unhappy with most annuities available through 403(b) plans. There are some notable exceptions. TIAA-CREF (Teachers Insurance Annuity Association—College Retirement Equity Fund), a major player in the 403(b) field, has recently begun offering a socially screened variable annuity called the Social Awareness Fund. The annuity is invested in a socially screened stock mutual fund. If you are a teacher, check to see if your employer sponsors a 403(b) through this company. State Bond Mutual (SBM) Life Insurance offers a fixed annuity that is about 75 percent invested in Ginnie Maes, with the balance in private mortgages and commercial paper. The company has promised to First Affirmative Financial Network, an association of social investment advisers now offering SBM's annuity, that it will maintain a socially responsible profile for this product. Pax World Fund, Dreyfus Third Century, and Calvert Social Investment Funds offer 403(b) plans. As of this writing, First Affirmative is developing a fixed annuity invested in ethical government agency securities which will qualify as a 403(b) investment. If you work for a nonprofit organization and your employer does not offer an SRI option, ask the plan manager to investigate one of the options just mentioned or inquire about other socially responsible products described in this book.

Retirement Plans for Self-Employed Persons and Owners of Small Companies

The most significant rule governing these retirement plans for the self-employed is that if your business has employees and you are setting aside company earnings for yourself, you must set up accounts for them (according to certain eligibility rules) and contribute the same percentage of their salaries as you do a percentage of your earnings.

For example, if you set aside 10 percent of your business's earnings, you must contribute an amount equal to 10 percent of employees' salaries to their accounts. You the employer can then deduct those contributions as a business expense. Generally speaking, you are required to make contributions only for eligible employees (employees who meet age and salary requirements) who have completed two years of service, although some employers choose to fund employees earlier in their employment.

Simplified Employee Pension IRA (SEP IRA)

Eligibility and Contribution Limits

SEP IRAs serve the needs of self-employed persons (including the part-time self-employed who report self-employment income to the IRS) and owners and employees of small companies.

As a self-employed person or small company owner, you may set aside about 15 percent of profit (according to a formula). You are allowed to set aside profits one year and not the next. (For example, if you have no business profits the year after you begin your plan, you do not have to set aside any money in your account that year or contribute to employee accounts.) A maximum of $30,000 can be set aside in any SEP IRA participant's (employer or employee) account.

Types of Investments

The same rules apply as for IRAs.

Withdrawal

Same as IRAs.

Other Rules, Features, and Strategies

Same as IRAs.

Ethical Considerations

Same as IRAs.

Money Purchase and Profit-Sharing Plans (Formerly Called Keogh Plans)

Eligibility and Contribution Limits

Any company or self-employed person (including the part-time self-employed) may set up one, or a combination, of these plans. Don't get caught up in guessing what is meant by each plan's title—just understand the general rules. Under a money purchase plan, you can set aside anywhere from 0 to 25 percent of earnings, up to a limit of $30,000 annually per participant. You inform the IRS of the level you choose, and the plan stays at that level until you arrange a change. The catch here is that once you have initiated a money purchase plan, you must fund it every year thereafter.

Profit-sharing plans, conversely, can be funded or not after the first year, entirely at your choice. However, the set-aside range is less than for money purchase plans—0 to 15 percent of profit—although the maximum per participant is still $30,000. You can change the percentage level of a profit-sharing plan without notifying the IRS, so there is less administrative overhead cost with a profit-sharing plan.

Some companies or self-employed individuals choose a third option—a combination money purchase/profit-sharing plan. Set-asides in the combination are limited to a maximum of 25 percent of earnings or $30,000 per participant, whichever is less. The maximum set-aside for the profit-sharing side of the combination is still 15 percent of profits. Thus 10 percent money purchase/15 percent profit-sharing is a popular combination. See the following for more on this strategy.

Types of Investments

Any investment eligible for an IRA is also eligible for one of these plans.

Withdrawal

Same rules as for IRAs.

Other Rules, Features, and Strategies
- You may still contribute the maximum allowable to an IRA in addition to your participation in a money purchase and/or profit-sharing plan.

- Probably the most common combination plan is a 10 percent money purchase/15 percent profit-sharing program. This mix gives you the flexibility to set aside the maximum 25 percent (again, up to $30,000 per participant), but only 10 percent of the set-aside is mandatory.
- You can save some money by running your plan through a mutual fund company as opposed to investing plan contributions in a self-directed investment portfolio. Mutual funds generally charge far less for administering pension plans than do private custodial services.
- Some employers start new employees on a money purchase and/or profit-sharing plan rather than waiting for the two-years-of-service eligibility requirement to kick in. Often this is done in lieu of a competitive salary, with the difference made up by plan contributions. As an employee, realize that you will probably not have full rights to this money if you leave the company for any reason in your first few years of service. When companies put new employees on a plan immediately after hiring, the rules change and what is called a *vesting* schedule comes into play. Vesting refers to the percentage of your plan account to which you are entitled. You could be as little as 20 percent vested after two years of service, and the company is not required to vest you 100 percent until you have been there six years.

Ethical Considerations

See discussion under IRA.

Other Retirement Plan Options Noted Briefly

Salary Reduction SEP IRAs (AKA SR SEP IRAs)

For employers who own small companies. An alternative to employer-sponsored plans detailed previously. Rules are similar to those for 401(k) plans. For more information, see IRS Publication 590.

Defined-Benefit Program

For business owners who want to set aside more than $30,000. The 1991 law allows over $108,000 to be set aside annually and is indexed for inflation for future years. Withdrawals can be arranged based on current earnings and IRS longevity tables to set up a predictable income for retirement. These plans are expensive to administer but worth it if they're a good fit with your circumstances. Consult a pension specialist if you want to know more about this option.

PLASTIC FANTASTIC—SRI CREDIT CARD OPTIONS

There are two truly socially responsible credit cards, available through Working Assets and Co-op America (the latter available to Co-op America members only—$20 annual membership). Contact information on these is provided at the end of this section.

Several organizations that themselves serve socially responsible functions offer credit cards. Your favorite environmental association, religious group, or fraternal organization may have already asked you if you would consider switching to its card. Before selecting one of these "affinity" cards, keep the following important fact in mind: Most such credit card services are offered in association with banks; even though a portion of the profits may be donated to a cause that you support, the overall credit operation may be financed by banking investment policies that you do not. For example, the National Wildlife Federation (NWF) receives a small percentage of the profits generated from its cards. But the operation is run by Maryland National Bank, which can afford to give NWF its little cut because Maryland National keeps the balance of the extremely profitable near 20 percent interest that it charges all of its customers (not to mention the several additional percentage points that it charges the merchant). Maryland National is not an SRI institution. It takes its credit card profits and invests them according to strictly financial criteria.

Socially responsible institutions like credit unions and minority-owned banks also offer credit cards. The card you obtain through a credit union will probably carry a considerably smaller interest rate than you can obtain from another institution because the credit union card is nonprofit. A credit union card is compromised ethically, however, by the fact that it too is run through a clearinghouse bank probably without regard for ethical criteria.

Resources

Co-op America
2100 M Street NW
Suite 403
Washington, D.C. 20063
(800) 424-2667

Working Assets
230 California Street
San Francisco, CA 94111
(415) 989-3200, (800) 533-3863

STARTING A SOCIALLY RESPONSIBLE INVESTMENT CLUB

Investment clubs are a way to play the SRI game with limited resources (although as you'll see in a moment, there are benefits to club membership to which any investor would want access). Some investments have fairly substantial "entry fees": corporate bonds are typically $1,000 a pop, as are unit investment trust opening investments; a diverse stock portfolio will cost in the low- to mid-five figures if the stocks are bought in round lots; and even many of the SRI mutual funds require opening investments of $1,000 to $2,500. If those numbers are currently out of reach for you, an alternative approach (among others that we have detailed throughout this book) is to gather a group of your friends, relatives, or work or volunteer organization associates and form a socially responsible investment club. If twenty of you throw in $50 per month, the club will have $1,000 per month to invest. With that kind of capital, the club can make some serious profits for everyone. Not that your club needs to operate at that level—even ten members putting up $10 per month is a worthwhile endeavor.

Clubs can also be a good source of emotional support for your investment decisions. First of all, a club can be a mechanism for educating its members. You can invite speakers to your meetings (many financial experts are glad to speak without charge), you can work as a group with a financial planner who will help educate you, and the group can subscribe to investment publications, including SRI advisory publications like *Clean Yield* and *Insight*.

One of the best things about investment clubs is the nonprofit organization designed to serve them—the National Association of Investors Clubs (NAIC), founded in 1951, which assists clubs both educationally and financially. The NAIC offers materials on subjects ranging from information on starting a club to methodologies used to research a stock. The association also sponsors investments in more than sixty stocks, allowing members to invest directly in those stocks without paying brokerage commissions. By cross-checking the NAIC list of stocks with lists of socially screened stocks in this book, your club members can put together a socially screened portfolio at a lower cost than any individual member could.

Traditionally typical clubs have between twelve and twenty members and meet monthly to review past investments and vote upon new ones. Members normally contribute $10 to $50 per month. The monthly contribution can vary, of course, with the new investment decided upon. We suggest that you recruit your membership from groups of like-minded people, since social criteria are among the issues to be decided upon. Possible sources might include your church or synagogue, a union, your place of employment, or a volunteer organization to which you belong.

For more information, contact National Association of Investors Clubs: 1515 East Eleven Mile Road, Royal Oak, MI 48067 (annual membership for individuals, $32; for investment clubs, $30 per club plus $9 per member).

PARKING MONEY WITHOUT A VALET

"Parking money" means keeping it in a safe, liquid place until the best opportunity to invest it more permanently presents itself. If you are house hunting, you will want to park enough money for the down payment and closing costs so you can leap when you find *the* house. If you have just inherited a large amount of cash, you will want to keep it in some high-interest parking places while you make up your mind on how to invest it long-term. In volatile economic times, many investors like to park their funds in cash equivalents like money market funds while waiting for the more lucrative markets to show some definite trends.

As we hope you realize by now, bank (or credit union) passbook savings accounts are not sensible parking places unless you absolutely have to have that deposit insurance to sleep at night. Most experienced investors choose to park in money market funds these days. You can access your money market deposits immediately with a check, you don't have to pay a commission to get your money in or out, and the rates are as high as you can get with that type of liquidity. Besides, money market funds are as safe historically as banks, even though the money is not insured.

If the amount you have to park is very large ($25,000 or more), you may get slightly better rates in a CD. This strategy only works, of course, if you know you will not be needing the money until the CD's term is up.

One other parking option offered at some banking institutions is the so-called liquid CD. These are as liquid as money market accounts but generally pay lower rates, the trade-off for their deposit insurance. One other drawback is that most liquid CDs have rules limiting the mobility of your money (high minimum balances, limited number of withdrawals per month, and so on).

TAX PLANNING (BEYOND CIRCLING APRIL 15 ON YOUR CALENDAR)

Given that you gotta pay 'em, the most positive way to view taxes is as another way to make money. Anything you can do to cut your tax bill is money in your pocket that would not otherwise be there. Now don't you feel better about taxes already?

Tax planning is vital to your overall financial planning. If you are in the higher tax brackets, you need to know at what point it is in your interest to be in tax-free investments or other tax-advantaged programs. Even if you are not in those brackets, it makes no sense to pay more taxes than the minimum the government is asking from you. And because the government has made it almost impenetrably complex to figure out what that minimum amount is, you will almost certainly need the help of a tax professional.

The authors of this book are not tax professionals. Nor do we intend this book to be a detailed tax planning guide. The following information is general and applies to most taxpayers. If you are in the higher tax brackets, you should consult with your broker and accountant, enrolled agent, or other tax adviser about possible tax-sheltering investments.

Some steps you can take to defer income:

- You can open an IRA, Keogh, or other similar tax-deferred account if you have not already done so. This will enable you to reduce adjusted gross income and generate tax-deferred interest income instead.

- You can contribute to a 401(k) or 403(b) plan if your employer offers one.
- You can purchase CDs that will mature in the next tax period.
- You can transfer money to your children under the Uniform Gifts to Minors Act (UGMA). You can still control the money by investing it for your children under their name and Social Security number and registering yourself as the custodian. The money will compound at a lower tax rate than yours, making UGMA accounts ideal for college planning. Specifically, for children fourteen years of age or younger, the first $500 of income is tax-free; the next $500 is taxed at 15 percent. Income over $1,000 is taxed at the parents' rate. Income for children over fourteen will be taxed at a rate typically lower than their parents' (depending upon how much income is generated by the account). Up to $10,000 a year per child can be given by each parent without gift tax.

Some steps you can take to increase your legal tax deductions (if you can itemize your deductions in a given year):

- Make your annual charitable donations no later than December 31 of the tax year.
- Pay both halves of your property tax payment before December 31 (if you pay property taxes directly instead of through an escrow account with the lender of your home loan).
- Give old clothing and other unwanted items to charity no later than December 31. Be sure to keep an itemized receipt of your donation (value the items at the price they would sell for in a thrift store).
- Try to lump medical payments within a single year because medical payments must exceed 7½ percent of your adjusted gross income to be deductible.

By the way, you may be wondering how to figure out whether a tax-free investment, which pays far lower returns than comparable taxables, makes sense in your bracket. The math is easy:

(Tax-free interest rate) divided by (one minus your tax bracket percentage) or

$$\frac{\text{Tax-free rate}}{1 - \text{tax bracket}}$$

Example: Barbara is considering purchasing a double tax-free municipal bond paying 8 percent. She is in the 28 percent federal and 7 percent state tax brackets, and since this bond is double tax-free, she will need to figure in both numbers. The calculation is

$$\frac{8\%}{1 - (28\% + 7\%)} \quad \text{or} \quad \frac{.08}{.65} \quad \text{or} \quad .1230769 \text{ (about 12.3\%)}$$

This means that in Barbara's tax situation, purchasing this 8 percent bond has the same financial impact as purchasing a 12.3 percent taxable investment and paying the 35 percent tax on her income from it. This bond sounds like an excellent deal for her, since 12.3 percent income investments of safety comparable to the muni bond are not to be found in the current, or foreseeable future, market.

The table below gives you a sense of what a tax-free investment can do for you if you are in the combined tax bracket shown.

Tax-Free Investing

Combined Federal and State Tax Rate	Tax-Free Yields						
	4.5%	5.5%	6.0%	6.5%	7.0%	7.5%	8.0%
	Equivalent Taxable Yields						
34.7%*	6.89%	8.42%	9.19%	9.95%	10.72%	11.49%	12.25%

* Brackets for 1990 federal and California taxable incomes (single return: $26,380–$47,050; Joint return: $52,760–$78,400).

FINDING A FINANCIAL ADVISER

Have you decided that you could use a little professional assistance with your investment decisions? Following are a few basics to consider in your search.

What's in a Name? How Financial Advice Is Packaged

Once you start looking for financial advice, you will notice that it is everywhere and that it comes in a variety of packages. Since there are no government licenses or certifications regulating who can offer this advice, anyone can legally call him- or herself a financial adviser or the analogous terms *financial planner, financial consultant, investment adviser,* and so on. You will also notice a number of insurance agents, real estate agents, tax professionals, attorneys, accountants, and stockbrokers advertising financial planning services. Some of them will be well qualified to help you, with a broad base of experience and knowledge behind their claims. Others are simply offering an auxiliary service to the captive audience that came to them for their primary expertise.

There are titles in the field with real meaning, however. None of the following are guarantees of the professional's competence and experience, so investigate thoroughly the background of any adviser you choose:

- **CFP (certified financial planner).** Someone who has completed six courses given by the College of Financial Planning in Denver, Colorado. Not associated with SEC licensing.
- **ChFC (chartered financial consultant).** A graduate of a program offered by the American College in Bryn Mawr, Pennsylvania. The program focuses on insurance-based financial planning. Also not associated with SEC process.
- **CPA (certified public accountant).** Someone who holds a state accounting license after having met educational and professional requirements. Usually has undergraduate and/or graduate degree in accounting or related curriculum. Many CPAs offer financial planning services but must be licensed separately by the SEC to sell investment products.
- **Enrolled agent.** Tax specialist who has passed an examination administered by the U.S. Department of the Treasury on applications and interpretations of the IRS Code (can alternatively qualify by having worked with IRS for five continuous years, applying code). Only professional besides CPAs and attorneys authorized to represent all classes of taxpayers before the IRS.
- **CLU (chartered life underwriter).** Insurance specialist who has completed an advanced program at the American College in Bryn Mawr.
- **LIC (licensed insurance counselor).** An insurance specialist who is licensed to charge fees for insurance advice, as opposed to making commission-based sales. LICs have also completed advanced study in insurance topics.
- **RIA (registered investment adviser).** Someone who is registered with both the state and the SEC to hold securities for another person and to charge fees for investment advice. A registration only, certifying that the registrant has met requirements; not a guarantee of competence.

- **Registered representative.** Someone licensed by the SEC to represent specific types of products (such as mutual funds, stocks, bonds, limited partnerships, and so on) and sell those products for a commission. There are no specific educational or background prerequisites for taking the licensing exam.

Titles other than the above are probably self-granted, which is not a slight on the person so titled. A self-titled person may be the most qualified financial adviser you could find. Again, a thorough investigation of your adviser's experience and track record is your best guarantee of quality.

Finding Good Advice on SRI

Does your adviser need to be an SRI specialist to advise you on SRI products? Not necessarily, but obviously it is preferable. An excellent place to start your search for an SRI specialist in your area is *A Guide to Social Investment Services,* published by the Social Investment Forum. The forum incorporated in 1985 as a national association of financial professionals, researchers, community organizations, publishers, and investors who wished to participate directly in developing the concept and practice of SRI. The guide lists the forum's professional members and the services they offer. (The authors are members of the Social Investment Forum, and Jack Brill is listed with those offering financial services.) You can contact the forum at Social Investment Forum: 430 First Avenue North, #290, Minneapolis, MN 55401; (612) 333-8338.

If you live in a smaller city or town, however, there may not be an SRI specialist in your area or even someone with a passing knowledge of SRI. At the end of this section we will show you how to make socially responsible investments with a broker who knows little or nothing about them.

Finding the Right Adviser for You, Part I—How Financial Planners Are Paid

Regardless of whether you plan to do your socially responsible investing through an SRI specialist or through a conventional adviser, there are several elements to consider in your search. The first concerns the way your adviser is compensated, because that may affect the type of advice you can expect to receive. There are essentially four ways for people in this business to make money.

1) *Commission-based advisers* get paid only when they sell a client a product (the commission is generally included in the sale price). Most advisers are compensated this way; the advice they offer is free. Obviously an honest commission-based adviser will only recommend buys that are in your interest. If your financial circumstances indicate that you should have all your money in a CD that pays no commission, that is the advice an honest adviser will dispense. Commissions paid to an honest, competent adviser is money well spent, because the long-term progress of solid investments will make that initial commission expense seem insignificant.

   Inevitably, of course, there are unscrupulous sorts who will always recommend a commission-paying product in circumstances where a no-commission product like a money market fund or CD would be more appropriate. Your best protection against this possibility is to know the basics of financial planning.

2) *Fee-based advisers* charge a basic or hourly fee for financial advice. They do not sell products, so they have no vested interest in the advice they give. That may make you feel more secure about their recommendations, but at $50 or $100 or more per hour, the price for your peace of mind can add up quickly.

3) *Combination fee/commission advisers* will typically charge you hourly for advice and then allow you to credit commissions owed for making recommended investments against your fee account. For example, such an adviser may charge you from $250 to $1,000 for putting together a personalized financial plan. If you make the investments recommended on the plan through the plan's designer, commissions are credited against the plan fee.

4) *Discount brokers* will purchase stocks and bonds (not mutual funds, which can't be discounted) for you at commissions of 30 to 70 percent less than regular commissioned, or "full-service," brokers. The trade-off is that they do not offer advice—they only buy and sell investments. It is also not uncommon for discount brokers to charge minimum fees that override their discounts on small transactions. Sophisticated investors who know what they want and who track their own investments keep these people in business, and rightly so. But most investors lack either the time or the knowledge to make effective use of these professionals. (Some full-service brokers discount off their published rates for stock and bond trans-

actions, although not to the level of discount brokers. Shop for a good balance of service and rates if you choose to go in this direction.)

Finding the Right Adviser for You, Part II—The Interview

Financial advisers expect to be interviewed by prospective clients. They know that if they are to be trusted with important decisions about your future, they will have to establish considerable trustworthiness with you. So don't be shy about interviewing an adviser, and don't feel obligated to buy anything just for taking up some of his or her time. It is also perfectly acceptable to walk into a financial adviser's office unannounced as long as you realize that the broker may be with a client and unable to see you. Here are some of the questions you will want to ask in your interviews, with explanatory notes where relevant:

- How are you paid?
- What is your training? (A college degree in business or finance? Licensing in real estate, law, insurance sales? A professional certification?)
- How many years have you been doing financial planning (not just stock or insurance sales)?
- How many middle-income clients do you have? (You don't want to get less attention than your planner's stable of wealthy investors.)
- Are you an official representative for any particular products? (You want an adviser who will offer you a broad array of products.)
- What is your expertise in taxes? insurance? employee benefits? (This is not a disqualifying question, but it will cue you as to what other sorts of advice you may need as a supplement.)
- Do you consider yourself a conservative or aggressive adviser?
- How long have you been involved with socially responsible investments? (Don't expect a ten-year-long track record here, but you may want a better answer than "Since I read it was a hot market in Tuesday's *Wall Street Journal*.")
- What percentage of your clients are involved with socially responsible investing?
- How do your social accounts perform relative to your nonscreened accounts? (There is no good reason for socially screened accounts performing more poorly than others.)
- Does your firm support socially responsible investing? (If not, why are you there?)

- How do you do your research on socially responsible investments?
- How will you ensure that my ethical investing criteria are met?
- What are your references?

After you finish your questioning, do not be alarmed when your prospective financial planner turns things around to ask you some detailed questions about your personal finances and goals. A professional needs to know if he or she can work with you, too. In fact, any financial planner licensed to sell investments for a commission must *by law* determine if you are qualified financially to take the financial risks involved with the investments offered to you and evaluate as well if you are qualified intellectually/emotionally to understand financial advice. Don't be embarrassed to admit, though, that your resources are meager. Most financial planners will be eager to establish a relationship with you just for the prospects of future investing.

Finding the Right Adviser for You, Part III—Better Living Through Chemistry

The personal chemistry between you and your financial adviser may turn out to be as important a qualifying factor as any of his or her paper credentials, so after each of your interviews, ask yourself how you felt with that individual. You're going to be sharing intimate details about your financial condition with your adviser. And there will be other frank talk, or at least there should be, because you want your adviser to feel comfortable enough to yank you back to reality when you are about to make a silly or panicky financial decision.

So put the same care into finding the right adviser for your financial needs as you would in finding a doctor or dentist for your family. In fact, you can almost consider your adviser part of your family's health care team, because financial health is an aspect of total health. If that sounds like stretching a point, consider the health consequences of the stress that can result from financial woes.

Keep in mind that you don't need to find a financial wizard out there. An adviser with average professional abilities will do, because the progress of an average investment portfolio will gain you the financial security you seek. It is far more important that your adviser be an exceptional *person*—someone with integrity and good interpersonal skills. The hand that holds yours when you make crucial financial decisions should be both warm and firm.

Red Flags

Advice you should avoid is often easy to spot. Among the telltale signs:

- **Brokers who make outlandish claims of above-market returns or other quick-and-easy money.** With rare exceptions, wealth is built through the patient accumulation of market-rate profits. Any adviser who tells you otherwise is taking you for a ride—to the soup lines.
- **Brokers who promise you that a particular investment "can't miss."** There are no sure things in investing.
- **Brokers who want to sell you something without a detailed knowledge of your financial circumstances.** No investment is right for everybody.
- **"Limited time offer" sales pitches.** No investment decision should be made without careful study and consideration.
- **"Churners and burners."** These are brokers who advise you to sell one investment and buy another with alarming frequency. The rapid-fire turnover in investments lines their pockets with commissions while devouring your profits or even principal. In most cases you should be paying commissions only on investments that you expect to hold a number of years.

Getting SRI Advice from a Non-SRI Adviser

Most of what you are looking for in a financial adviser is the same whether or not the adviser knows SRI. You can do the rough sorting of investments for their social worthiness by using resources listed in this book—from there, any competent financial adviser can evaluate a vehicle's worthiness as an investment. In any event, though, you will need to be more actively involved with the social screening of investments than you would with an SRI specialist.

One situation that you do want to avoid is using an adviser who is antagonistic to the principles of SRI. That could be the source of a number of problems, not least of which is the introduction of some toxic elements into your professional relationship.

DEALING WITH AN INHERITANCE AND OTHER WINDFALLS

Inheritances, despite their ability to enhance your financial security, have a curious tendency to undermine the security of your emotional state. If you are naive about sophisticated financial dealings, you may find the lucrative investment portfolio you inherited from Grandpa Louie upsetting, as if you were suddenly made responsible for a dozen small children screaming for your attention. As a socially responsible investor, you may find yourself the embarrassed owner of Louie's extensive holdings of Thermonuclear Unlimited stock. Yet because Gramps was so dear to you, you might feel uncomfortable about selling those shares, as if doing so would destroy the little bit of worldly presence remaining of the guy.

The first step with any inheritance is to understand exactly what you have. And what you have will almost certainly have its debit as well as credit side in the form of legal expenses, inheritance taxes, and possibly other obligations attached to those assets. You need to be brought up to speed on those debits and obligations, because you may need to liquidate part of the inheritance to handle them. The deceased's attorney and accountant should be able to help you here; perhaps relatives can, also.

Your financial adviser can evaluate assets and give you a complete rundown on their quantity and value. If this is the first time you are using an adviser, you will find out right away just how good a communicator and educator this person is, because the evaluation should come wrapped in a first lesson on financial planning. Do not make any decisions until you understand what you are doing, and do not commit anything to long-term investments until you have allowed for your financial obligations as an heir.

Dealing with an inheritance in a socially responsible manner is one circumstance where your ethics can cost you some real money. Grandpa Louie may have been a sweet, loving man, but a lot of sweet, loving men and women invest without any regard for social implications. That leaves you with the difficult choice of churning and burning your own investments at great cost in commissions or compromising your ethics in this exceptional case.

If you decide to liquidate part of the inheritance for ethical reasons, be sure to park it in a high-interest account like an SRI money market fund until you select a new home for it. But try to find that new home ASAP. Remember, you paid some commission for your principles. You can minimize that loss by finding an ethical investment right away that will earn you market-level returns.

Windfalls other than inheritances are a terrific opportunity to put a substantial amount of money to work compounding interest. After you have allowed for taxes and other obligations, put your new money on the job immediately.

Before we leave this topic, a true story will illustrate the importance of dealing with your inheritance or windfall not only responsibly but also *promptly:*

Hal Brill, who did much of the investment research for this book, recently began working with a client named Janna. Janna, now twenty-five and a single parent, inherited about $170,000 some five years ago. Although socially conscious, Janna knew little about investing and nothing about SRI. Figuring she should do something sophisticated with that much money, she enlisted a mainstream broker to compile a portfolio for her. He selected some blue-chip corporate bonds—not socially screened, of course. Uncomfortable with the social implications of her new investment, Janna instructed the broker to sell the bonds just days later. Since she was still unsure of the SRI alternative and more nervous about the high-finance world than ever, she neglected to follow through with another financial plan. Instead she parented her child full-time, living off the inherited principal. By the time she found Hal, she was down to her last $15,000. With Hal's advice that money has been invested in an SRI income vehicle. The tragedy, of which she is now well aware, is what could have been. "If I had met you earlier," she told Hal, "I would still have all of my inheritance, I would be making a comfortable living, and I would feel good about myself. Please dedicate part of your book to me so others won't make the same mistake." Thanks for sharing your story, Janna. This section is dedicated to you.

HOME OWNERSHIP AND OTHER REAL ESTATE

If you are interested in direct ownership of real estate investments (as opposed to REITs and limited partnerships), we suggest you consult other sources that specialize in this topic. But because real estate ownership, particularly home ownership, is a major consideration in most financial planning, we will cover it here in a general way.

Owning Your Own Home

You have probably heard it said in many places and by many people that owning your own home is the best investment you will ever make.

Those places and people were probably all in California before 1990—when residential real estate appreciated at a dizzying rate. One of the authors of this book saw his home triple in value in nine years.

But recently the market has actually backslid a little. Many experts call this a "correction"—by that they mean that the market is simply returning to a more appropriate level after overshooting buyers' ability and willingness to pay. Other experts blame a general economic slowdown in the United States for the decline. Whoever is right, the market reversal only confirms that there are no sure things in investing, including home ownership.

The prevailing "wisdom" about home ownership is based on several factors, including supply and demand (everyone needs a place to live, and there are only so many of them) and the fact that the interest you pay on your mortgage is tax-deductible. "Leverage" is another plus—you can usually finance 80 percent or more of your purchase at interest rates that are low compared with other financing (especially when you figure in the tax break). You can also increase the value of your property with home improvements.

All these things are true. Now, consider the "buts." If the neighborhood deteriorates, so may the value of your house. Yes, your mortgage interest and property taxes are deductible, but even considering those, your house payment may be greater than you would pay in the rental market. Besides, not everyone needs that tax break. And if you rented, the cash that would otherwise go into your down payment on your home is money you could invest. The income that cash could produce has to be figured into the renting vs. owning equation. So does the fact that your landlord will probably handle landscaping and maintenance expenses; some landlords will pay for upgrades and utilities as well. When you own, all property expenses are out of your pocket. Finally, houses are an illiquid investment—in a slow market you could be waiting for half a year or more to find the right buyer.

So should you rent or buy? For a strictly financial answer, start by comparing your after-tax costs of buying to what you would pay in rent, then calculate the other factors mentioned already. But if you are like most Americans, this is not strictly a financial question, is it? Owning a home is part of the American dream; it gives you a greater sense of control of your life (at least until "they" start jacking up the property taxes), and there is a lot more immediate pleasure in home ownership than in buying shares in a mutual fund. In a normal real estate market, even with all the caveats we spoilsports have mentioned, you will do at least okay.

If you do decide to buy a home, you can increase your chances of realizing a good profit on your home if

- **you buy a fixer-upper that is priced like one.** Assuming you can do the repairs yourself, the costs of repairs should be significantly less than the price difference between yours and a comparable "turn-key" home. If not, the fixer-upper is overpriced.
- **you buy one of the lesser houses in the neighborhood.** The fancier homes in the area will pull up the value of yours; conversely, the best house in the neighborhood will be pulled down in value by the others.
- **you plan to keep the house for three years or more.** Transaction costs in real estate are expensive and will swallow your profits if you pay them too often.

The Ethics of Real Estate Investments

Direct ownership of investment properties has been the foundation of many fortunes. There are significant risks involved, too. It is far beyond our expertise to advise you in this area, other than to highlight some of the ethical issues involved.

We'll start with commercial property (that is, buildings and other property leased to businesses). First off, think of all those "For Rent" signs that you have seen in vacant offices and storefronts for months on end. Each one represents a building owner sweating bullets. Generally, when you buy a commercial building, you are planning for a positive cash flow. At worst you need enough lease income to cover your loan payments if you are going to be able to hang on to the building. So commercial tenants are an extremely valuable commodity to you.

Now, suppose the only business that has offered to lease your long vacant front space is Toxic Waste Dumping Lobbyist Associates. Well, you get the idea.

Vacancies are not quite as serious in residential real estate, because the demand situation is different. (Good times or bad, everyone needs a home—our deplorable homeless situation notwithstanding—but in bad times fewer will open new businesses, and some will close existing ones.) You can usually find a tenant for your vacant apartment by dropping the rent a few bucks a month. But what if Mr. Benton down the hall just lost his job and can't pay rent at all? Are you going to toss him and his young family of four to the wolves while you look for a paying tenant? And how do you feel about rent control? Being a landlord can do funny things to your political sympathies, especially if regular rent increases are the only way you can keep your building or feed *your* family.

SAVINGS—FINDING THAT LITTLE EXTRA

"You can't squeeze blood from a turnip," goes the old saying. You *can* usually find extra cash in your budget to feed your financial plan, however, because cash is flowing through the system constantly. Even if your cash flow is not nearly what you would like it to be, the fact that there is a flow means you should be able to divert at least a trickle toward your future security.

We won't kid you. Saving regularly may mean making some sacrifices in your consuming habits. If you habitually spend all that you make, clearly you will have to cut back somewhere to make regular investments—but not necessarily as severely as you might fear. Smart shopping can cut your expenses considerably. Understanding the various attributes and options available to you in some products will help you avoid paying for features you do not need. Making the suggested changes in the following areas will help you win the battle of the budget and create your own "peace dividend":

■ **Insurance.** Not understanding insurance in the least, most of us usually nod absently to the questions our insurance agent peppers us with and then pay the fattened premiums without further reflection. The little bit of research it would take to buy more wisely would probably pay us far more per hour than we earn at our regular employment. We have already covered consumer issues in health and life insurance. But auto insurance is another area where you might save a great deal if you better understood the product and your needs. Higher deductibles where feasible will lower your premiums, as will dropping unnecessary coverages, as will shopping around. Ask your financial adviser to help you revise your coverage, or ask your insurance agent directly.

■ **Home repairs.** Many of your most frequent home repairs require only a minimum of skills and tools. Fixing leaky faucets, painting, repairing malfunctioning toilets, even building fences don't necessarily require the skills of an expert. Get an easy-to-follow comprehensive home maintenance book and discover how dexterous you really are.

- **Large purchases (automobiles, furniture, appliances, electronic gear, and so on).** One key here is to buy used instead of new, the consumer's equivalent of recycling. Three-year-old cars, for example, are said to give the best balance of price and remaining life. You should research the makes with good repair records. Reconditioned appliances, electronic gear, and previously owned furniture may serve you as well as new items. You can also save money by buying through discount mail-order houses.
- **Other family expenses.** Shop around, shop sales, and keep to your budgets. Aim to buy quality on sale rather than buying low-quality items at retail.
- **Buying on credit.** If you can't pay cash for an item, you can't afford it. We have already shown you how to use your credit cards to make money. Other than that, you should see your credit cards strictly as a convenience to avoid carrying cash, plus as an emergency source of quick cash. Don't let credit card issuers and other consumer loan hawkers steal away your financial plan. Effectively, their "borrow, borrow, borrow" campaigns are bidding to do just that. (Obviously this advice does not apply to home or car purchases. It doesn't necessarily apply to other large purchases, either. When the washing machine dies, you need to replace it. But in general, use credit only when you absolutely need to, and in those cases, shop for the best terms. Your financial adviser should be able to help you with the specifics.)

Once you have started the habit of saving money, the challenge is, first, to keep it up, and, second, not to dip into the till. Regard your savings as sacrosanct—a fixed expense like your rent and utility bills with serious consequences if it goes unpaid. A little sacrifice and discipline now will pay substantial dividends down the line.

SAMPLE PORTFOLIOS FOR VARIOUS STAGES OF LIFE AND LIFE CIRCUMSTANCES

Do you want to test yourself on how well you now understand socially responsible financial planning? In this section we have fictionalized some actual financial planning case histories. These anecdotes cover a variety of stages of life, but they do not necessarily fall neatly into the classic molds because of little circumstantial twists. We've followed each anecdotal description with an analysis of the problem and a recommended solution. To check your knowledge, don't read the analysis and recommendations right away. Do your own, and then check to see how you did.

Case History One

DESCRIPTION: Marty Jenkins, 30, is single, with no children. He makes $30,000 annually as an editor for a local magazine and has health insurance through his work. He has habitually spent most of what he earns, although he does have $8,000 in a savings account (which he has recently contemplated putting toward a new sports car). He has never invested a cent. He wandered into an SRI financial planner's office after seeing a brokerage firm's television commercial that chastised someone in just about his circumstances for running his financial life like a kid.

ANALYSIS: He's never planned for retirement and now is when he should start to give compounding interest time to build an estate. Viewed as emergency cash instead of as a luxury automobile, his liquid reserves are adequate, although barely. Given his past history, it is important that he develop an investing habit right now. He can bolster his reserves later. Having never established good financial habits, he would benefit from an imposed discipline like a check withdrawal system. He's also getting killed at tax time as a single employed person without children. Being young and on a decent career path, he can afford to take some calculated risk.

SOLUTION: First, the planner convinces Marty to move his $8,000 savings account to a double tax-free money market fund where it will earn higher and more socially responsible returns than at the bank, and where it can serve as an emergency account. (After being enlightened about the benefits of compounding interest—and sobered by the reality of compounding inflation—Marty agrees that the car can wait.) Marty will also do most of his banking through the money market account, paying his large expenses (rent, car payment, taxes, and so forth) with money market checks while earning high, tax-free interest until those payments come due.

Next, Marty and his planner determine that he can find $100 a month to invest by spending less here and there. Their ultimate goal is to invest for Marty's retirement in an SRI growth mutual fund earning an average annual return of 12 percent. However, the fund selected requires an opening investment of $1,000 (although subsequent investments through the fund's check debit plans can be as little as $50 a month.) Since Marty does not have a spare $1,000 to open the account, he starts his investing through a Ginnie Mae fund—returning an average of 9 percent per year—which allows opening investments of $25 minimum through its check debit plan. He opens with $100 and arranges the check debit at $100 a month. One further refinement: By setting up the account as an IRA, the money he invests each year can be excluded from income tax and the earnings will compound tax-deferred.

As soon as he has accumulated $1,000 in the Ginnie Mae fund (between seventeen and eighteen months), he will switch the money into the growth fund but maintain the IRA account structure (easily arranged with the fund). Through the growth fund's check debit, he will continue to invest $100 a month. As with the Ginnie Mae fund, the new mutual funds are now doing his investing for him.

The long-term payoff for all these little maneuvers? After thirty-five years, when he reaches normal retirement age, Marty will have about $580,000—assuming he never puts in more than the $100 a month.

Case History Two

DESCRIPTION: Barbara and Cecil Rollins have been investing for most of their working lives, but without an adviser. They're frustrated now because their financial package has become so complex that they have no idea how it all works together, or even if it works. They're also concerned about the social implications of their investments, having just read an article on SRI.

Here are the details: Barbara is thirty-eight, a part-time social worker. Cecil is forty-five and manages a local sales office for a tire company. They have two kids, ages eight and twelve. Over the fifteen years of their marriage, Barbara and Cecil have amassed $100,000 in investment value between IRAs, his 401(k) at work, Stride Rite stock, and a small rental property (market value: $29,000), which an older couple (retired on a fixed income) has occupied for the last ten years. The Rollinses have a total of fifteen investments, all of them but the 401(k) in medium-, high-, or very high-risk stock funds (the 401(k) is in a money market fund). Although they have plenty of investments, they've never planned directly for their kids' college educations.

ANALYSIS: First, the Rollinses are overdiversified. Second, their portfolio is much too risky, especially for their stage of life—except for the 401(k), that is, which is too conservatively invested. Third, holding a single stock makes no sense from a diversification standpoint, even if it's a good, ethical company's stock.

SOLUTION: The overall goal is get the Rollinses down to a more manageable ten investments, reduce the portfolio's risk balance, and get some sensible education planning going for the kids. The action plan also includes selling the Stride-Rite at a profit; the adviser counsels them to honor their concerns for their tenants and hold on to the rental.

When the restructuring is complete, the portfolio includes two Uniform Gifts to Minors Act accounts invested in socially responsible conservative growth funds. These, of course, are designed to finance the kids' college educations (income funds simply won't compound powerfully enough, considering the age of the Rollins kids). The 401(k) is also redirected into a socially responsible conservative growth fund. The overall portfolio, all socially responsible, ends up being 19 percent in very low-risk investments (including a Ginnie Mae mutual fund and an SRI bond fund), 20 percent in low-risk investments (including an SRI balanced fund and an ethical limited partnership), 32 percent in medium-risk investments (including an SRI growth fund and another, slightly riskier ethical limited partnership), and 29 percent in a high-risk investment (the rental). The overall return for the portfolio projects 10 to 12 percent. Not counting the $10,000 in the kids' accounts (which should more than cover their college expenses after compounding) or the $10,000 plus $2,500 a year in Cecil's 401(k), the Rollinses should have at least $538,200 in investment value when Cecil retires in twenty years. His 401(k) will add another $224,781.

Case History Three

DESCRIPTION: Julio and Maria Hernandez, both fifty years old, teach at the same university. Their only child, a daughter, is married. The Hernandezes live well but spend virtually all they make. They pay 28 percent in income taxes each year and have saved nothing; suddenly they are agonizing over the prospects of a desperate retirement, which they face no later than age sixty-five because of university regulations. Admittedly a bit leery of the investing world, they know they have to do something to secure their future. They are aware of socially responsible investing because the university offers SRI options through its 403(b) plan, even though they have never participated.

ANALYSIS: Because of their preretirement stage of life, the Hernandezes need to be quite protective of whatever principal they can generate for investment. They need to do something immediately, with only fifteen years of compounding time left before they reach the university's mandatory retirement age.

SOLUTION: Following the advice of an SRI broker referred to them by another professor, Julio and Maria rebudget their life-style to include $500 a month contributed to the university's salary reduction plan. Because of their stage of life and a low emotional tolerance for risk, the adviser suggests conservative investments returning about 9 percent per year. Julio's $250 a month contribution goes into a socially screened bond portfolio offered through a variable life annuity. Maria's $250 a month goes into a Ginnie Mae mutual fund. After fifteen years they will have amassed $192,015 in the 403(b)s. They will also have saved $25,200 in income taxes.

Case History Four

DESCRIPTION: Harold and Betty Greene are retired, quite comfortably, in fact—living entirely off their investment income, they are still in the 28 percent tax bracket. A former high school principal, Harold, seventy-three, contributed heavily to his 403(b) plan from the first year it was offered (1974) and also receives state teacher's retirement pay. Betty, sixty-four, had maximally funded an IRA during her years as a real estate salesperson.

Following tips that he picked out from the newspaper from time to time, Harold had directed their investing without ever consulting an expert about their circumstances. Until recently he was proud of what he had to show for his efforts: a total portfolio value of $325,000, including five taxable money market funds, T-bills, stock in a South African–based gold mining company (worth $20,000), Exxon stock (worth $30,000), seven growth and aggressive growth mutual funds, four different government bond mutual funds (mainly Ginnie Maes), and a limited partnership invested in a seniors housing project. About 40 percent of the total value is in aggressive growth investments begun twelve years earlier.

Although both are lifelong peace activists, neither Harold nor Betty had given much thought to socially responsible investing until the

Exxon *Valdez* disaster. Thus motivated, Harold studied SRI thoroughly and went to an SRI professional.

ANALYSIS: Not only is the Greenes' portfolio out of step with their social values, but it is overdiversified as a whole, almost impossible to track because of the complexity, and far too risky for their stage of life. The stock holdings are underdiversified. The Greenes are also paying out far too much in taxes.

SOLUTION: Following their planner's recommendation, the Greenes sell the stocks at a profit; the $50,000 raised is invested in a double tax-free municipal bond fund with dividends reinvested. This solves their tax problem and makes the portfolio much more ethical. They then pull out of the five unscreened money market funds and get into two ethical money market funds—one a double tax-free that is part of the muni bond fund's family. This gives them access to the bond fund money by check (the other money market fund is a broadly screened taxable). The seven unscreened stock fund investments are sold, with the proceeds invested in two broadly screened conservative growth funds. The partnership and government bond funds, both ethically okay, are left intact.

KEEPING TRACK

Financial plans are like cars and human beings in that they need occasional check-ups and maintenance. Sometimes your goals change. Sometimes circumstances interrupt your plans. Sometimes investments don't perform as expected.

Financial plans are also like cars in that failure to maintain them can get quite expensive. If an investment isn't performing as you wish after you've given it every reasonable chance, you'll want to replace it with a more hopeful prospect—and the sooner the better. You also want to make sure that you keep tabs on everything you own. As your investments get increasingly diverse, there is a chance that you could actually forget you have some of them, particularly if for some reason—like a change of address—their mail stops coming to you.

Keeping track of your financial plan and the investments it includes is not difficult. We suggest the following simple routine: *Once per year* update your net worth statement (Worksheet II).

Once every six months review your goals (Worksheet IV) and revise if necessary, and chart your investments on the worksheet that follows.

Worksheet V: Investment Tracker

Investment				
Date Purchased				
$				
Date/Value				
Date/Value				
Date/Value				
Date/Value				
Date/Value				
Date/Value				
Date Sold				
$				

Investment				
Date Purchased				
$				
Date/Value				
Date/Value				
Date/Value				
Date/Value				
Date/Value				
Date/Value				
Date Sold				
$				

Of course, if you are working with a financial planner, you can ask him/her to do the tracking and keep you abreast. But not all planners, even those whose advice and other service is impeccable, will keep up-to-date on your file without periodic reminders from you. Good advisers are not always good detail persons. So you ought to check in with your planner every six months or so, even if all your investments are in long-term products.

Frankly our advice is to keep your own records whether or not you have a planner (although your planner can supply the details). These are your investments, after all, and you are the one who will pay if something gets lost in the shuffle. Make it fun, though—think of it like checking the sports pages to see how your home team is doing. Your investments are kind of a team, anyway, working to win in your behalf. But unlike sports teams, which have to win championships to make the big money, your investments only have to finish in the middle of the pack to make your family's future a whole lot more secure. (Actually, since the advent of obese sports television contracts and millionaire salaries for mediocre ball players, that's true of some sports teams, too. Thanks, guys, for spoiling a perfectly good analogy.) By the way, computer users: ask your local computer dealer to recommend a user-friendly software package that will further simplify the tracking process.

STAYING ON TRACK—THE *INVESTING FROM THE HEART* ACTION PLAN

We began this book by making two "outrageous" claims: first, that you can secure your financial future with socially responsible money management and investing without sacrificing market-level returns; and second, that doing so will help secure the future of the planet as well.

Of course, now you can see as well as we can that we were not being outrageous at all. We in SRI can match the financial worthiness of our vehicles with anybody's. And when we join in spirit with other like-minded consumers, bank depositors, and investors to use all our dollars ethically, our collective power is profound. Recall from earlier pages that socially responsible investors (and divestors) share much of the credit for the current dismantling of South Africa's apartheid; several major corporations now submit to annual environmental audits and others have, for example, ceased animal testing because of the efforts of committed shareholders; flexing the financial muscle represented by their vast investment accounts, managers of large SRI mutual funds are influencing corporations to improve their social behavior in myriad ways; and finally, socially responsible bank depositors are financing the "advantaging" of disadvantaged communities through their devotion to socially responsible financial institutions.

It is our hope that you take *Investing from the Heart* to heart. To make it even easier for you to do that, we have provided the following reminders about some key money management, investing, and investor activist strategies covered in the book. If you follow these steps, not only will your life begin to work financially, but life for *everybody* will be all the richer.

- Keep only enough money in the bank to handle your day-to-day bills.
- Do your banking at a regular credit union, community development credit union, minority-owned bank, or socially responsible bank.
- Keep other liquid reserves in a socially responsible money market account. Pay your large bills (house payments, car payments, and so on) from this account.
- Only use credit cards to make purchases you can pay for when the credit card bill comes due. Never pay consumer interest except in emergencies and for large purchases like house and car payments. Use only socially responsible credit cards.
- Set financial goals and make specific plans to meet them.
- Start investing in your living estate now, so the magic of compound interest has maximum time to do its stuff.
- Dollar cost average when you invest in stocks and stock and bond mutual funds.
- Contribute to tax-advantaged retirement plans to the maximum amount available.
- Select only socially responsible options from the plans available at your work and for your personal retirement plans as well. If your employer does not offer SRI options, request that the company begin doing so.
- If others depend on your wage-earning ability, be sure to carry adequate life insurance until you have established a living estate. Select a socially responsible plan if there is one to fit your needs.
- Carry adequate medical insurance at all times.
- Keep three to six months' living expenses in liquid assets such as socially responsible money market and mutual funds.
- If you want professional assistance in your financial planning, use the services of an SRI professional if possible. At the very least, cross-check the advice of a non-SRI professional with advice given in this book and other SRI resources.
- Invest regularly. One excellent method is to authorize a socially responsible mutual fund to debit your checking account monthly. If

yours is a stock mutual fund, this method will build dollar cost averaging into your investing.

- Expand your socially responsible investment options with investments in mortgage-backed securities and municipal bonds and mutual funds invested in these vehicles.
- If you do not have the capital for a mutual fund's minimum opening investment, see if the fund allows you to open with less through a check debit plan.
- If you only have a small amount to invest each month, join or start an investment club.
- Be sure to include some growth investments such as stocks in your financial plan even if you are near retirement or retired. These are necessary to protect against the effects of inflation.
- If you own stock, inspect your proxy statements carefully for opportunities to vote for socially responsible shareholders' resolutions.
- Diversify your investments for protection against losses.
- If you can afford to accept below-market returns with some of your investments in exchange for a high social impact, consider loaning money to a community development loan fund.
- Set up Uniform Gifts Act to Minors accounts for your children's education planning.
- Reduce debt as you near retirement.
- Never invest in anything you don't fully understand or with someone you don't fully know or trust.
- Track your investments.
- Shop as well as invest responsibly.
- Tell your friends and relatives about socially responsible investing.

Appendix A

LISTS OF COMPANIES BY SOCIAL CATEGORY

The lists in this appendix should be regarded only as starting places from which to evaluate a company's social worthiness. Obviously some lists are provided to alert you to companies with potential SRI value, and others alert you to companies you may want to exclude from your investment planning. Remember also that many of these lists change constantly, so you should confirm whether or not the information is still current for any company you are evaluating. One other note: A few companies listed may be privately held, that is, not available for public investment.

South Africa

Corporate connections with South Africa are constantly in flux, and additional information becomes available almost daily. The following list is presumed accurate as of the publication date of the cited source. However, before taking final divestment or other action, readers should seek confirmation from the companies themselves about their current involvement in South Africa, especially where a trust responsibility is concerned.

Note also that this list only concerns companies with a direct involvement in South Africa. It does not include companies with strictly financial involvement through bank loans or underwriting; companies involved mainly through licensing, franchising, distribution, or other sales arrangements; companies involved through international affiliates in South Africa; or companies once involved that no longer do business there. The best source of that information is the *Unified List of United States Companies Doing Business in South Africa* (*current edition*), available through The Africa Fund: 198 Broadway, New York, NY 10038; (212) 962-1210.

The Domini 400, a socially responsible stock index, offers a list of four hundred companies screened for "operations in, or substantial involvement with, South Africa," among other social criteria. This list may be found in appendix B on page 369.

U.S. Publicly Traded Companies with Direct Investment or Employees in South Africa

AM International
Abbott Laboratories
Air Express International
Albany International
Allergan
Amerford International
American Cyanamid
Arvin Industries
Avery-Dennison
Baker Hughes
Bandag
Beckman Instruments
Borden
Brilund Ltd.
Bristol-Myers Squibb
CBI Industries
Cascade
Caterpillar
Chevron
Colgate-Palmolive
Crown Cork & Seal
Deere
Donaldson
Dresser Industries Inc.
Du Pont
Echlin
Ferror
Franklin Electric
Gillette
Grey Advertising
Harnischfeger Industries
Harsco
Ingersoll-Rand
International Flavors & Fragrances
International Paper

Interpublic Group of Companies
Johnson & Johnson
Kellogg
Kimberly-Clark
Lawter International
Lilly, Eli
Loctite
Lubrizol
MacDermid
Mine Safety Appliances
Minnesota Mining & Manufacturing
Molex
Monsanto
Nalco Chemical
Olin
Parker Hannifin
Pfizer
Phelps Dodge
Premark International
Quaker Chemical
Raytheon
Reader's Digest Association
Robertson-Ceco
Schering-Plough
Schlumberger
Standard Commercial
Terex
Texaco
Tinker
Tokheim
Twin Disc
USG
Union Camp
Union Carbide
United Technologies

UpJohn   Wear
Warner-Lambert   Wynn's International

Source: Franklin Research & Development Corporation, *Investing in a Better World* (March 15, 1991): 711 Atlantic Avenue, 5th Floor, Boston, MA 02111; (617) 423-6655. Reprinted with permission.

Labor and Equal Opportunity

The Best Companies for Women

American Express/Shearson Lehman	Home Box Office
AT&T	Honeywell
Avon Products	International Business Machines
Barrios Technology	Levi Strauss
Bidermann Industries	Lotus
CBS	Manufacturers Hanover Trust
Children's Television Workshop	Merck
Citizens and Southern	Mountain Bell
Cognos	Mount Carmel Health
Conran Stores	Neiman-Marcus
The Denver Post	Northwestern Bell
Digital Equipment Corporation	Payless Cashways
Drake Business Schools	PepsiCo
Federal Express	Pitney Bowes
Fidelity Bank	Procter & Gamble
First Atlanta	Recognition Equipment
Gannett	Restaurant Enterprises Group
General Mills	Rowland Company, The
Grey Advertising	Saks Fifth Avenue
GTE	Salomon Brothers
Hallmark	Simon & Schuster
Hearst Trade Books	Southern New England Telephone
Herman Miller	Syntex
Hewitt Associates	Time
Hewlett-Packard	U.S. West Direct

Source: *The Best Companies for Women,* copyright © 1988 by Baila Zeitz and Lorraine Dusky, Simon and Schuster: Rockefeller Center, 1230 Avenue of the Americas, New York, NY 10020.

The Fifty Best Places for Blacks to Work

BEST OF THE BEST

AT&T	Anheuser-Busch
Amtrak	Atlantic Richfield

Avon Products
Chase Manhattan
Chrysler
Coca-Cola
Coors
Eastman Kodak
Equitable
Exxon
Federal Express
Ford
Gannett
General Mills
General Motors
Hallmark
IBM
Johnson & Johnson

Kellogg
Kraft
McDonald's
Merck
N. C. Mutual
J. C. Penney
PepsiCo
Philip Morris Cos.
Port Authority (NY/NJ)
Procter & Gamble
Ryder
J. E. Seagram & Sons
Soft Sheen Products
U.S. Armed Forces
Xerox

COMPANIES TO WATCH IN THE NINETIES

(". . . companies whose efforts are headed in the right direction, but [which] have a ways to go [with regards to] numbers of black employees and managers . . .")

Aetna Life
American Airlines
Apple Computer
Bristol-Myers Squibb
Chevron
Digital
General Electric

Hewlett-Packard
Inner City Broadcasting
Mobil
H. J. Russell
Sears
Stroh Brewery
Time

Source: *Black Enterprise* magazine (February 1989). Copyright 1989 by The Earl G. Graves Publishing Co., Inc., 130 Fifth Avenue, New York, NY 10011; all rights reserved.

Seventy-five Best Companies for Working Mothers

Aetna Life & Casualty
Allstate Insurance
America West Airlines
American Bankers Insurance Group
American Express
AT&T
Arthur Andersen & Co.
Apple Computer
Atlantic Richfield Company

BE&K
Bellcore (Bell Communications Research)
Boston's Beth Israel Hospital
Bureau of National Affairs, The
Leo Burnett
CMP Publications
Campbell Soup Company
Champion International

Citicorp (Citibank)
Consolidated Edison Company of New
York
Corning
Digital Equipment
Dominion Bankshares
Dow Chemical
Du Pont
Fel-Pro
Gannett
Genentech
General Mills
Grieco Bros., Inc. (Southwick)
Group 243
G. T. Water Products
Hallmark
John Hancock Mutual Life Insurance
Hechinger
Hewitt Associates
Hill, Holliday, Connors, Cosmopulos
Hoffmann-La Roche
Home Box Office
IDS Financial Services
INDEECO
International Business Machines
Johnson & Johnson
S. C. Johnson & Sons
Lancaster Laboratories
Lincoln National
Little Tikes

Lost Arrow (Patagonia)
Lourdes Hospital
MNC Financial
Marriott
Catherine McAuley Health System
Merck
Minnesota Mining & Manufacturing
Morrison & Foerster
NCNB
Official Airline Guides
Oracle
Phoenix Mutual Life Insurance
Pitney Bowes
Procter & Gamble
SAS Institute
South Shore Bank
Steelcase
Stride Rite
Syntex
Tenneco
Trammell Crow Company
Travelers Corporation, The
United States Hosiery
Unum
U.S. West
Warner-Lambert
Wegmans
Wells Fargo & Company
Xerox

Companies Responding to AIDS: Voluntary Signees of the "Ten Principles for the Workplace"

THE PRINCIPLES

1) People with AIDS or HIV (Human Immunodeficiency Virus) infection are entitled to the same rights and opportunities as people with other serious or life-threatening illnesses.

2) Employment policies must, at a minimum, comply with federal, state, and local laws and regulations.

3) Employment policies should be based on the scientific and epidemiological evidence that people with AIDS or HIV infection do not pose a risk of transmission of the virus to co-workers through ordinary workplace contact.

4) The highest levels of management and union leadership should unequivocally endorse nondiscriminatory employment policies and educational programs about AIDS.

5) Employers and unions should communicate their support of these policies to workers in simple, clear, and unambiguous terms.

6) Employers should provide employees with sensitive, accurate, and up-to-date education about risk reduction in their personal lives.

7) Employers have a duty to protect the confidentiality of employees' medical information.

8) To prevent work disruption and rejection by co-workers of an employee with AIDS or HIV infection, employers and unions should undertake education for all employees before such an incident occurs and as needed thereafter.

9) Employers should not require HIV screening as part of general preemployment or workplace physical examinations.

10) In those special occupational settings where there may be a potential risk of exposure to HIV (for example, in health care, where workers may be exposed to blood or blood products), employers should provide specific, ongoing education and training, as well as the necessary equipment, to reinforce appropriate infection control procedures and ensure that they are implemented.

CORPORATE AND SMALL BUSINESS SIGNEES
(AS OF AUGUST 15, 1991)

Abt Associates
Aetna Life Insurance
Alan Emery Consulting
Allstate Insurance
American Telephone and Telegraph Company
Archie Comics
Atlantic Industries
Atlantic Magazine
Bankers Trust Company
Birch & Davis Associates
Burroughs Wellcome Company

Chemical Bank
Chevron Corporation
CIBA-GEIGY Corporation
Coastal Training Institute
Digital Equipment Corporation
Dow Jones
Du Pont
ETR Associates
EduCare Associates
Equicor Health Plan
The Equitable
Ethicon

Franklin Research & Development
Corporation
General Electric
Girard Video
Glaxo
Good Money Publications
Hoffmann-La Roche
Howard J. Rubenstein Associates
International Business Machines
ITT
Johnson & Johnson
League of Resident Theatres
Levi Strauss
Levine, Huntley, Schmidt, and Beaver
Lola Restaurant
The Mercantile and General
Reinsurance Company
Merck
Metropolitan Life Insurance
Midwest Title Guarantee Company of
Florida
Mobil
Morgan Guaranty Trust Company of
New York
Ms. Magazine
National Association of Public
Television Stations
Norton Company, Ogilvy and Mather
Advertising
Ortho Pharmaceutical Corporation
Outreach

Pacific Bell
Philip Morris Management
Corporation
Playboy Enterprises
Princeton Project Resources
The Principal Financial Group
Progressive Asset Management
The Prudential Insurance Company
Sassy Magazine
Schering-Plough Corporation
Shubert Organization, The
SmithKline Beckman Corporation
Squibb Corporation
Swing Shift
Syntex Corporation
Tennessee Department of Health and
Environment
Time
Times Mirror
Transamerica Life Companies
U.S. News & World Report
Union Carbide Corporation
United Jersey Banks
Warner-Lambert
Wells Fargo Bank
Whole Wheat 'n Wild Berrys
Restaurant
WNET—Public Television
Xerox

Source: Citizens Commission on AIDS for New York City and Northern New Jersey, 121 Avenue of the Americas, 6th Floor, New York, NY 10013; (212) 925-5290. Reprinted with permission.

The One Hundred Best Companies to Work for in America (Composite and Sublists)

Advanced Micro Devices
Analog Devices
Anheuser-Busch
Apple Computer
Armstrong World Industries
Atlantic Richfield
Baxter Travenol

Bell Laboratories
Leo Burnett
Celestial Seasonings
Citicorp
Control Data
Trammell Crow
CRS/Sirrine

Cummins Engine
Dana
Dayton Hudson
Deere
Delta Air Lines
Digital Equipment
Donnelley
Doyle Dane Bernbach
Du Pont
Eastman Kodak
A. G. Edwards
Electro Scientific
Erie Insurance
Exxon
Federal Express
Fisher-Price Toys
H. B. Fuller
General Electric
General Mills
Goldman Sachs
Gore
Hallmark Cards
H. J. Heinz
Hewitt Associates
Hewlett-Packard
Inland Steel
Intel
IBM
Johnson & Johnson
Johnson Wax
Knight-Ridder
Kollmorgen
Levi Strauss
Liebert
Linnton Plywood
Los Angeles Dodgers
Lowe's
Marion Labs
Mary Kay Cosmetics
Maytag
McCormick
Merle Norman Cosmetics
Herman Miller

Minnesota Mining & Manufacturing
Moog
J. P. Morgan
Nissan
Nordstrom
Northrop
Northwestern Mutual Life
Nucor
Odetics
Olga
J. C. Penney
Physio-Control
Pitney Bowes
Polaroid
Preston Trucking
Procter & Gamble
Publix
Quad/Graphics
Rainier Bancorporation
Random House
Raychem
Reader's Digest
Recreational Equipment
Remington Products
ROLM
Ryder
Saga
Security Pacific Bank
Shell Oil
Southern California Edison
Springs
Steelcase
Tandem Computers
Tandy
Tektronix
Tenneco
Time
Viking Freight System
Wal-Mart
Westin Hotels
Weyerhaeuser
Worthington Industries

Best Places for Women to Work

Citicorp
Control Data
Doyle Dane Bernbach
Federal Express
Hallmark Cards
IBM
Levi Strauss

Mary Kay Cosmetics
Nordstrom
Northwestern Mutual Life
J. C. Penney
Recreational Equipment
Security Pacific Bank
Time

Best Places for Blacks to Work

Cummins Engine
Federal Express
General Electric
Hewlett-Packard
IBM

Levi Strauss
Los Angeles Dodgers
Polaroid
Time

Most Strongly Unionized

Anheuser-Busch
Cummins Engine
Dana
Deere

General Electric
Inland Steel
Maytag

Where Employees Own a Large Piece of the Company

Federal Express
Hallmark Cards
Linnton Plywood

Lowe's
Publix Super Markets
Quad/Graphics

The Ten Best Companies to Work for in America

Bell Laboratories
Trammell Crow
Delta Air Lines
Federal Express
Goldman Sachs

Hallmark Cards
Hewlett-Packard
IBM
Pitney Bowes
Time

. . . And Five Runners-Up

Herman Miller
Minnesota Mining & Manufacturing
J. P. Morgan

Northwestern Mutual Life
Publix Super Markets

Source: *The 100 Best Companies to Work for in America.* A Signet Book, New American Library, April 1985.

National Boycotts Sanctioned by the AFL-CIO

Note: All data in parentheses indicates the stock exchange, stock name, and stock symbol for those companies on the AFL/CIO list that are publicly traded. Researched by the authors.

Ace Drill Corporation

Brown & Sharpe Mfg. Co. (NYSE, Brown & Sharpe—BNS)

Brown Corporation

California Table Grapes

Daily News, New York—newspaper publication

Garment Corporation of America—A1 Manufacturing

Greyhound Lines, Incorporated (NYSE, Greyhound Dial—G)

Guild Wineries & Distilleries

Holly Farms (NYSE, Holly Farms—HFF)

International Paper Company (NYSE—IP)

Kawasaki Rolling Stock, U.S.A.

Krueger International, Incorporated

Louisiana-Pacific Corp. (NYSE—LPX)

Mohawk Liqueur Corporation

R. J. Reynolds Tobacco Co. (NYSE, RJR Nabisco—RJR)

Rome Cable Corporation

Shell Oil Company (NYSE, Royal Dutch Petroleum—RD)

Silo, Inc.

United States Playing Card Co. (OTC, Jesup Group—JGRP)

Source: *Label Letter* (March/April 1991), Union Label and Service Trades Department, AFL-CIO: 815—16th Street, NW, Washington, D.C. 20006.

Companies with More Than 20 Percent Employee Ownership and Their Level of Employee Participation

COMPANY	% EMPLOYEE OWNED	LEVEL OF EMPLOYEE PARTICIPATION
American Recreation Centers	22%	High
American West Airlines	26%	High
Apple Computer	24%	High
Applied Power	34%	No info
Ashland Oil	23%	Low
Baker, Michael	70%	High
Carter Hawley Hale	40%	Medium
Crown Crafts	23%	Medium

E-Systems	22%	Medium
Enron	20%	Medium
Figgie International	21%	Low
FMC	29%	Low
Grumman	37%	Medium
Interregional Corp.	37%	Medium
Kerr Glass	33%	Low
Kroger	34%	Medium
Lillian Vernon Corporation	28%	Medium
Lowe's	31%	High
Louisiana General Services	32%	Low
McCormick	24%	Medium
McDonnell Douglas	30%	Medium
Miller, Herman	20%	High
Modine	20%	No info
Oregon Metallurgical	62%	No info
Oregon Steel	32%	High
Pacific Enterprises	22%	Low
Penney, J. C.	25%	Medium
Phillips Petroleum	25%	Low
Piper Jaffrey	46%	Medium
Preston Trucking	20%	Medium
Procter & Gamble	20%	Medium
Products Research & Chemicals	20%	No info
Ruddick Corp.	41%	Medium
Sears	21%	Low
Stanley Works	28%	Medium
Sterling Chemical	48%	High
Tandy Crafts	44%	Medium
USG	26%	Low
Weirton Steel	80%	High

Source: Franklin's Insight study, *The ESOP Effect,* August 1990. Reprinted with permission.

Includes shares owned by employees outside the ESOP.

Employee participation includes programs that give workers greater control over their jobs, improve communication between management and employees, or involve employees more deeply in corporate decision making. Firms that have long histories of broad worker participation receive a "high" rating; those that have at least some form of worker participation are rated "medium"; those with no apparent forms of participation (other than the ESOP) are rated "low."

The Environment (Including Energy Production)

Alternative Energy Companies

SOLAR ENERGY

Alpha Solarco   Satco Power
Applied Solar   Solar Electric Engineering
Energy Conversion   Sunlight Technologies
FAFCO   THT

GEOTHERMAL

California Energy   Magma Power

ALTERNATIVE AND COGENERATION

Bonneville Pacific   Environmental Power

WASTE TO ENERGY

Wheelabrator Technology   Zurn Industries

Source: Compiled by the authors, January 1991.

Natural Gas Companies

Anadarko   Energen
Apache   Equitable Resources
Arkal   NICOR
Atlanta Gas   New Jersey Resources
Bay State Gas   Oneok
Brooklyn Union Gas   Oryx Energy
Burlington Resources   Pacific Enterprises
Connecticut Energy   Peoples Energy
Diversified Energy   Washington Gas & Light
Energas

Source: Compiled by the authors, January 1991.

Recycling Companies

Astec   Compliance Recycling
CPAC   Diversified Industries
Commercial Metals   Gencor

IMCO Recycling
Ogden
Ogden Projects
Prab Robots
Proler International

Reserve Industries
Roanoke Electric
Scope Industries
Wellman

Source: Compiled by the authors, January 1991.

Water Supply, Treatment, and Equipment Companies

Autotrol
Badger Meter
Betz Laboratories
Calgon Carbon
California Water Service
Connecticut Water
Consumers Water
Davis Water & Waste
E'town
Enviropact
ESSEF
Franklin Electric
Geraghty & Miller
Groundwater Technology
Gundle Environmental
GWC
Hydraulic
Ionics
IWC Resources

Keystone International
Lindsay Manufacturing
Metcalf & Eddy
Millipore
Milton Roy
Nalco Chemical
NCH
Osmonics
Pennsylvania Enterprise
Philadelphia Suburban
SJW
Smith (A. O.)
Southern California Water
Southwest Water
United Water Resources
Univar
Valley Industries
Watts Industries
Western Canada Water

Source: Compiled by the authors, January 1991.

Waste Management Companies

Advanced Precision Technology
Andura
Athey Products
Browning-Ferris Industries*
CECO Filters Inc.
Chambers Development
Chemical Waste Management*
Davis Water & Waste Industries
Eastern Environmental Services
ECI Environmental
Environmental Monitoring & Testing

EnviroSource Inc.
F & E Resource Systems Technology
GZA Geo Environmental Services
Handex Environmental Recovery
Hunter Industrial Facilities
Industrial Services of America
Integrated Waste Services
Kimmins Environmental Service
Laidlaw
Met-Pro
Metcalf & Eddy

Micro-Ergics
Mid-American Waste Systems
Midwesco Filter Resources
NDB Environmental
OFRA Corp. of America
Ogden Projects
Quadrex
Resource Recycling Technologies
Riedel Environmental Technologies
Safe-Waste Systems
Safety-Kleen

Sanifill
Scope Industries
Sparta Surgical
Stevenson Environmental Services
Techops Landauer
Tetra Technologies
Tri-R Systems
USA Waste Services
Versar
Waste Management*

* These companies have been the subject of public investigation and litigation concerning their environmental records and/or business practices in the recent past. These problems may have been resolved since the publication of this book.

Source: Compiled by the authors, January 1991.

United States Public Utilities Operating Nuclear Power Plants (as of December 31, 1990)

Alabama Power
Arizona Public Service
Arkansas Power & Light
Baltimore Gas & Electric
Boston Edison
Carolina Power & Light
Cleveland Electric Illuminating
Commonwealth Edison
Connecticut Light & Power
Consolidated Edison
Consumers Power
Detroit Edison
Duke Power
Duquesne Light
Florida Power & Light
Florida Power
Georgia Power
Gulf States Utilities
Houston Lighting & Power
Illinois Power
Indiana/Michigan Electric
Iowa Electric Light & Power
Jersey Central Power & Light
Kansas City Power & Light
Kansas Gas & Electric

Louisiana Power & Light
Maine Yankee Atomic Power
Metropolitan Edison
Nebraska Public Power District
New England Power
New York Power Authority
Niagara Mohawk Power
North Carolina Electric Membership
North Carolina Municipal Power
Northern States Power
Ohio Edison
Omaha Public Power District
Pacific Gas & Electric
Pennsylvania Power & Light
Philadelphia Electric
Portland General Electric
Public Service Electric & Gas
Rochester Gas & Electric
Sacramento Municipal Utility District
San Diego Gas and Electric
South Carolina Electric & Gas
Southern California Edison
System Energy Resources
Tennessee Valley Authority
Union Electric

Vermont Yankee Nuclear Power   Wisconsin Electric Power
Virginia Electric Power   Wisconsin Public Service
Washington Public Power System

Source: *Commercial Nuclear Power 1990:* Energy Information Administration, Office of Coal, Nuclear, Electric, and Alternative Fuels, U.S. Department of Energy, Washington, D.C. 20585.

Top Fifty Manufacturing Companies Releasing Toxic Chemicals

Numbers in parentheses indicate rank order.

AAC Holdings (24)

Agrimont (50)

Air Products and Chemicals (38)

Allied Signal (8)

AMAX (7)

American Cyanamid (3)

Amoco (31)

ASARCO (9)

ATOCHEM (19)

Avtex Fibers (25)

BASF Corporation (16)

BP America (5)

Chevron (34)

Chrysler (49)

CIBA-GEIGY Corporation (39)

Coastal (35)

DOE Run (36)

Dow Chemical (29)

Du Pont (1)

Eastman Kodak (14)

Eli Lilly (46)

First Mississippi (32)

Ford (28)

Freeport McMoran (6)

General Motors (11)

General Electric (21)

Georgia Pacific (37)

Hercules (47)

Hoechst Celanese (20)

ICI American (48)

Inland Steel (18)

International Paper (33)

James River (44)

Kaiser Aluminum (43)

LTV (30)

Minnesota Mining and Manufacturing (13)

Monsanto (2)

National Steel (17)

Occidental Petroleum (12)

Pfizer (40)

Phelps Dodge (15)

PPG Industries (41)

SCM Chemicals (42)

Shell Oil (4)

Sterling Chemical (22)

Texaco (45)

Unocal (23)

USX (27)

Vulcan Chemicals (10)

Westvaco (26)

Source: Based on Environmental Protection Agency 1988 data as supplied by "Manufacturing Pollution: A Survey of the Nation's Toxic Polluters," Citizen Action/Citizens Fund, May 1990: 1300 Connecticut Avenue NW, Washington, D.C. 20036.

Pollution Control Equipment Companies

Advance Ross   American Ecology

Air & Water Technologies   American Pacific

American Plastics & Chemicals
Badger Meter
Betz Laboratories
Canonie Environmental Service
Ceco Filters
Chambers Development
Clean Harbors
Commercial Metals
Commodore Environmental Services
Compuchen
Control Resource Industries
Corning
Duratek
Ecolab
Ecology & Environment
Emcon Associates
Engineered Support Systems
Environment One
Enviropact
Envirosafe Services
Epolin
Farr
Gorman-Rupp
Groundwater Technologies
Harding Associates
Health-Chem
Industrial Services of America
Insituform Gulf South
International Recovery

International Technology
IRT
JWP
Krug International
Lakeland Industries
Lindsay Manufacturing
Memtek
Na-Tec Resources
Nalco Chemical
Noxso Corp.
On-Site Toxic Control
Pacific Nuclear Systems
Pall
Peerless Manufacturing
Pinnacle Environmental
Ply-Gen Industries
Pure Tech International
RCM Technologies
Radon Testing Corporation of America
Reuters
Rollins Environmental Services
Somerset Group
Techops Landauer
Texal International
Thermo Instrument Systems
Trenstech Industries
Wheelabrator Technology
Zurn Industries

Source: Compiled by the authors, January 1991.

Weapons Involvement

Top One Hundred Defense Department Contractors—Fiscal Year 1990

Numbers in parentheses indicate rank order.

Aerospace (43)
Allied Signal (28)
Ameranda Hess (96)
Amoco (72)
Arvin Industries (93)
Astronautics Corp. America (79)

ARCO Products (47)
AT&T (23)
Avondale Industries (34)
Bath Holding (27)
Black & Decker (81)
Boeing (11)

Bollinger Machine (80)
Bolt Beranek & Newman (92)
CAE Industries (74)
CFM International (54)
Chevron (73)
Chrysler (65)
Coastal (51)
Computer Sciences (53)
Contel (62)
Control Data (64)
Crowley Maritime (94)
CRS Sirrine Metcalf & Eddy JV (84)
CSX (63)
Day & Zimmerman (95)
Digital Equipment (88)
Duchossois (82)
Dyncorp (39)
E-Systems (41)
Eastman Kodak (98)
Eaton (77)
Emerson Electric (67)
Exxon (42)
Federal Express, Pan American, Northwest Airlines, et al. (60)
FMC (30)
Ford (25)
Forstmann Little & Co. (55)
Foundation Health (36)
Gencorp (21)
General Dynamics (2)
General Electric (3)
General Electric PLC (57)
General Instruments (99)
General Motors (4)
Goodyear Tire & Rubber (99)
Grumman (9)
GTE (17)
Harris (71)
Harsco (75)
Hercules (38)
Hewlett Packard (89)
Honeywell (15)
IBM (18)

International Marine Carrier (68)
International Shipholding (87)
ITT (24)
Johns Hopkins University (48)
Johnson Controls (59)
Kaman (70)
Litton Industries (14)
Lockheed (6)
Logicon (86)
Loral (31)
LTV (20)
Martin Marietta (7)
Massachusetts Institute of Technology (40)
McDonnell Douglas (1)
McDonnell Douglas/General Dynamics JV (33)
MIP Instandsetzungsbetric (91)
Mitre (44)
Mobil (61)
Morrison Knudsen (50)
Motorola (45)
Northrop (26)
Olin (32)
Oshkosh Truck (58)
Penn Central (49)
Peterson Builders (66)
Philip Morris (90)
Philips Gloeilampenfabricken (56)
Ram System Gmbh (100)
Raytheon (5)
Rockwell (13)
Rolls-Royce (83)
Royal Dutch Shell (46)
Science Applications Int'l. (37)
Sequa (69)
Talley Industries (97)
Teledyne (35)
Tenneco (10)
Texas Instruments (29)
Textron (19)
Thiokol (52)
TRW (22)

Unisys (16)
United Technologies (8)
Varian Associates (78)

Westinghouse (12)
Westmark Systems (76)
World Rosen Key JV (85)

Source: Department of Defense, 100 Companies Receiving the Largest Dollar Volume of Prime Contract Awards, fiscal year 1990.

Top One Hundred Nuclear Weapons Contractors (Fiscal Year 1988–1989)

Numbers in parentheses indicate rank order.

Allied Signal (12)
Analytic Sciences Corp. (82)
Applied Technology Div. (60)
Astronautics Corp. of America (98)
AT&T (9)
B. F. Goodrich (50)
Boeing (13)
British Aerospace, P.L.C. (47)
Brunswick (68)
C.A.E. Industries, Ltd. (91)
Canadian Commercial (52)
Charles Stark Draper Lab (38)
Colt Industries (87)
Computer Sciences (41)
Construcciones Aeronauticas, SA (85)
Control Data (61)
Costruzioni Aero (75)
Curtiss Wright (73)
Delta Industries (81)
Dover (93)
Earth Technology (99)
Eaton (35)
EG&G (11)
Electro-Methods (92)
Emerson Electric (40)
Fabrique National Herstal, SA (83)
Fairchild Industries (67)
Farley Industries (76)
Figgie International (53)
Flight Safety International (84)
FMC (39)
Ford Motor (57)

Forster Enterprises (71)
Forstmann, Little & Co. (74)
Gencorp (23)
General Dynamics (3)
General Electric (4)
General Motors (15)
General Instrument (42)
Goodyear Tire & Rubber (89)
Grumman (24)
GTE (63)
Gull (72)
H.A.C. (96)
Harris (78)
Harsco (19)
Hayes International (69)
Heney Group (77)
Hercules (29)
Holmes & Narver (44)
Honeywell (33)
IBM (16)
Israel Military Industries (49)
ITT (20)
Korean Air Lines Col, Ltd. (80)
Litton Industries (28)
Lockheed (8)
Logicon (70)
Loral (26)
LTV (18)
MDTT (JV) (65)
Martin Baker Aircraft Co., Ltd. (95)
Martin Marietta (7)
Mason & Hanger-Silas Mason (32)

McDonnell Douglas (1)
Ministry of Defense, U.K. (66)
Motorola (94)
Nav Con Defense Elect. (48)
Northrop (36)
Pan Am (43)
Parker-Hannifin (58)
Penn Central (37)
Philips Gloeilampen Fabrieken (54)
Quintron (90)
Raytheon (14)
Rockwell International (10)
Rohr Industries (51)
Rolls-Royce, P.L.C. (25)
Sargent Fletcher (56)
Schlumberger, Ltd. (97)
Scientific Atlanta (100)
Sequa (45)

Singer (31)
Slocomb J. T. (79)
Smith Industries (34)
Sparton (64)
Sundstrand (55)
Teledyne (46)
Texas Instruments (27)
Textron (30)
Thiokol (22)
TRW (21)
Unisys (17)
United Technologies (6)
United Industrial (62)
University of California (5)
Wackenhut Services (59)
Westinghouse Electric (2)
Westmark Systems (88)
Williams International (86)

Source: Nuclear Free America: 325 E. 25th Street, Baltimore, MD 21218; From Department of Energy fiscal year 1988 and Department of Defense fiscal year 1989 data.

Firearms and Ammunition Companies

Allied Research
American Body Armor & Equipment International
BEI Electronics
Ballistic Recovery Systems
Coleman
Colt Industries
Explosive Fabricators
Feather Industries

International Hitech Group
Megebar
Mining Services International
North American HiTech Group
Piranha International
Remington Arms
Sturm Ruger & Co.
Wildley

Source: Compiled by the authors, January 1991.

Product Integrity and Corporate Citizenship

Tobacco Companies

American Brands
American Filtrona
American Maize-Products
American Tobacco

Brooke Group Ltd.
Carolina Leaf Tobacco
Conwood Co., L.P.
Culbro

Dibrell Brothers
Loews
Lorillard
Maclin-Zimmer-McGill Tobacco
Philip Morris
RJR Nabisco
Southern Processors

Southwestern Tobacco
Standard Commercial Tobacco
Swisher JNO & Son
J. P. Taylor Co.
Tobacco Processors
Universal Corporation
UST

Source: Compiled by the authors, January 1991.

Alcohol Companies—Brewers and Distillers

Aquasciences International
American Brands
Anheuser-Busch
Bacardi
Jim Beam Brands
Boulder Brewing
Brown & Forman
Buffalo Don's Artesian Wells
Cananda/Gua Wine Co.
Chalone
Conrotto Winery
Adolph Coors
Erly Industries
Falstaff Brewing
Genesee
Glenmore Distilleries

G. Heileman Brewing Co.
Ingredient Tech
McCormick Distilling
Metatec
Midwest Grains Products
Miller Brewing
Millipore
Nutri Bevco
Pavichevich Brewing
Penwest Ltd.
Philip Morris
Pureco Energy
Seagram & Sons
W.I.N.E.
Wine Society of America
UST

Source: Compiled by the authors, January 1991.

Gambling: Gaming, Casinos, and Racing Associations

American Equine Products
Aztar
Bally Manufacturing
Brainerd International
Caesar's World
Churchill Downs
Circus-Circus Enterprises
Cloverleaf Kennel Club
Fairgrounds
James H. Fors & Associates

Golden Nugget
Hilton Hotels
Hollywood Park Realty Enterprises
International Game Technology
International Sensor Technologies
International Speedway Car
Jack Pot Enterprises
Mid-State Raceway
Mile High Kennel Clubs
Multnomah Kennel Clubs

Promus Companies   Turf Paradise
Santa Anita Realty (R.E.I.T.)   Universal Capital
Show Boat
Stinson & Farr Thoroughbred
Management

Source: Compiled by the authors, January 1991.

Appendix B

SOCIALLY RESPONSIBLE STOCK INDEXES

The following lists are the SRI publication *Good Money*'s socially screened alternatives to the Dow Jones Industrial and Utility Averages.

Good Money Industrial Average (GMIA)

Ametek
Consolidated Papers
Cross, A. T.
Cummins Engine
Dayton Hudson
Digital Equipment
Walt Disney
First Virginia Banks
Flight Safety International
H. B. Fuller
Hartmarx
Hershey Foods
Johnson & Johnson
Maytag
McDonald's

MCI Communications
Melville Corp.
Meredith Corp.
Herman Miller
Minnesota Mining & Manufacturing
Pitney Bowes
Polaroid
Rouse Company
Snap-on Tools
Stride Rite
Volvo A.B.
Wang Laboratories
Washington Post
Worthington Industries
Zurn Industries

Good Money Utility Average (GMUA)

Citizens Utilities
Consolidated Natural Gas
Hawaiian Electric
Idaho Power

Kansas Power & Light
LG&E Energy
Magma Power
Montana Power

Oklahoma Gas & Electric
Orange & Rockland Utilities
Otter Tail Power Utilities
Southwest Gas

South West Public Service
TECO Energy
United Water Resources

Source: *Good Money* (January/February 1991). Good Money Publications, Inc.: P.O. Box 363, Worcester, VT 05682. Reprinted with permission.

Domini Social Index 400

The following is Kinder, Lydenberg, Domini & Co.'s socially screened alternative to Standard & Poor's 500 Index.

* Acme-Cleveland
Acuson
Advanced Micro Devices
* Aetna Life & Casualty
Affiliated Publications
Ahmanson, A. H.
* Air Products & Chemicals
Airborne Freight
Alberto-Culver
Albertson's
* Alco Standard
Alexander & Alexander Services
Allwaste
* Aluminum Co. of America
Amdahl
America West Airlines
American Express
* American General
American Greetings
American International Group
American Stores
American Water Works
Ameritech
AmeriTrust
* Amoco
* AMP
AMR Corp
Anadarko Petroleum
Analog Devices
Angelica
* Apache
Apogee Enterprises
Apple Computer

Applied Materials
Archer Daniels Midland
ARCO Chemical
* Arkla
ASK Computer Systems
Atlanta Gas
* Atlantic Richfield
Autodesk
Automatic Data Processing
Avnet
* Avon Products
Baldor Electric
Bank One Corp
Bank of Boston
BankAmerica
* Bankers Trust New York
* Barnett Banks
Bassett Furniture
Battle Mountain Gold
* Baxter International
Becton Dickinson
* Bell Atlantic
* BellSouth
Bemis
Ben & Jerry's
* Beneficial
Bergen Brunswig
Betz Laboratories
Biomet
* Block, H & R
* Bob Evans
* Briggs & Stratton
* Brooklyn Union Gas

* Brown Group
Burlington Resources
* Cabot
Calgon Carbon
California Energy
Campbell Soup
Capital Cities/ABC
* Capital Holding
Carolina Freight
* CBS
Centex
Chambers Development
* Chemical Banking
* Chubb
* Church & Dwight
* CIGNA
Cincinnati Milacron
Cintas
Circuit City Stores
Citizens Utilities
Claire's Stores
CLARCOR
Clark Equipment
* Clorox
* Coca-Cola
Comcast
Commerce Clearing House
Community Psychiatric Centers
Compaq Computer
Computer Associates Int'l
Connecticut Energy
Consolidated Freightways
* Consolidated Natural Gas
Consolidated Papers
* Consolidated Rail
Continental
Cooper Tire & Rubber
* Corning
* CPC International
CPI
Cross, A. T.
Cross & Trecker
* CSX

Cummins Engine
Cyprus Minerals
* Dana
* Dayton Hudson
* Delta Airlines
* Deluxe
Digital Equipment
Dillard Department Stores
Dime Savings Bank, NY
Dionex
Disney, Walt
* Diversified Energies
Dollar General
Donnelley, R. R., & Sons
* Dow Jones
DSC Communications
Echo Bay Mines Ltd.
Edwards, A. G.
Egghead
* Energen
* Enron Corporation
* Equitable Resources
Fastenal
Fedders
Federal Express
Federal Home Loan Mortgage
* Federal National Mortgage
Association
Federal-Mogul
* First Chicago
* First Fidelity Bancorp
* First Wachovia
Fleetwood Enterprises
* Fleet/Norstar Financial
* Fleming
Forest Laboratories
* Fuller, H. B.
* Gannett
Gap
* GATX
GEICO
* General Cinema
* General Mills

General Re
* General Signal
Genuine Parts
* Gerber Products
* Giant Food
Gibson Greetings
Golden West Financial
* Goulds Pumps
Graco
Grainger, W. W.
Great Atlantic & Pacific Tea
* Great Western Financial
Groundwater Technology
* Handelman
* Hannaford Brothers
Harland, John H.
Harman International
Hartford Steam Boiler Insurance
* Hartmarx
Hasbro
Hechinger
* Heinz, H. J.
Helmerich & Payne
* Hershey Foods
Hillenbrand Industries
Home Depot
HON Industries
* Household International
Hubbell
* Huffy
Humana
Hunt Manufacturing
* Idaho Power
Illinois Tool Works
INB Financial
Inland Steel Industries
Intel
International Dairy Queen
Ionics
Isco
James River Corp of Virginia
Jefferson-Pilot
Johnson Products

* Jostens
* Kansas Power & Light
Kaufman & Broad Home
Kelley Services
King World Productions
* Knight-Ridder
Kroger
K mart
Land's End
Lawson Products
Lee Enterprises
Leggett & Platt
Lillian Vernon
Limited, The
* Lincoln National
Liz Claiborne
Longs Drug Stores
Lotus Development
Louisiana Land & Exploration
* Louisville Gas & Electric
* Lowe's
Luby's Cafeterias
Magma Power
Manor Care
Marriott
* Marsh & McLennan
Mattel
May Department Stores
* Maytag
McCaw Cellular Communications
* McDonald's
* McGraw-Hill
MCI Communications
* McKesson
* Mead
Measurex Corp
Medco Containment Services
* Media General
Medtronic
Mellon Bank
Melville
Mercantile Stores
* Merck

Meredith
* Merrill Lynch
Micron Technology
Microsoft
Miller, Herman
* Millipore
Modine Manufacturing
Monarch Machine Tool
* Moore
Morgan, J. P.
Morrison
Morton International
Mylan Laboratories
National Education Corporation
* National Medical Enterprises
* National Services Industries
NBD BANCORP
NCR
Neutrogena
New England Business Service
New England Critical Care
* New York Times
Newell
* NICOR
Nike
Nordson
Nordstrom
* Norfolk Southern
Northern Telecom Ltd.
Northwestern Public Service
Norwest Corp
* Nucor
NWNL
Omnicom Group
Oneida Ltd.
ONEOK
Oryx Energy
Oshkosh B'Gosh
PACCAR
* Pacific Enterprises
* Pacific Telesis Group
* Penney, J. C.
* Pennzoil

* Peoples Energy
Pep Boys, The
* PepsiCo
Petrie Stores
Phillips-Van Heusen
Piper, Jaffray & Hopwood
* Pitney Bowes
* PNC Financial
* Polaroid
Potomac Electric Power
* Premier Industrial
Price
Primerica
* Procter & Gamble
* Quaker Oats
* Ralston Purina
Reebok International
Roadway Services
* Rochester Telephone
Rouse
Rowan
* Rubbermaid
* Russell
Ryan's Family Steakhouse
* Ryder System
SAFECO
* Safety-Kleen
Santa Fe Pacific
Sara Lee
Scott Paper
Sealed Air
* Sears Roebuck
* Security Pacific
Service Corp International
Shared Medical Systems
Shaw Industries
Shawmut National
Sherwin-Williams
Sigma-Aldrich
Skyline
Smith, A. O.
* Smucker, J. M.
Snap-On Tools

Software Toolworks
* Southern New England Telecor
Southwest Airlines
* Southwestern Bell
Spec's Music
Springs Industries
* SPX
* Square D
St. Jude Medical
St. Paul Companies
Standard Register
Stanhome
* Stanley Works
Stratus Computer
* Stride Rite
Stryker
Student Loan Marketing
* Sun Company
Sun Microsystems
* Sun Trust Banks
Super Valu Stores
SYSCO
* Tambrands
Tandem Computers
Tandy
TCBY
Tektronix
* Telephone & Data Systems
Tele-Communications
* Tennant
Thermo Electron
Thermo Instrument Systems
Thomas Industries
Thomas & Betts
Times Mirror
TJ International
TJX Companies
Tonka
Tootsie Roll
Torchmark

Toro
Toys ''R'' Us
* Transamerica
Travelers
U.S. West
UAL
United Telecommunications
UNUM
* USF&G
USLIFE
U.S. Healthcare
* V. F. Corp
Value Line
Van Dorn
Vermont Financial Services
Viacom
* Walgreen
Wallace Computer Services
Wal-Mart
Wang Laboratories
Washington Gas & Light
Washington Post
Watts Industries
Wellman
Wells Fargo
Wesco Financial
* Westvaco
Wetterau
* Whirlpool
Whitman
Williams Companies
* Woolworth, F. W.
Worthington Industries
* Wrigley, Wm.
* Xerox
Yellow Freight Systems
Zenith Electronics

* Companies offering a dividend reinvestment program.
Source: Copyright 1991 by Kinder, Lydenberg, Domini & Co., Inc.
        Reprinted courtesy of Kinder, Lydenberg, Domini & Co., Inc.: Dana Street, Cambridge,
        MA 02138.

Appendix C

HIGH SOCIAL IMPACT INVESTING

The following organizations help investors locate appropriate ways to make investments that have a high social impact. Some of the groups focus on local communities; others address national or global needs. Depending upon the needs of their borrowers, some of the investment opportunities developed by these organizations offer below market returns and are aimed at investors, both individual and institutional, who take this factor into account. Others offer more competitive returns. Some of the resources listed are community development loan funds, which have been described in more detail in chapter 1. See also chapter 1's section on socially responsible banking, since many of these institutions offer high social impact investments in local (and usually disadvantaged) communities as well.

Resource Organizations

Accion International
130 Prospect Street
Cambridge, MA 02139-9794
(617) 492-4930
Operates the Bridge Fund, a loan program for microenterprises located in Latin America.

Catalyst
64 Main Street, 2nd Floor
Montpelier, VT 05602
(802) 223-7943
A quarterly newsletter on new initiatives and investment opportunities in the emerging alternative economy. Focuses on small businesses, worker-ownership, co-ops, re-

sponsive (and usually small) publicly traded companies, nonprofit organizations, local currency, appropriate technology, and alternative banks.

Community Economic & Ecological Development Institute (CEED)
1807 Second Street, Studio #2
Santa Fe, NM 87501
(505) 986-1401
Focuses on ecologically sustainable community development. Operates the Right Livelihood Revolving Loan Fund, which finances environmentally sound businesses.

Campaign for Human Development
1312 Massachusetts Avenue NW
Washington, D.C. 20005
(202) 659-6650
An action and education program sponsored and funded by the Catholic Bishops of the United States. Operates the Economic Development Loan Program, which provides loans and loan guarantees to worker-owned or community-owned business in areas of poverty.

Common Good Loan Fund
1320 Fenwick Lane #600
Silver Springs, MD 20910
(301) 565-0053

Interfaith Center on Corporate Responsibility
475 Riverside Drive, Room 566
New York, NY 10115
(212) 870-2295
This organization, first discussed in chapter 1 in connection with shareholder activism, also serves as a clearinghouse to increase the flow of capital from religious institutions into community economic development.

Source: *A Socially Responsible Planning Guide,* Co-Op America, 1990.

Community Development Loan Funds

National Association of Community Development Loan Funds (NACDLF)
P.O. Box 40085
Philadelphia, PA 19106-5085
(215) 923-4754
An association of thirty-nine member funds that borrow capital and relend it on affordable terms to benefit communities. Develop housing, employment opportunities, and other resources and services for low-income people. Check to see if the NACDLF knows of a fund in your community.

The following funds are members of the NACDLF. Most serve the community, city, or state in which they are located. Funds preceded by an asterisk serve a wider region or are national in scope.

*ANAWIM Fund of the Midwest
517 W. 7th
P.O. Box 4022
Davenport, IA 52808
(319) 324-6632

**Association for Regional Agriculture
Building the Local Economy
(ARABLE)**
1175 Charnelton Street
Eugene, OR 97401
(503) 485-7630

Boston Community Loan Fund
30 Germania Street
Jamaica Plain, MA 02130
(617) 522-6768

**Capital District Community Loan
Fund**
33 Clinton Avenue
Albany, NY 12210
(518) 436-8586

*Cascadia Revolving Fund
4649 Sunnyside North, Suite 348
Seattle, WA 98103
(206) 547-5183

***Catherine McAuley Housing
Foundation**
1601 Milwaukee, 5th Floor
Denver, CO 80206
(303) 393-3906

**Catskill Mountain Housing
Development Corporation Revolving
Loan Fund**
P.O. Box 473
Catskill, NY 12414
(518) 943-6700

**Common Wealth Revolving Loan
Fund**
1221 Elm Street
Youngstown, OH 44505
(216) 744-2667

*Cooperative Fund of New England
108 Kenyon Street
Hartford, CT 06105
(203) 523-4305

Cornerstone Loan Fund
P.O. Box 8974
Cincinnati, OH 45208
(513) 871-3899

***Delaware Valley Community
Reinvestment Fund**
924 Cherry Street
Philadelphia, PA 19107
(215) 925-1130

Enterprise Loan Fund
218 W. Saratoga Street, 3rd Floor
Baltimore, MD 21202
(301) 727-8535

**Federation of Appalachian Housing
Enterprises**
Drawer B
Berea, KY 40403
(606) 986-2321

*Fund for an OPEN Society
311 S. Juniper Street, Suite 400
Philadelphia, PA 19107-5804
(215) 735-6915

**Greater New Haven Community
Loan Fund**
5 Elm Street
New Haven, CT 06510
(203) 789-8690

**Human/Economic Appalachian
Development Community Loan
Fund**
P.O. Box 504
Berea, KY 40403
(606) 986-1651

***Industrial Cooperative Association
Revolving Loan Fund**
58 Day Street #203
Somerville, MA 02144
(617) 629-2700

***Institute for Community
Economics Revolving Loan Fund**
57 School Street
Springfield, MA 01105
(413) 746-8660

Koinonia Fund for Humanity
Koinonia Partners
Route 2
Americus, GA 31709-9986
(912) 924-0391

The Lakota Fund
P.O. Box 340
Kyle, SD 57752
(605) 455-2500

*Leviticus 25:23 Alternative Fund
Mariandale Center
P.O. Box 1200
Ossining, NY 10562
(914) 941-9422

Low Income Housing Fund
605 Market Street, Suite 709
San Francisco, CA 94105
(415) 777-9804

*McAuley Institute
8300 Colesville Road, Suite 310
Silver Spring, MD 20910
(301) 588-8110

Michigan Housing Trust Fund
3401 East Saginaw, Suite 212
Lansing, MI 48912
(517) 336-9919

**MICRO Industry Credit Rural
Organization**
802 East 46th Street
Tucson, AZ 85713
(602) 622-3553

**Montreal Community Loan
Association**
3609 boul. St. Laurent, 3è étage
Montreal, PQ, Canada H2X 2V5
(514) 849-3271

**New Hampshire Community Loan
Fund**
Box 666
Concord, NH 03302
(603) 224-6669

New Jersey Community Loan Fund
126 North Montgomery Street
Trenton, NJ 08608
(609) 989-7766

**New Mexico Community
Development Loan Fund**
P.O. Box 4979
Albuquerque, NM 87196
(505) 243-3196

Non-Profit Facilities Fund
12 West 31st Street
New York, NY 10001
(212) 868-6710

***Northcountry Cooperative
Development Fund**
2129-A Riverside Avenue
Minneapolis, MN 55454
(612) 371-0325

**Northern California Community
Loan Fund**
14 Precita Avenue
San Francisco, CA 94110
(415) 285-3909

***Rural Community Assistance
Corporation**
2125 19th Street, Suite 203
Sacramento, CA 95818
(916) 447-2854

***Southeastern Reinvestment
Ventures, Inc.**
159 Ralph McGill Boulevard, N.E.,
Room 410
Atlanta, GA 30308
(404) 659-0002, ext. 240

Vermont Community Loan Fund
P.O. Box 827
Montpelier, VT 05601
(802) 223-1448

**Washington Area Community
Investment Fund**
2201 P Street NW
Washington, D.C. 20037
(202) 462-4727

**Wisconsin Partnership for Housing
Development**
152 W Wisconsin, Suite 200
Milwaukee, WI 53203
(414) 223-2740

Worcester Community Loan Fund
P.O. Box 271, Mid Town Mall
Worcester, MA 01614
(508) 799-6106

Source: National Association of Community Development Loan Funds, November 1990.

Appendix D

RESOURCES

Banking
See listings on page 17.

Brokerage Houses

Progressive Asset Management
1814 Franklin Street, Suite 710
Oakland, CA 94612
(415) 834-3722

Financial Professionals

Clean Yield Asset Management, Inc.
224 State Street
Portsmouth, NH 03801
(603) 436-0820

Ethical Investments, Inc.
430 First Avenue North, Suite 214
Minneapolis, MN 55401
(612) 339-3939

First Affirmative Financial Network
1040 South 8th Street, Suite 200
Colorado Springs, CO 80906
(800) 422-7284
A nationwide network of financial professionals.

Franklin Research & Development Co.
711 Atlantic Avenue, 5th Floor
Boston, MA 02111
(617) 423-6655

Kinder, Lydenberg, Domini & Co.
7 Dana Street
Cambridge, MA 02138
(617) 547-7479

Social Investment Forum
430 First Avenue North, Suite 290
Minneapolis, MN 55401
(612) 333-8338
National professional association of socially responsible financial professionals, research and community organizations, and publishers dedicated to developing concepts and practices of SRI. Publishes Social Investment Services: A Guide to A Guide to Forum Members, *available for purchase by the public.*

The Social Responsibility Investment Group Inc.
The Chandler Building, Suite 622
127 Peachtree Street, N.E.
Atlanta, GA 30303
(404) 577-3635

Trust and Asset Account Management

U.S. Trust Company
30140 Court Street
Boston, MA 02109
(617) 726-7244

General Information Providers

Co-op America
2100 M Street NW, Suite 403
Washington, D.C. 20063
(800) 424-2667 or (202) 872-5307
Publish A Socially Responsible Financial Planning Guide *and the quarterly* Building Economic Alternatives, *which provide information and updates about a wide range of social investing opportunities.*

Council on Economic Priorities
30 Irving Place
New York, NY 10003
(800) 822-6435
A nonprofit public research organization. Publishes research reports, Shopping for a Better World, *and* The Better World Investment Guide.

Data Center
464—19th Street
Oakland, CA 94612
(415) 835-4692
Clipping and research service. Excellent source of information of a broad range of progressive issues addressed by social investing. Publishes monthly Corporate Responsibilities Monitor.

The Ethical Investment Research Service
401 Bondway Business Center
71 Bondway
London, SW8 1SQ
United Kingdom

Interfaith Center on Corporate Responsibility
475 Riverside Drive, Room 566
New York, NY 10115
(212) 870-2295
A working coalition of religious organizations addressing corporate social responsibility issues. Developing shareholder resolutions for presentation at corporate annual meetings is one of its primary activities. Publishes The Corporate Examiner.

Investor Responsibility Research Center
1755 Massachusetts Avenue NW
Washington, D.C. 20036
(202) 939-6500
A primary information source on the interface of social and public policy issues with the corporate community. Maintains Social Issues Service on shareholder resolutions; South Africa Review Service; and Corporate Governance Service. Financed mainly by annual subscriptions.

Kinder, Lydenberg, Domini & Co.
(See Financial Professionals page 379 for contact information.)
Developed and maintains the Domini 400, a socially responsible stock index. Publishes Update, *describing the index and each company's performance for the prior month.*

Insurance

Consumers United Insurance Co.
2100 M Street NW
Washington, D.C. 20063
(800) 422-6653
Offers health, term life, annuity, and Medicare supplement policies for members of Co-op America, National Organization for Women, and other groups. Will offer group insurance to employee groups. If you are not a member of a group associated with Consumers United, you may join Co-op America for $20 annually.

First Affirmative Financial Network
1040 South 8th Street, Suite 200
Colorado Springs, CO 80906
(800) 422-7284
Developing term life, universal life, whole life and annuity products for ethical investors. Anticipates first availability as October 1991.

Investment Newsletters

Catalyst
P.O. Box 1308
Montpelier, VT 05601
(802) 223-7943
A quarterly on new initiatives and investment opportunities in the emerging alternative economy. Focuses on small businesses, worker-ownership, co-ops, responsive (and usually small) publicly traded companies, nonprofit organizations, local currency, appropriate technology, and alternative banks.

The Clean Yield
Box 1880
Greensboro Bend, VT 05842
(802) 533-7178

Franklin's Insight: The Advisory Letter for Concerned Investors
711 Atlantic Avenue, 5th Floor
Boston, MA 02111
(617) 423-6655

Good Money
P.O. Box 363 B
Calais State Road
Worcester, VT 05682
(800) 223-3911

Organizational Resources on Specific Issues Addressed By SRI

See listings under appropriate topic headings in chapter 2.

Publications

Business Ethics Magazine
1107 Hazeltine Boulevard, Suite 530
Chaska, MN 55318
(612) 448-8864

Co-op America Quarterly
2100 M Street NW, Suite 403
Washington, D.C. 20063
(800) 424-2667

The Corporate Examiner
Interfaith Center on Corporate Responsibility
475 Riverside Drive, Room 566
New York, NY 10115
(212) 870-2293

Directory of Socially Responsible Investments
666 Broadway, 5th Floor
New York, NY 10012
(212) 529-5300

National Boycott Newsletter
6506-28th Avenue, NE
Seattle, WA 98115
(206) 523-0421

Shopping for a Better World
Council on Economic Priorities
30 Irving Place
New York, NY 10003
(212) 420-1133
An excellent one-stop guide to socially responsible companies and products.

Appendix E

GLOSSARY OF MONEY MANAGEMENT, INVESTMENT, AND SOCIALLY RESPONSIBLE INVESTMENT TERMS

AGGRESSIVE GROWTH FUND. A mutual fund concentrating on the stocks of young, still developing companies and other smaller companies. Such funds offer high growth potential as a trade-off for high risk.

AGGRESSIVE INVESTMENT STRATEGY. An investment strategy aimed at maximizing the returns on a portfolio. The trade-offs for high-return potential include high risk to principal and, in some cases, illiquidity as well.

ANNUAL REPORT. The formal financial statement that a corporation is required to issue yearly to shareholders. The annual report lists assets, liabilities, earnings, profit performance, and the like and usually includes textual items intended to show the corporation in the best possible light.

ANNUITY. An investment vehicle offered by insurance companies that returns a payment stream to the investor generated by the insurance company's investment of the investor's principal. Annuities are popular retirement planning vehicles because they permit tax-deferred compounding. Most companies offer several different types of annuities and a multiplicity of options within each type.

APPRECIATION. Increase in the value of an investment or other asset.

ASSET. That which is owned or receivable.

BALANCED FUND. A mutual fund invested mainly in high-quality bonds and high-dividend-paying, "blue chip" stocks. These are conservative investments designed to combine the fixed income attributes of bonds with the appreciation potential of stocks with minimal risk to the investor's principal. Balanced funds maintain a ratio of about 60 to 40 percent stocks to bonds.

BEAR. One who predicts the stock market will decline. The opposite of a **BULL.**

BEAR MARKET. A stock market on the decline.

BIG BOARD. The New York Stock Exchange.

BLOCK. A large number—generally ten thousand or more—of common stock shares either held or transacted.

BLUE CHIP STOCK. The stock of large, well-established corporations with long, stable histories of success. These are conservative investments with little chance of spectacular growth but highly probable continued moderate growth. Their stability is such that they can be expected to pay dividends even during an economic downturn.

BOND. A loan to a corporation, government agency, financial institution, or municipality for an agreed-upon term (period of time). In return for the loan, the issuer promises to repay the principal amount borrowed when the term expires plus a percentage of the amount borrowed as interest. See also **CORPORATE BOND, MUNICIPAL BOND.**

BOND FUND. A mutual fund primarily invested in bonds.

BROKER. An agent who transacts purchases and sales of securities, property, and/or other investments at the request of his/her clients. Brokers are generally compensated by commission—in other words, a percentage of the transaction amount.

BROKER/DEALER. See **DEALER.**

BULL. One who predicts that the stock market will rise. The opposite of a **BEAR.**

BULL MARKET. A stock market on the rise.

CALLABLE. A bond that is callable may be redeemed—or "called"—by the issuing corporation or municipality under specified conditions before its maturity date. Preferred stock shares may also be called if such terms are specified upon purchase. See also **MATURITY.**

CAPITAL GAIN. The personal income you earn when you sell an asset such as stock or real estate. The amount of the gain is the difference between the price you originally paid for the asset and the price you sold it at. Currently the IRS taxes capital gains at the same rate you pay on your ordinary income. Prior to the mid-1980s tax reform, capital gains were taxed at a lesser rate.

CAPITALIZATION. 1. The total investment of the owners in a business enterprise. 2. The total value of the various securities issued by a company, including bonds, common and preferred stock, debentures, and surplus. See also **BONDS, COMMON STOCK, DEBENTURE, PREFERRED STOCK, SURPLUS.**

CERTIFICATE OF DEPOSIT (CD). An investment account deposited at a financial institution for a specified term in return for a guaranteed interest rate. Interest rates significantly exceed those paid on regular passbook deposit accounts and carry the same deposit insurance as regular accounts. CDs can range in term from a few months to several years; generally, the longer the term, the higher the interest rate. At the end of the term, the investor may withdraw both the principal and accrued interest. Early withdrawals carry interest penalties.

CLOSED-END MUTUAL FUND. A mutual fund that issues a fixed number of shares, closing the fund to new investors once those shares are sold.

COLLATERAL. Assets pledged by a borrower to secure the repayment of a loan.

COLLECTIBLES. Tangible goods purchased for their investment value. Examples include fine art, stamps, baseball cards, rare coins, antiques, and rare books.

COMMERCIAL PAPER. Short-term (generally thirty-, sixty-, or ninety-day) loans to corporations. Corporations sell these issues to meet short-term cash needs (for example, a major toy company may issue commercial paper every spring to build up its inventories for the Christmas season). Commercial paper is considered a very safe investment and is a primary instrument of money market funds. It is usually sold in $100,000 denominations.

COMMISSION. The fee you pay to a financial agent such as a stockbroker for arranging a purchase or sale. The fee is generally based on a percentage of the transaction amount.

COMMON STOCK. Shares of ownership in a corporation that carries voting rights and a claim on corporate profits and assets after preferred stockholders and debt holders are paid. Common stock is the most standard form of individual stock investment. See also **PREFERRED STOCK.**

COMMUNITY DEVELOPMENT CREDIT UNION (CDCU). A credit union established to serve the needs of a low-income community. CDCUs are favored by many ethical investors for both daily banking and bank-related investments such as certificates of deposit. See **CREDIT UNION.**

COMMUNITY DEVELOPMENT LOAN FUND (CDLF). See **REVOLVING LOAN FUND.**

COMPOUNDING. Occurs when investment earnings are reinvested, thereby becoming principal and themselves earning interest.

CONSERVATIVE INVESTMENT STRATEGY. An investment strategy aimed at preserving principal. A conservative portfolio sacrifices growth potential for safety and liquidity. Conservative investments range from government-guaranteed and insured investments (virtually no risk to principal, no growth potential) to blue-chip stocks (some risk to principal as a trade-off for modest growth potential). See also **AGGRESSIVE INVESTMENT STRATEGY.**

CONVERTIBLE BOND FUND. A mutual fund that invests in convertible bonds. See also **CONVERTIBLE SECURITIES.**

CONVERTIBLE SECURITIES. Generally, bonds (convertible bonds) or preferred stock (convertible preferred stock) that can be converted into common stock at the holder's discretion (although the rate of exchange is fixed and conversion rights may be limited to a defined period of time).

CORPORATE BOND. A loan to a corporation from an investor, usually for a fixed term at a fixed interest rate. At the end of the term, the bondholder's principal is refunded. Interest is usually paid biannually. See also **BOND.**

CORPORATE BOND FUND. A mutual fund that invests in corporate bonds. See also **CORPORATE BOND.**

CORPORATION. A form of business organization in which the total worth of the business is divided into equal units of ownership called shares of stock. Stockholders enjoy limited liability—that is, their financial liability cannot exceed the amount they have invested. Corporations are allowed to raise capital by selling stock to the public (known as "going public" or "taking the company public").

CREDIT UNION. A nonprofit financial institution owned by its depositors and existing solely for the needs of its depositors. Money pooled through members' deposits is loaned back to members. Interest rates paid on deposits generally exceed those paid at other banking institutions, and rates charged for loans are lower than at other banking institutions. Credit unions are a favored form of banking for ethical investors because the institution's primary income is generated through loans to members rather than investments in unscreened portfolios. See also **COMMUNITY DEVELOPMENT CREDIT UNION.**

CUSTODIAN. A company responsible for holding the assets and/or portfolio of investments of another entity (such as a mutual fund company or a pension fund). Also, an individual responsible for overseeing another individual's investments (a parent overseeing a minor child's account, for example).

DEALER. An individual or company in the securities business acting as a principal (in other words, the actual purchaser or seller) rather than as an agent for another. Dealers normally buy securities for their own account and then sell from that inventory to their customers; their hope, of course, is that they will sell the securities for more than they paid for them. The same individual or company may act as a broker or dealer at different times but must disclose to customers securities purchased as a principal. Such individuals are often called *broker/dealers.* See also **BROKER.**

DEBENTURE. A promissory note backed by the general credit of the company. Normally debentures are not secured by any form of lien on specific property.

DEPRECIATION. A bookkeeping procedure for tax purposes, depreciation allows an owner to write off the decrease in value of a work or investment-related tangible asset (such as a building or equipment) over its useful life. Depreciation takes into account such factors as decay, wear and tear, decline in price, and so forth. It does not represent an actual cash outlay or money designed for that purpose.

DISCOUNT. The amount under par value by which a bond or preferred stock may sell on the secondary market. See also **PREMIUM, PAR VALUE.**

DIVERSIFICATION. The strategy of making a variety of investments to minimize the investor's risk (in other words, to counteract the danger of putting all one's eggs in one basket). Investment portfolios are typically diversified by investment category—for instance, an investor may seek appreciation through a combination of common stocks, real estate investments, and precious metals investments. Investors further minimize risk by diversifying *within* a category (for example, an investor in common stocks will own several at any one time and may further diversify in terms of industry, growth stock vs. blue chip, economic factors such as dependence on consumer borrowing, and so on).

DIVIDEND. A dividend is the proportion of a company's earnings that are distributed to its stockholders.

DIVIDEND REINVESTMENT PLAN. A stockholders' feature offered by some publicly traded companies, allowing stockholders to reinvest dividends in lieu of receiving cash dividend payments. Dividends are wholly applied to stock purchase—no commissions are charged and purchase may include fractional shares. Dividend reinvestment plans—when available—save commission costs, eliminate considerations of lot size, and build dollar cost averaging into the investment. These plans are arranged directly with the company, not through a broker, after the initial purchase of stock.

DOLLAR COST AVERAGING. An investment strategy that involves investing fixed dollar amounts at regular intervals. When you dollar cost average, your money will buy fewer shares when the price is high and more when the price is low. Your hope is that your average cost per share will end up being lower than if you had purchased the same number of shares at one time. There is an excellent chance of that happening if you apply the strategy to stock investments, because the historical upward trend of the stock market increases the probability of buying bargain-priced shares over time.

DOW JONES INDUSTRIAL AVERAGE. See **STOCK AVERAGES.**

ECONOMIC RISK. The potential loss of investment value due to economic conditions. The most common form of economic risk in the United States is inflationary risk. A currency devaluation is another example of economic risk. See also **INFLATION, RISK.**

EQUITY. In non–real estate investing, the ownership interests of common and preferred stockholders in a company. In real estate, the value of a property over and above what is owed in mortgages, liens, and other charges against the property.

EQUITY FUND. A mutual fund invested mainly in common stocks. A growth fund, for instance, is a type of equity fund.

FACE VALUE. See **PAR VALUE.**

FEDERAL DEPOSIT INSURANCE. The federal government's insurance—for all depositors at member banks, savings & loan institutions, and credit unions—against loss of principal and interest (up to a specified amount).

FIDUCIARY. Someone who is legally entrusted with the power to act for another in financial matters.

FIXED INCOME FUNDS. Any mutual fund type invested primarily in bonds and money market instruments.

GOVERNMENT BONDS. Debt obligations of the U.S. government. Government bonds are generally considered the safest bonds available. However, some government bonds are considered safer than others for technical reasons, with those considered less safe paying correspondingly higher returns. See also **BOND.**

GOVERNMENT FUND (taxable). A mutual fund invested in U.S. Treasury and/or U.S. government agency bonds.

GOVERNMENT FUND (tax-free). A mutual fund invested in municipal bonds. See also **MUNICIPAL BOND.**

GROWTH. See **APPRECIATION.**

GROWTH FUND. A mutual fund primarily invested in growth stocks. Growth fund investors accept above-average risk as a trade-off for above-average potential returns. See also **GROWTH STOCKS.**

GROWTH AND INCOME FUND. A mutual fund invested in conservative stocks and convertible corporate bonds to produce both long-term growth and income. See also **CONVERTIBLE SECURITIES.**

GROWTH STOCKS. The stocks of companies that are reinvesting all or most of their earnings into research and/or development (as opposed to paying high dividends to shareholders) in anticipation of highly profitable returns on their projects. These stocks have above average appreciation potential but also above average risk because they perform with more volatility than the stocks of mature companies that have passed their rapid growth phase.

INCOME FUND. A stock mutual fund that has as its primary objective current income as opposed to appreciation. As such, the fund's managers will select stocks with a history of market stability and high dividend payments that predicts a similar or better performance in the future. Most income funds also include bonds and money market instruments.

INDEX FUNDS. A mutual fund invested in the stocks of a popular stock index (for example, the Standard & Poor 500) with the goal of replicating the index's performance.

INDIVIDUAL RETIREMENT ACCOUNT (IRA). Current IRS rules allow individuals under age 70½ to contribute up to $2,000 annually to an account in a bank or in an investment and have that money excluded from taxable income. Additionally, the money compounds tax-deferred in the account until withdrawn.

INFLATION. A raise in general price levels (and thus a decrease in purchasing power). Inflation is caused by an increase in the supply of money and credit relative to the supply of available goods and services—not by a change in those goods or services.

INFLATION HEDGE. An investment considered likely to at least keep pace in value with the rate of inflation. A classic example is real estate. See also **INFLATION.**

INITIAL OFFERING. The original sale of a corporation's securities (which at this point are called *new issues*). Also called *primary offering* or *primary distribution*.

INSTITUTIONAL INVESTOR. An organization that manages substantial investing assets, in many cases for others. Examples include mutual funds and other investment companies, pension funds, banks, and insurance companies.

INTEREST. The return for lending money or the charge for borrowing it. Lending-type investments such as certificates of deposit, bonds, and mortgage-backed securities pay interest.

INTERNATIONAL FUND. A mutual fund invested in securities outside the United States.

INVESTMENT. The use of money with the expectation that it will make more money either through income or appreciation or both.

ISSUE. 1. (noun) A security. 2. (verb) The act of distributing securities by the securities' originator. See also **SECURITY.**

JUNK BOND. A corporate bond issued by a corporation carrying a quality rating of BB (Standard & Poor's) or Ba (Moody's) or less. Corporations receive these low-quality ratings because they are considered to be too indebted to be fully creditworthy or because they have a past record of defaulting on debt, or both. Corporations issuing junk bonds attract investors by offering higher rates of interest on their bonds than are offered on "investment-grade" bonds (bonds rated higher than BB or Ba). See also **CORPORATE BOND** and **QUALITY RATINGS.**

LEVERAGE. Borrowing money to make possible a larger investment than would be possible with cash alone. Real estate, whether residential or commercial, is typically leveraged with a 20 to 25 percent cash down payment and the rest mortgaged. Those corporations that choose to leverage themselves typically do so by issuing corporate bonds or preferred stock. See also **CORPORATE BOND** and **PREFERRED STOCK.**

LIABILITY. That which is owed.

LIMITED PARTNERSHIP. A business enterprise composed of one or more *general partners,* who make the business decisions and assume unlimited liability for those decisions, and several (sometimes hundreds of) *limited partners,* who do not participate in decision making and whose liability is normally limited to the amount of their investment.

LIQUID ASSET. A cash asset or an asset that can be quickly converted into cash (generally in one to seven working days).

LIQUIDITY. For an individual, liquidity means having enough readily available assets to meet financial needs without having to sell off long-term investments or nonliquid assets such as a personal residence. For a business, liquidity means having adequate assets on hand to meet financial obligations without having to sell off productive assets such as equipment or buildings. Regarding investments, liquidity means the relative ease and speed with which the investment can be converted to cash. See also **LIQUID ASSET.**

LISTED STOCK. A stock listed on a stock exchange, as opposed to being sold over the counter. See **OVER-THE-COUNTER.**

LOAD. A load is the sales fee, or commission, charged by many mutual funds. Those funds that do not charge a sales fee are called "no-load" funds.

MARKET VALUE. The price that investors will pay for a security or other investment at the current time. Market value in investing is determined by the laws of supply and demand and by risk/reward trade-offs.

MATURITY. The date when the principal paid on a bond or certificate of deposit is repaid to the investor.

MONEY MARKET MUTUAL FUND. A mutual fund invested solely in "money market instruments"—typically short-term debt instruments such as treasury bills, commercial paper, and certificates of deposit. An SRI money market fund would not invest in treasury bills and would invest only in debt instruments issued by socially screened corporations and banking institutions.

MUNICIPAL BOND. A bond issued by a city or state. The interest earned on municipal bonds ("munis") is not taxed by the federal government. If you are a resident of the state issuing the bond, you will not pay state taxes on the interest, either. (These bonds are often advertised as "double tax-free." Some cities issue "triple tax-free bonds"—free of city taxes as well.) Municipal bonds are popular investments with many ethical investors because they generally finance projects with high social value such as school construction, libraries, parks, roads, and other infrastructure needs.

MUTUAL FUND. A company that pools investors' money to assemble a portfolio of securities (usually common stocks and bonds) and/or other investment vehicles and then manages that portfolio for the investors. Mutual funds are available in a variety of investment vehicles, including stocks, corporate bonds, municipal bonds, money market instruments, mortgage-backed securities, other government bonds, precious metals, and the like. Some mutual funds concentrate on a single area, such as stocks. Others (for example, balanced funds) may combine two or more vehicles.

NASDAQ. National Association of Securities Dealers Automated Quotations, an automated information-reporting network that provides brokers and dealers with price quotations of securities traded over the counter. The business pages of large newspapers often print NASDAQ listings. See also **OVER-THE-COUNTER.**

NET ASSET VALUE. In mutual funds, the net value (the gross value minus liabilities, expenses, taxes, and required reserves) of the investments held by the fund at yesterday's closing price divided by the number of shares outstanding that day. In other words, "net asset value" really means the net asset value *per share.*

NET WORTH. Total value of assets (that which is owned) minus total liabilities (that which is owed). Applies to both businesses and individuals.

NEW ISSUE. A stock or bond sold by the issuer (whether a corporation, government agency, or municipality) for the first time. See also **SECONDARY MARKET.**

OPEN-END MUTUAL FUND. A mutual fund authorized to issue an unlimited number of shares. When new investors buy into the fund, new shares are created and the new money is invested in additional securities that meet the fund's guidelines. Nearly all mutual funds are open-end.

OVER-THE-COUNTER. A market system for stocks not listed on stock exchanges. Unlike a stock exchange, the over-the-counter market is not an actual place. Instead it is a network of broker-dealers who conduct trades according to specific guidelines established by the SEC. Many stocks of small, newer companies are traded this way; some stocks of major corporations and public utilities are only sold over-the-counter as well. With the major corporation and large utility stocks notable exceptions, most stocks sold over-the-counter have insufficient shares outstanding, stockholders, or earnings to qualify for listing on an exchange.

PAPER PROFIT. Profit that is unrealized because the investment has not yet been sold. If you buy a stock at $40 a share and it rises to $80 a share, your paper profit of 100 percent will become a realized profit if you can sell the stock at $80.

PAR VALUE. An arbitrary dollar value assigned to newly issued shares of stock. Also, a synonym for **FACE VALUE,** the principal amount of a bond.

PENNY STOCKS. Highly speculative stocks selling for a few dollars a share or less. The companies issuing penny stocks usually have no track record and may not have even manufactured their product yet.

POINT. 1. A percentage point (of interest, commission, and so on). 2. Regarding shares of stock, a point means $1. That is, a stock that rises two points has risen $2 in price per share. Note, however, that if a market average like the Dow Jones Industrial Average rises from, for example, 2501.25 to 2502.25, it will also be described as gaining a point, which in such an average is not the same as $1.

PORTFOLIO. The total group of investments held by an individual or organization.

PREFERRED STOCK. Ownership shares in a corporation that pay a fixed or stated dividend. Preferred stockholders have a priority claim on company earnings and assets over that of common stockholders (but after bond- and other secured debt holders). Preferred stock, unlike common stock, does not usually carry voting rights. See also **COMMON STOCK.**

PREMIUM. The amount over par value by which a bond or preferred stock may sell on the secondary market. Regarding a new issue of stock or bonds, the amount by which the market price exceeds the original selling price. See also **DISCOUNT.**

PRICE-EARNINGS RATIO (P/E RATIO). A popular indicator of stock value determined by dividing the price of a share by the corporation's annual earnings per share. If the stock is selling for $40 a share and the company's annual earnings are at $5 a share, the P/E ratio is 8 to 1 (usually spoken of as 8).

PRIMARY OFFERING or **PRIMARY DISTRIBUTION.** See **INITIAL OFFERING.**

PRINCIPAL. 1. The original amount of money invested or lent (not including profit and/or interest). 2. The person actually buying or selling an investment, as opposed to the broker acting in his/her behalf. A dealer buying or selling for his/her own account is also known as the principal in the transaction.

PRIVATELY HELD STOCK. Stock held by the owners of a corporation that has not *gone public*—stock, in other words, that is not for sale on the public markets.

PROFIT-TAKING. Realizing a profit by selling a stock that has appreciated in value since purchased. Profit taking is frequently offered as the explanation when a surging market suddenly takes a downturn.

PROSPECTUS. For securities, the legal document describing a company in detail as well as the security it is offering. For mutual funds, the legal document describing the fund in detail, including its goals and strategies. (A socially screened mutual fund's prospectus will describe its social criteria as well.) In each of these cases, prospectuses are due each new investor.

PROXY. A stockholder's written authorization to have someone else vote his/her shares as instructed at a shareholder's meeting.

PROXY STATEMENT. Information that the SEC requires a company to give to stockholders in advance of an annual shareholder's meeting and as a prerequisite to soliciting proxies. The statement lists the date, time, and place of the meeting and the matters to be decided there. Accompanying the statement is a card, the submitting of which instructs the corporation how the stockholder wishes to vote. (The stockholder also has the option of attending the meeting and voting there.) See also **PROXY** and **SHAREHOLDER RESOLUTION.**

PUBLIC UTILITY. Actually a private company granted a monopoly by a municipal government and regulated by that government for the purpose of providing a needed service or product to the general population. Public utilities provide water, gas, electricity, and the like to municipalities.

QUALITY RATINGS. Independent evaluation of the safety of a bond investment, performed by a number of financial service organizations, of which Moody's and Standard & Poor's are the best known. (Federal bonds are not rated, although the creditworthiness of the United States is assumed to be top grade. In rare instances corporate or municipal issues may go unrated as well because the issuer does not wish to pay the rating fee.) Because of the risk/reward trade-off, the highest-rated bonds offer the lowest interest rates. Similarly, lower-rated bonds must offer higher rates to entice investors.

QUOTATION (QUOTE). Report of the highest bid to buy and the lowest offer to sell a security at a given time.

REAL ESTATE INVESTMENT TRUST (REIT). A company that pools money from investors (much like a mutual fund) and invests it in real estate, mortgages, and mortgage-backed securities. REIT shares are traded publicly like stocks.

REAL RETURN. The difference between the annual return on the investment and the annual rate of inflation. Also referred to as the *real interest rate.*

RETURN. The earnings on an investment. Return may be expressed as a percentage (*rate of return*) or as a ratio of earnings to the original investment. A *net return* is the earnings minus sales charges (commissions) or other investment costs. See also **YIELD.**

REVOLVING LOAN FUND. A nonprofit corporation that pools loans from institutional and individual lenders at below market rates, then lends the funds to various projects that meet its community-based goals. As a rule, these loans are made at below market rates to projects unable to obtain capital through traditional lending sources. Some revolving loan funds also provide technical assistance that helps ensure the success of the project and thus the repayment of the loan. See also **COMMUNITY DEVELOPMENT LOAN FUNDS.**

RISK. The potential loss of money from the investment. The two most common types of risk are *risk of principal* and *inflationary risk.* Risk of principal means the potential loss of part (or, in extreme cases, all) of the money invested. Inflationary risk, a form of economic risk, usually occurs with investments having little or no risk of principal and fixed rate of return. The risk with these investments is that the rate of inflation may exceed the rate of return on the investment, resulting in a loss of purchasing power. As a rule, the greater the potential risk an investment carries, the greater the potential returns. See also **ECONOMIC RISK** and **REAL RETURN.**

SALES CHARGE. See **COMMISSION.**

SECONDARY MARKET. The sale or purchase of any investment that is not a primary (or new) issue. See also **NEW ISSUE.**

SECTOR FUND. A mutual fund invested primarily in a narrow segment of the economy such as a particular industry. Hi-Tech funds are sector funds invested in high technology companies.

SECURITIES AND EXCHANGE COMMISSION (SEC). A federal agency that supervises the exchange of securities and takes action against companies and individuals who defraud investors.

SECURITY. Generally refers to stock and bond investments. Technically, a security is written evidence of ownership or creditorship. Stock and bond certificates are examples of securities in this meaning of the term.

SHAREHOLDER ACTIVIST. A stockholder (or one with access to a sympathetic stockholder) who uses the shareholder resolution process to influence a corporation to cease what he/she perceives as socially destructive behavior and/or initiate what he/she perceives as socially constructive behavior. See also **SHAREHOLDER RESOLUTION.**

SHAREHOLDER RESOLUTION. Corporate policy proposal put to a vote of the total shareholder body through the proxy mechanism. Shareholders meeting certain requirements set by the SEC may draft their own resolutions. Because shareholder resolutions are listed on the proxy statement, they represent an important strategy for compelling corporate management and other shareholders to consider the company's social responsibilities.

SOCIAL SCREEN. A social criterion applied to an investment or group of investments, such as a screen for business involvement in South Africa. Social screening is performed mainly by socially responsible investment organizations such as mutual funds and other socially responsible research organizations. Some social screens are negative in form, such as those designed to assure investors that a company or investment product is *not* involved with an objectionable practice or product. Positive screens select investments for their socially constructive impact.

SOCIALLY RESPONSIBLE INVESTMENT. "The channeling of personal, community, or workplace capital toward just, peaceful, healthy, environmentally sound purposes and away from destructive uses." (Source: The Social Investment Forum.)

SPECULATOR. One who is willing to risk much more than the average investor for the chance at much greater than average gains.

STANDARD & POOR'S 500. See **STOCK AVERAGES.**

STOCK. Shares of ownership in a corporation. There are two kinds of stock, common and preferred, each with its own rights and investment attributes. See **COMMON STOCK** and **PREFERRED STOCK.**

STOCK AVERAGES. Methodologies designed to simulate the total performance of stocks listed on exchanges. For example, the Dow Jones Industrial Average totals the price of thirty stocks and then adjusts that figure to compensate for such factors as past stock splits and dividends paid. The Standard & Poor's 500, which tracks five hundred stocks, is another popular stock average. The Domini 400 is an average derived from the performance of four hundred socially screened stocks.

STOCK EXCHANGES. The physical places where the buy and sell orders for stock are processed. The purpose of stock exchanges and the over-the-counter market is to provide an orderly marketplace where stockholders can instantly find buyers (and buyers can find sellers) and where transactions can be recorded such that new owners of stock will be recognized as bona fide stock holders. Each exchange trades only those stocks that are listed on it. Many stocks are listed on more than one exchange. The New York Stock Exchange (the largest) and the American Stock Exchange are the two national stock exchanges. The Pacific Stock Exchange and Midwest Stock Exchange are examples of major regional stock exchanges. See also **OVER-THE-COUNTER.**

STOCK SPLIT. Corporations sometimes *split* the price of their stock to make it more affordable to the small investor. The split is accomplished by increasing the number of shares outstanding and giving shareholders an appropriate amount of new stock. For example, a two-for-one split of a $100 stock would leave current shareholders with twice as many shares, while cutting the price of the stock to $50.

STREET NAME. Securities held in the name of the broker instead of his/her customer are said to be held in street name.

SURPLUS. Assets over and above a company's capital stock value—in other words, over and above all shares of ownership (common and preferred stock).

TAX-ADVANTAGED INVESTMENT. Investments that include a tax savings through deferral, decrease, or elimination of taxable income. Federal, state, and local governments create tax advantages to encourage investments in areas of social need. For example, IRAs allow tax-deferred retirement planning, reducing the government's need to provide for indigent retirees. Tax-free municipal bonds encourage investment in needed public projects.

TREASURY BILL (AKA T-BILL). A short-term (thirteen-, twenty-six-, or fifty-two-week) federal government debt investment sold at a discount from its face value. The income to the investor is the difference between the purchase price and the face value. Most ethical investors avoid treasury bills because the U.S. Treasury funds the activities of the government indiscriminately.

TREASURY BOND (AKA T-BOND). Long-term (ten- to thirty-year) federal debt investments sold by the U.S. Treasury at face value and paying a stated interest rate on a semiannual basis. They are also sold in auctions at more or less than the face value, depending upon whether or not the stated interest rate is a good deal in the current interest rate climate. Avoided by most ethical investors (see **TREASURY BILL**).

TREASURY NOTE (AKA T-NOTE). These are like treasury bonds in every way except that their maturities vary from two to ten years. See **TREASURY BOND**.

UNIT INVESTMENT TRUST. A company that invests in a fixed portfolio of securities and then sells shares of the portfolio to the public.

U.S. AGENCY OBLIGATIONS. Bonds issued by U.S. government agencies and backed by either the agency itself or the government as a whole. Although not socially screened per se, many U.S. agency obligations are considered by social investors to be ethical investments because of the nature of the agency's activity. For example, the Government National Mortgage Association ("Ginnie Mae") supports affordable housing; the Student Loan Marketing Association ("Sally Mae") supports affordable education. See also **BOND, GOVERNMENT BONDS**.

VENTURE CAPITAL. High-risk investments providing start-up capital for new business ventures. In exchange for their faith, venture capitalists generally demand either a high rate of return, a percentage of future business, shares of private stock, or some combination of the above. Venture capital is frequently raised through private limited partnerships (limited partnerships restricted to investors who can meet a high-level net worth requirement).

VOLATILITY. The relative degree of fluctuation in price, value, and/or interest rate of a particular investment.

YIELD. 1. In general, the annual rate of return on an income-producing investment (such as bonds or dividend-paying stocks). Expressed as a percentage (calculation: annual interest or dividends divided by amount of initial investment). 2. Three types of yields, with important differences, may be mentioned in discussions regarding bonds. The *stated* or *coupon yield*—sometimes simply spoken of as the bond's *coupon*—is the percentage of the bond's par value that the bond issuer promises to pay annually. Example: Small International, Inc., issues a $1,000 bond stating an interest rate of 7.5 percent. The *current yield* is the coupon in dollars divided by the bond's current price on the secondary market. Example: Small International's bond now sells discounted on the secondary market for $600. Thus $75 (the coupon) divided by $600 (current price) equals a current yield of 12.5 percent. If the bond had been purchased at a premium price of $1,050, the current yield would be 7.14 percent. The *yield to maturity,* sometimes called *basis yield,* is the rate of return on a bond held to maturity. This is a complicated calculation involving stated yield; the difference between the current bond price and the amount the bond will be redeemed for at maturity (in our example, the Small International bond bought for $600 will be redeemed for $1,000); and accrued interest (accumulated interest, including interest earned on interest). Your broker should be able to provide you with the yield to maturity of any bond you are considering. See also MATURITY and PAR VALUE.

ZERO COUPON BOND. Bonds that make no actual interest payments but instead are bought at a discount off their face value and then redeemed for the full face value when they mature. Because the interest earned is not paid out until maturity, it compounds until that time. See also COMPOUNDING.

Bibliography

Corporate Responsibility

Booth, Helen, and Kenneth Bertsch. *The MacBride Principles and U.S. Companies in Northern Ireland.* Washington, D.C.: Investor Responsibility Research Center, 1989.

Corson, Ben, et al., and The Council on Economic Priorities. *Shopping for a Better World.* New York: Council on Economic Priorities, 1990.

Freudberg, David. *The Corporate Conscience.* New York: Amacom, 1986.

Lydenberg, Steven, Alice Tepper-Marlin, Sean O'Brien Strub, and The Council on Economic Priorities. *Rating America's Corporate Conscience.* New York: Addison-Wesley, 1986.

General Financial Planning

Bloch, H. I., and Grace Lichtenstein. *Inside Real Estate.* New York: Grove Weidenfeld, 1987.

Gardiner, Robert. *The Dean Witter Guide to Personal Investing.* New York: New American Library, 1988.
Quite simply, this is the most lucid, friendly beginner's book on financial planning that we've seen.

Hardy, C. Colburn. *Dun & Bradstreet's Guide to Your Investments.* New York: Harper & Row, 1983.

Van Caspel, Venita. *Money Dynamics for the New Economy.* New York: Simon & Schuster, 1986.

Van Caspel, Venita. *The Power of Money Dynamics.* Reston: Reston Publishing Company, 1983.

General Financial Topics

Bruck, Connie. *The Predators' Ball.* New York: Simon & Schuster, 1988. *This version of the Michael Milken/Drexel Burnham story reads like good fiction.*

Engel, Louis. *How to Buy Stocks.* New York: Bantam Books, 1962.
*This is a classic, a most readable introduction to the inner workings of the
stock market and stocks in general.*
Lee, Susan. *ABZ's of Economics.* New York: Poseidon Press, 1987.
Mayer, Martin. *Markets.* New York: W. W. Norton, 1988.
Mayer, Martin. *Wall Street.* New York: Collier Books, 1955.

Social Issues
Brown, Lester. *Building a Sustainable Society.* New York: W. W. Norton,
1981.
Brown, Lester. *State of the World, 1989.* New York: W. W. Norton, 1989.
Cohen, Gary, and John O'Conner, eds. *Fighting Toxics.* Washington, D.C.:
Island Press, 1990.
Cook, Fred. *The Great Energy Scam.* New York: Macmillan, 1982.
Earthworks Group. *50 Simple Things You Can Do to Save the Earth.* Berke-
ley: Earthworks Press, 1989.
Goldsmith, Edward, et al. *Imperiled Planet.* Cambridge, Mass.: MIT Press,
1990.
Lappé, Frances Moor, and Joseph Collins. *Food First.* Boston: Houghton
Mifflin, 1977.
McGovern, James. *The Oil Game.* New York: Viking Press, 1981.
Oelhaf, Robert C. *Organic Agriculture.* Montclair: Allanheld, Osmun & Co.,
1978.
*A dispassionate, objective comparison of organic and conventional agri-
culture methods, this is a superior resource on the interdependence of
chemical fertilization and chemical pesticides, as well as the economic
factors that perpetuate it.*
Smith, Hedrick. *The Power Game.* New York: Random House, 1988.
*The basis for the PBS television series, this is a seminal resource on the
military-industrial and other government-industrial complexes.*
Tompkins, Peter, and Christopher Bird. *The Secret Life of Plants.* New York:
Harper & Row, 1973.
Tompkins, Peter, and Christopher Bird. *Secrets of the Soil.* New York: Harper
& Row, 1989.

Socially Responsible Money Management and Investing

Alperson, Myra, et al. *The Better World Investment Guide.* New York:
Prentice-Hall Press, 1991.
Co-op America. *A Socially Responsible Financial Planning Guide.* Wash-
ington, D.C.: Co-op America, 1988.
Bruyn, Severyn T. *The Field of Social Investment.* Cambridge: Cambridge
University Press, 1987.

Intended apparently for economic and business scholars, this is a comprehensive academic treatment of SRI theory and practice.

Domini, Amy, and Peter Kinder. *Ethical Investing.* New York: Addison-Wesley, 1986.

This was the first comprehensive book on the field and is still one of the best.

Judd, Elizabeth. *Investing with a Social Conscience.* New York: Pharos Books, 1990.

Meeker-Lowry, Susan. *Economics as if the Earth Really Mattered.* Philadelphia: New Society Publishers, 1988.

Index